118 Springer Series in Solid-State Sciences
Edited by Peter Fulde

Springer Series in Solid-State Sciences

Editors: M. Cardona P. Fulde K. von Klitzing H.-J. Queisser

Managing Editor: H. K.V. Lotsch

Volumes 1–89 are listed at the end of the book

90 **Earlier and Recent Aspects of Superconductivity**
Editors: J. G. Bednorz and K. A. Müller

91 **Electronic Properties of Conjugated Polymers III** Basic Models and Applications
Editors: H. Kuzmany, M. Mehring, and S. Roth

92 **Physics and Engineering Applications of Magnetism** Editors: Y. Ishikawa and N. Miura

93 **Quasicrystals** Editors: T. Fujiwara and T. Ogawa

94 **Electronic Conduction in Oxides**
By N. Tsuda, K. Nasu, A. Yanase, and K. Siratori

95 **Electronic Materials**
A New Era in Materials Science
Editors: J. R. Chelikowsky and A. Franciosi

96 **Electron Liquids** By A. Isihara

97 **Localization and Confinement of Electrons in Semiconductors**
Editors: F. Kuchar, H. Heinrich, and G. Bauer

98 **Magnetism and the Electronic Structure of Crystals** By V. A. Gubanov, A. I. Liechtenstein, and A.V. Postnikov

99 **Electronic Properties of High-T_c Superconductors and Related Compounds**
Editors: H. Kuzmany, M. Mehring, and J. Fink

100 **Electron Correlations in Molecules and Solids** 2nd Edition By P. Fulde

101 **High Magnetic Fields in Semiconductor Physics III** Quantum Hall Effect, Transport and Optics By G. Landwehr

102 **Conjugated Conducting Polymers**
Editor: H. Kiess

103 **Molecular Dynamics Simulations**
Editor: F. Yonezawa

104 **Products of Random Matrices** in Statistical Physics By A. Crisanti, G. Paladin, and A. Vulpiani

105 **Self-Trapped Excitons** By K. S. Song and R. T. Williams

106 **Physics of High-Temperature Superconductors** Editors: S. Maekawa and M. Sato

107 **Electronic Properties of Polymers**
Orientation and Dimensionality of Conjugated Systems Editors: H. Kuzmany, M. Mehring, and S. Roth

108 **Site Symmetry in Crystals** Theory and Applications By R. A. Evarestov and V. P. Smirnov

109 **Transport Phenomena in Mesoscopic Systems** Editors: H. Fukuyama and T. Ando

110 **Symmetry and Optical Phenomena in Superlattices and Other Heterostructures**
By E. L. Ivchenko and G. E. Pikus

111 **Low-Dimensional Electronic Systems**
New Concepts
Editors: G. Bauer, F. Kuchar, and H. Heinrich

112 **Phonon Scattering in Condensed Matter VII**
Editors: M. Meissner and R. O. Pohl

113 **Electronic Properties of High-T_c Superconductors**
Editors: H. Kuzmany, M. Mehring, and J. Fink

114 **Interatomic Potential and Structural Stability**
Editors: K. Terakura and H. Akai

115 **Ultrafast Spectroscopy of Semiconductors and Semiconductor Nanostructures** By J. Shah

116 **Electron Spectrum of Gapless Semiconductors**
By J. M. Tsidilovski

117 **Electronic Properties of Fullerenes**
Editors: H. Kuzmany, J. Fink, M. Mehring, and S. Roth

118 **Correlation Effects in Low-Dimensional Electron Systems**
Editors: A. Okiji and N. Kawakami

A. Okiji N. Kawakami (Eds.)

Correlation Effects in Low-Dimensional Electron Systems

Proceedings of the
16th Taniguchi Symposium
Kashikojima, Japan, October 25-29, 1993

With 37 Figures

Springer-Verlag
Berlin Heidelberg New York
London Paris Tokyo
Hong Kong Barcelona
Budapest

Professor Ayao Okiji

Department of Applied Physics, Osaka University
Suita, Osaka 565, Japan

Professor Norio Kawakami

Yukawa Institute for Theoretical Physics, Kyoto University
Sakyo-ku, Kyoto 606, Japan

Series Editors:

Professor Dr., Dres. h. c. Manuel Cardona
Professor Dr., Dr. h. c. Peter Fulde
Professor Dr., Dr. h. c. Klaus von Klitzing
Professor Dr., Dres. h. c. Hans-Joachim Queisser

Max-Planck-Institut für Festkörperforschung, Heisenbergstrasse 1
D-70569 Stuttgart, Germany

Managing Editor:

Dr. Helmut K.V. Lotsch

Springer-Verlag, Tiergartenstrasse 17, D-69121 Heidelberg, Germany

ISBN-13:978-3-642-85131-5 e-ISBN-13:978-3-642-85129-2
DOI: 10.1007/978-3-642-85129-2

Library of Congress Cataloging-in-Publication Data. Correlation effects in low-dimensional electron systems: proceedings of the 16th Taniguchi symposium, Kashikojima, Japan, October 25-29, 1993/A. Okiji, N. Kawakami (eds.) p. cm. — (Springer series in solid-state sciences; 118) Includes bibliographical references.
ISBN-13:978-3-642-85131-5 1. Condensed matter—Congresses. 2. Quantum Hall effect—Congresses. I. Okiji, A. (Ayao), 1934- . II. Kawakami, N. (Norio), 1958- . III. Taniguchi International Symposium on the Theory of Condensed Matter (16th : 1994 : Kashikojima, Japan) IV. Series. QC173.45.C67 1994 530.4'1—dc20 94-8947

Softcover reprint of the hardcover 1st edition 1994

Typesetting: Camera ready copy from the authors/editors
SPIN: 10130823 54/3140 – 5 4 3 2 1 0 – Printed on acid-free paper

Preface

This volume contains the proceeding of the 16th Taniguchi International Symposium on the Theory of Condensed Matter, which was held during October 25–29, 1993, at the Shima-Kanko Hotel, Kashikojima, Mie, Japan. The subject of the symposium was "Correlation Effects in Low-Dimensional Electron Systems". In particular, the selection of papers emphasized exact treatments of the electron-correlation problem.

In the last decade our knowledge of low-dimensional correlated electron systems has become precise and rich, following recent progress in analytical and numerical methods. In particular, extensive studies based on exactly solvable models and conformal-field theory have revealed various new aspects of integrable systems in one dimension. Furthermore, numerical approaches have clarified the key features of correlation effects in both one- and two-dimensional electron systems. In this volume, various topics in low-dimensional correlated electron systems are reported by leading physicists. Parts I–III are devoted to exact results in one dimension, and contain reports on new developments of the exact solution, conformal-field-theoretical approaches, application of quantum groups, and numerical diagonalization techniques. In Part IV, the edge states of the fractional quantum Hall effect are treated by field theoretical methods as well as by numerical diagonalization. In Part V, various key properties, as clarified by numerical approaches, are reported for two-dimensional highly correlated electron systems.

The symposium was fully supported financially by the Taniguchi Foundation established in 1929 by Mr. Toyosaburo Taniguchi from his personal funds. The purpose of the Foundation is to promote research in the field of basic science thus contributing to the development of industry and economy, and also to promote mutual understanding and international friendship especially among young researchers. The Taniguchi Foundation sponsors 18 international symposia on selected topics every year. The concept of these symposia is unique in that participants, both Japanese and from abroad, are limited in number to small discussion groups, and live together, although for a short period, as a close-knit community.

Fruitful discussions held during the symposium have certainly encouraged participants, and will hopefully stimulate this field of physics. On behalf of the participants we would like to express our sincere thanks to Mr. Toyosaburo Taniguchi and the Taniguchi Foundation.

Osaka
Kyoto
January 1994

Ayao Okiji
Norio Kawakami

Contents

Part III Phase Diagram for 1D Correlated Electron Systems

Part IV Edge States of the Quantum Hall Effect

Part V 2D Correlated Electrons

Participants in the Taniguchi International Symposium

October 25–29, 1993 (at Kashikojima, Japan)

Seated (*left to right*)
F.D.M. Haldane, B. Sakita, B.S. Shastry, V.E. Korepin, A. Okiji, J. Kanamori, M. Imada, A. Muramatsu, T.M. Rice, I. Affleck

Second row (*left to right*)
N. Kawakami, D. Yoshioka, K. Ueda, M. Ogata, N. Furukawa, T. Nishino, Y. Kuramoto, K. Vacek, S. Suga

Third row (*left to right*)
E. Hanamura, S. Hikami, N. Nagaosa, H. Fukuyama, S. Sorella

Part I

Quantum $1/r^2$ Models

Physics of the Ideal Semion Gas: Spinons and Quantum Symmetries of the Integrable Haldane-Shastry Spin Chain

F.D.M. Haldane

Department of Physics, Princeton University, Princeton, NJ 08544, USA

Abstract. Various aspects of the Haldane-Shastry spin chain with $1/r^2$ exchange, and its various generalizations, are reviewed, with emphasis on its Yangian quantum group structure, and the interpretation of the model as the generalization of an ideal gas (of "spinons") to the case of *fractional statistics*. Some recent results on its dynamical correlation function are discussed, and conjectured extensions of these remarkably simple results to the $SU(n)$ model, and to the related Calogero-Sutherland model with integer coupling constant, are presented.

1. Introduction

Most of our understanding of quantum many-body physics is based on solvable models – the ideal Bose and Fermi gases, and harmonic oscillators. These are the paradigms, solvable in full detail, and treatments of more complicated, interacting systems usually aim to find a description of the system that is close to one of these paradigms. There are very few other fully-solvable models; the only general class of non-trivial solvable quantum models are the integrable one-dimensional models that can be traced back to the $S = 1/2$ Heisenberg chain solved by Bethe [1] in 1931. Starting with Bethe, there has been success in calculating the energy eigenvalues and thermodynamics of these models, and the algebraic structures that make them solvable have been identified as the quantum Yang-Baxter equation and "quantum groups", but progress in explicit calculation of their correlation functions has so far been very limited.

Bethe's model and its generalizations generally involve *contact* or *delta-function* interactions. A different class of integrable models based on *inverse-square* interactions was introduced around 1970 by Calogero [2] and Sutherland [3]. However, this class of models involved only gases of spinless impenetrable particles, and did not have the richness and lattice generalizations of the Bethe family. On the other hand, certain (static) correlations functions were explicitly found [3]. A few years ago, spin-chain relatives of these models were discovered (the "Haldane-Shastry" model [4, 5]), and it was subsequently found how to put internal spin degrees of freedom [6, 7, 8, 9] into the Calogero-Sutherland model (CSM).

In this paper, I will focus on the inverse-square-exchange spin-chain, and its *trigonometric* ($1/\sin^2$) variants [4, 5], mentioning in passing also its *hyperbolic* ($1/\sinh^2$) and *elliptic* ($1/\mathrm{sn}^2$) variants [10], which provide a continuous interpolation to Bethe's model. The trigonometric models are remarkable because they are explicitly solvable in much greater detail than Bethe's model, and share many of the characteristics of the ideal gas, but with non-standard or *fractional* statistics, which can occur in spatial dimensions below three. In particular, it seems that the action of the physically-relevant local operators (such as the spin operator on a given site of the spin chain) on their ground states excites

Springer Series in Solid-State Sciences, Vol. 118
Correlation Effects in Low-Dimensional Electron Systems
Editors: A. Okiji and N. Kawakami

only a *finite* number of elementary excitations. This parallels the action of a one-particle operator on an ideal gas ground state, and greatly simplifies the calculation of correlation functions in terms of the "form factors", the matrix elements of local operators between the ground state of the system and eigenstates characterized by finite numbers of elementary excitations.

In the *rational* limit (pure $1/r^2$ interactions), it has proved possible to obtain simple explicit closed-form expressions for the *thermodynamic potentials* [11, 12], and very recently, for *ground-state dynamical correlation functions* [13], which are again simple, but non-trivial. I believe that these models will finally be solved in full detail, approaching that with which the ideal gas can be solved, including, for example, full correlation functions at all finite temperatures. This should provide the first fully-developed extension to the standard paradigms.

In this presentation, Section (2) will provide a self-contained introduction to the Yangian "quantum group" [14, 15, 16] symmetries of these models; I will mainly restrict the discussion to the $S = 1/2$ spin-chain models with $SU(2)$ symmetry, and try to avoid the more esoteric mathematical characterizations (Hopf algebras, coproducts, *etc.*), disguising them in a more pedestrian physicists' terminology. Section (3) describes the application of this to the eigenspectrum of the trigonometric models, and Section (4) describes the recent remarkably simple results [13] for some dynamical correlations of the $S = 1/2$ chain (and the $\lambda = 2$ CSM), and presents new conjectured (and certainly correct) generalizations to the $SU(n)$ version of the chain, and to the CSM with arbitrary integer coupling constant λ.

2. The Yangian "Quantum Group" and Integrable Heisenberg Chains

In this section, I will present a self-contained outline of the Yangian "quantum group" algebra, and its application to the Haldane-Shastry model (HSM) and its variants, pointing out a number of open questions. The HSM contains a novel realization of the Yangian algebra, originally characterized by Drinfeld [14] in the context of the algebraic Bethe Ansatz [17], but a mathematical structure in its own right. I will not give a comprehensive history of this elegant mathematical edifice; a concise account by Kirillov and Reshetikin [18] references many of the original works. An account of the representation theory of the Yangian is given by Chari and Pressley [19].

The original integrable model is the $S = 1/2$ Heisenberg chain, solved by Bethe in 1931 [1]:

$$H = \sum_i P_{i\,i+1}, \quad P_{ij} = \tfrac{1}{2} + 2\vec{S}_i \cdot \vec{S}_j. \tag{2.1}$$

The integrability of Bethe's model derives from an underlying *"quantum group"* algebra called the $Y(gl_2)$ *Yangian* [14]: let $\mathbf{T}(u)$ be a 2×2 matrix with non-commuting operator-valued entries that depend on a "spectral parameter" u, act in the Hilbert space of the spin chain, and obey the algebra

$$\mathbf{R}^{12}\mathbf{T}^1\mathbf{T}^2 = \mathbf{T}^2\mathbf{T}^1\mathbf{R}^{12}, \tag{2.2}$$

where $\mathbf{T}^1$ is the 4×4 operator-valued matrix $(\mathbf{T}(u_1)\otimes\mathbf{1})$, $\mathbf{T}^2$ is $(\mathbf{1}\otimes\mathbf{T}(u_2))$, and $\mathbf{R}^{12}(u_1, u_2)$ is a 4×4 c-number matrix defined in the direct product space $V^1 \otimes V^2$ of two 2×2 c-number matrices. Consistency of the algebra requires that $\mathbf{R}$ satisfies the *quantum Yang-Baxter equation* (QYBE) $\mathbf{R}^{12}\mathbf{R}^{13}\mathbf{R}^{23} = \mathbf{R}^{23}\mathbf{R}^{13}\mathbf{R}^{12}$ in $V_1 \otimes V_2 \otimes V_3$. The Yangian $Y(gl_2)$ corresponds to the *rational* QYBE solution

$$\mathbf{R}^{12} = (u_1 - u_2)\mathbf{1} + h\mathbf{P}^{12}; \tag{2.3}$$

here $\mathbf{P}^{12}$ is the exchange matrix where (for c-number matrices) $\mathbf{P}^{12}(\mathbf{A}\otimes\mathbf{B})\mathbf{P}^{12} = (\mathbf{B}\otimes\mathbf{A})$, and h is the *quantum parameter*, which in this case sets the scale of the spectral parameter.

The Yangian algebra can be written as the commutation relation

$$(u - v)[T^{\alpha\beta}(u), T^{\gamma\delta}(v)] = h(T^{\gamma\beta}(v)T^{\alpha\delta}(u) - T^{\gamma\beta}(u)T^{\alpha\delta}(v)). \tag{2.4}$$

It is commutative in the "classical" limit $h \to 0$. The algebra implies that

$$[t(u), t(v)] = 0, \quad t(u) \equiv \mathrm{Tr}[\mathbf{T}(u)], \tag{2.5}$$

so the $t(u)$ are an infinite set of commuting operators. Another consequence is that $[\mathbf{T}(u), \mathrm{Det_Q}|T(v)|] = 0$, where $\mathrm{Det_Q}|T(v)|$ is the *quantum determinant* [20]

$$\mathrm{Det_Q}|T(u)| = T^{11}(u)T^{22}(u-h) - T^{12}(u)T^{21}(u-h). \tag{2.6}$$

The quantum determinant commutes with all elements of the algebra and is analogous to a Casimir operator. In the "classical limit" $h \to 0$, the quantum determinant reduces to the usual determinant of c-number-matrices.

It is consistent to impose the asymptotic condition $\mathbf{T}(u) \to \mathbf{1}$ for $u \to \infty$. An asymptotic expansion can then be defined:

$$\mathbf{T}(u) = \phi(u)\left(\mathbf{1} + \frac{h}{u}\left(\vec{J}_0\cdot\vec{\sigma} + \sum_{n=1}^{\infty}\frac{(\vec{J}_n\cdot\vec{\sigma} + J_n^0\mathbf{1})}{u^n}\right)\right); \quad \phi(u) = \left(1 + \sum_{n=1}^{\infty}\frac{a_n}{u^n}\right), \tag{2.7}$$

where $\vec{\sigma}$ are Pauli matrices, and $\{a_n\}$ are the commuting generators of an infinite set of Abelian subalgebras of $Y(gl_2)$: $[a_m, a_n] = [\vec{J}_m, a_n] = [J_m^0, a_n] = 0$; they alone determine the quantum determinant:

$$\mathrm{Det_Q}|T(u)| = \phi(u)\phi(u-h). \tag{2.8}$$

The infinite-dimensional non-Abelian $Y(sl_2)$ Yangian subalgebra, is completely generated by $\vec{J}_0$, the generator of the sl_2 Lie algebra (of $SU(2)$ generators, so $[J_0^a, J_n^b] = i\epsilon^{abc}J_n^c$) and the additional generator $\vec{J}_1$. All other operators can be expressed in terms of these: for example,

$$J_1^0 = \tfrac{1}{2}h\vec{J}_0\cdot\vec{J}_0; \quad J_2^0 = h\vec{J}_0\cdot\vec{J}_1; \quad \dots \quad , \tag{2.9a}$$

$$\vec{J}_2 = -i\vec{J}_1\times\vec{J}_1 + \tfrac{1}{2}h^2(\vec{J}_0\cdot\vec{J}_0)\vec{J}_0; \quad \dots \quad . \tag{2.9b}$$

Note that, as a consequence of the Yang-Baxter relation, $[J_m^0, J_n^0] = 0$; also $[\vec{J}_0, J_n^0] = 0$, but $[\vec{J}_1, J_n^0] \neq 0$.

The requirement that $\mathbf{T}(u)$ obeys the $Y(gl_2)$ algebra imposes consistency conditions on the $Y(sl_2)$ generators; the first non-trivial condition is the *Serre relation*

$$[J_1^a, J_2^b] = [J_2^a, J_1^b]. \tag{2.10}$$

In fact, this relation is sufficient to ensure that $\vec{J}_1$ generates $Y(sl_2)$, and can be used to compute the value of h^2; either sign of $h = \pm\sqrt{h^2}$ can be consistently chosen. If the "classical limit" $h \to 0$ is taken, the $\vec{J}_n$ obey the infinite-dimensional Lie algebra $\widehat{sl}_{2+}$:

$$[J_m^a, J_n^b] = i\epsilon^{abc}J_{m+n}^c, \quad m, n \geq 0. \tag{2.11}$$

(A familiar example of this algebra is the subalgebra of any $SU(2)$ Kac-Moody algebra defined by its non-negative modes).

The *fundamental representation* of $Y(gl_2)$ (called the "evaluation" in the mathematical literature [14, 19]) is

$$\mathrm{T}_1(u) = 1 + \frac{h\mathrm{P}_1}{u}; \quad \mathrm{P}_1 = \tfrac{1}{2}1 + \vec{\sigma}\cdot\vec{S}_1 , \tag{2.12}$$

where $\vec{S}_1$ is a $S = 1/2$ spin. In this irreducible representation, $\vec{J}_1 = 0$, and the quantum determinant is $(u+h)/u$.

If $\mathrm{T}(u)$ is a representation of $Y(gl_2)$, so is $f(u)\mathrm{T}(u-a)$, where $f(u)$ is a c-number function ($f(u) \to 1$ as $u \to \infty$) and a is a spectral parameter shift. The corresponding change in $Y(sl_2)$ is $\vec{J}_1 \to \vec{J}_1 + a\vec{J}_0$.

Like Lie algebras, "quantum group" algebras have a fundamental property that allows larger representations to be constructed from tensor-products of smaller representations (the "coproduct"); however, unlike Lie algebras, the tensor-product operation is "non-cocommutative", which means that the result of a tensor product depends on the *order* in which it is carried out. In the case of $Y(gl_2)$, if $\mathrm{T}_1(u)$ is a representation acting on a Hilbert space $\mathcal{H}_1$ and $\mathrm{T}_2(u)$ is another representation acting on another Hilbert space $\mathcal{H}_2$, then the matrix product

$$\mathrm{T}(u) = \mathrm{T}_1(u)\mathrm{T}_2(u) \tag{2.13}$$

is also a representation, acting on a Hilbert space $\mathcal{H}_1 \otimes \mathcal{H}_2$, and

$$\mathrm{Det_Q}|T(u)| = (\mathrm{Det_Q}|T_1(u)|)(\mathrm{Det_Q}|T_2(u)|). \tag{2.14}$$

For the $Y(sl_2)$ subalgebra, the tensor product representation is

$$\vec{J}_0 = \vec{J}_0(1) + \vec{J}_0(2) \tag{2.15a}$$

(the usual "law of addition of angular momentum") and

$$\vec{J}_1 = \vec{J}_1(1) + \vec{J}_1(2) + ih\vec{J}_0(1) \times \vec{J}_0(2). \tag{2.15b}$$

Note that reversing the ordering of $\vec{J}_0(1)$ and $\vec{J}_0(2)$ defines an alternative tensor product with $h \to -h$; either definition can be used, which is why only h^2 is fixed by the Serre relation. In $Y(gl_2)$ the two alternative tensor products are right and left matrix multiplication.

A (Yangian) *highest-weight state* (YHWS), is a state that, for all u, is annihilated by $T^{21}(u)$, and is an eigenstate of $T^{11}(u)$ and $T^{22}(u)$ with eigenvalues $\phi_1^+(u)$ and $\phi_2^+(u)$, where

$$(\mathrm{Det_Q}|T(u)|)|\mathrm{YHWS}\rangle = \phi_1^+(u)\phi_2^+(u-h)|\mathrm{YHWS}\rangle. \tag{2.16}$$

Every finite-dimensional representation has at least one highest-weight state. It is *irreducible* if there is *only* one YHWS; conversely, an irreducible representation can be derived from each YHWS contained in a reducible representation. To each such YHWS can be associated an invariant subspace of all the states that can be generated from it by successive action of elements of the algebra. In contrast to what happens in the unitary representation theory of Lie groups and their associated Lie algebras, the representation of $Y(gl_2)$ within this invariant subspace may still be reducible, as the subspace may contain *other* YHWS (and their invariant subspaces) which are eigenstates of the quantum determinant with the same eigenvalue.

An irreducible representation containing (only) a given YHWS is obtained by constructing its invariant subspace, and then projecting out the invariant subspaces of all other YHWS contained within it. This is not necessary in the important special case of *full reducibility*, associated with "quantum symmetries", when the invariant subspaces associated with each independent YHWS are always irreducible, and orthogonal to each other. Reducibility without *full* reducibility, where the YHWS $|2\rangle$ is contained within the invariant subspace generated by YHWS $|1\rangle$, means that there are non-vanishing matrix elements of the form $\langle 2|f(\mathbf{T}(u))|1\rangle$, but all matrix elements of the form $\langle 1|f(\mathbf{T}(u))|2\rangle$ vanish.

The representation theory of $Y(sl_2)$ has been described in detail in [19]. The basic theorem is that all finite-dimensional irreducible representations are isomorphic to an irreducible representation derived from the maximal-J_0^z YHWS of some tensor-product of fundamental representations. This means that the eigenvalues of $T^{11}(u)$ and $T^{22}(u)$ acting on a YHWS which generates a finite-dimensional invariant subspace have the form

$$\phi_1^+(u) = f(u)P^+(u+h); \quad \phi_2^+(u) = f(u)P^+(u), \tag{2.17}$$

where $P^+(u)$ is a finite-dimensional polynomial called the *Drinfeld polynomial.* Let

$$P^+(u) = \prod_{n\geq 1}\prod_{i=1}^{N_n}\left(\prod_{\nu=0}^{n-1}\left(u - u_i^{(n)} + \nu h\right)\right), \tag{2.18}$$

be the unique decomposition of the roots of $P^+(u)$ into "strings" of consecutive roots with spacing h, obtained by successively factoring out strings, starting with the longest ones present (this fixes how multiple roots are treated). The dimension and $SU(2)$-representation content of the irreducible representation of $Y(gl_2)$ derived from the YHWS can immediately be identified from this form: it is equivalent to a tensor product of $SU(2)$ representations where $S = n/2$ occurs N_n times, with a total dimension $(2^{N_1})(3^{N_2})\ldots((k+1)^{N_k})$, where the YHWS has $J_0^z = k/2 = \sum_n nN_n/2$. If only 1-strings are present in the Drinfeld polynomial of a YHWS, the invariant subspace generated from it must be irreducible.

The *Algebraic Bethe Ansatz* [17] is the solution to the problem of finding the eigenvectors of the commuting operators $t(u)$ within an irreducible finite-dimensional representation of $Y(gl_2)$: the eigenstates with eigenvalue $t_\nu(u)$ which are $SU(2)$ highest-weight states with $J_0^z = (N/2) - M_\nu$, are given by

$$|\nu\rangle = \left(\prod_{i=1}^{M_\nu} T^{12}(v_i^{(\nu)})\right)|\mathrm{YHWS}\rangle\ , \tag{2.19}$$

where the $M^{(\nu)}$ "rapidities" $\{v_i^{(\nu)}\}$ are solutions of the Bethe Ansatz equations (BAE)

$$\phi_1^+(u)Q_\nu(u-h) + \phi_2^+(u)Q_\nu(u+h) = t_\nu(u)Q_\nu(u) \tag{2.20}$$

and $Q_\nu(u)$ is the polynomial

$$Q_\nu(u) = \prod_{i=1}^{M_\nu}(u - v_i^{(\nu)}). \tag{2.21}$$

The eigenstates of $t(u)$ form an orthogonal basis only if $[(\vec{J}_0\cdot\vec{J}_1),(\vec{J}_0\cdot\vec{J}_1)^\dagger] = 0$.

The Bethe model on a periodic chain of N sites with $S = 1/2$ spins corresponds to a $Y(gl_2)$ representation that is a simple tensor product of fundamental representations with no spectral parameter shifts:

$$\mathbf{T}(u) = \mathbf{T}_1(u)\mathbf{T}_2(u)\ldots\mathbf{T}_N(u). \tag{2.22}$$

It is irreducible, with $\phi_1^+(u) = ((u+h)/u)^N$ and $\phi_2^+(u) = 1$. The limit as $u \to 0$ of $u^N t(u)$ is $h^N \exp iK$, where $\exp iK$ is the lattice translation operator, and we may write

$$u^N t(u) = h^N e^{iK} \exp\left(\sum_{n\geq 1} u^n H_{n+1}\right), \tag{2.23}$$

where $\{H_n\}$ are a set of local Hamiltonians. Then $[H_m, H_n] = [H_m, \vec{J}_0] = [H_m, \exp iK] = 0$, and $H \propto H_2$. The $Y(sl_2)$ generator may be written

$$\vec{J}_1 = ih\sum_{i<j} \vec{S}_i \times \vec{S}_j. \tag{2.24}$$

In the conventional Bethe parameterization, u is the rapidity parameter if the choice $h = i$ is made; in this case $\vec{J}_1$ is Hermitian.

For finite N, $[\exp iK, \vec{J}_1] \neq 0$, and $[H_m, \vec{J}_1] \neq 0$. However, in the *thermodynamic limit* $N \to \infty$, the model acquires a full non-Abelian *quantum symmetry*: the Yangian generator $\vec{J}_1$ now *commutes* with the lattice translation operator, and with all the H_n. The infinite-dimensional representation of the Yangian becomes *fully reducible* into orthogonal subspaces in which the H_n are diagonal. Furthermore, the antiferromagnetic ground state becomes the *unique* state that is singlet under the action of the Yangian. In the thermodynamic limit a spectral parameter "boost" operator B [21], where $[\vec{J}_1, B] = h\vec{J}_0$, is given by

$$B = \sum_i (i + \tfrac{1}{2}) P_{ii+1}. \tag{2.25}$$

This operator, which is incompatible with periodic boundary conditions, also generates the family of commuting Hamiltonians with quantum symmetry from the lattice translation operator: $[B, \exp iK] = hH_2 \exp iK$, $[B, H_n] = nhH_{n+1}$.

The most general representation of the $Y(sl_2)$ Yangian based on a simple tensor product of fundamental representations (including spectral parameter shifts) is

$$\vec{J}_1 = \sum_i \gamma_i \vec{S}_i + \tfrac{1}{2}\sum_{i\neq j} w_{ij}\ \vec{S}_i \times \vec{S}_j\ , \tag{2.26}$$

where $w_{ij} = ih\,\mathrm{sgn}(i-j)$, and the $\{\gamma_i\}$ are arbitrary. It is an instructive exercise to substitute a form of this type into the Serre relation, *without* making any assumptions about w_{ij}, except that it can generally be taken to be odd. The result [15] is that this is a representation of the Yangian generator if

$$w_{ij}w_{jk} + w_{jk}w_{ki} + w_{ki}w_{ij} = h^2, \quad i \neq j \neq k. \tag{2.27}$$

The general solution of this is

$$w_{ij} = ih\left(\frac{z_i + z_j}{z_i - z_j}\right), \tag{2.28}$$

where $\{z_i\}$ are a set of N arbitrary, but distinct, complex parameters. The simple tensor-product representation is recovered in a limit where $|z_i/z_j| \to \infty$ for $i > j$.

Note that $Y(sl_2)$ representations with an invariant subspace where groups of $2S$ consecutive $S = 1/2$ spins remain in the symmetric spin-S state can be obtained from the simple tensor product by suitably choosing the associated spectral parameter shifts γ_i to be a $2S$-string, and this is the basis for the extension of Bethe's model to chains of spins with $S > 1/2$ [22]. However, this requires the stronger property $(w_{ij})^2 + h^2 = 0$, which eliminates more general representations based on (2.26) and (2.28) for such spins.

The $Y(sl_2)$ representation based on (2.28) is evidently derived from a more general class of representations of $Y(gl_2)$ than those obtained by a simple tensor product. Instead, following Ref.[16], they can be obtained as follows. Let $\{\gamma_i\}$ and $\{z_i\}$ be variables that commute with each other, and with the spin variables, and let K_{ij} be a permutation operator that permutes their labels: $K_{ij}\gamma_i = \gamma_j K_{ij}$, *etc.* These permutation operators do *not* affect the spins: $[K_{ij}, \vec{S}_k] = 0$. Now consider the quantities

$$\hat{\gamma}_i = \gamma_i + h \sum_{j(\neq i)}{}' \left(\frac{z_i\theta(j-i) + z_j\theta(i-j)}{z_i - z_j} \right) K_{ij}. \tag{2.29}$$

These variables commute, and obey a "degenerate affine Hecke algebra" [16]:

$$K_{ii+1}\hat{\gamma}_i - \hat{\gamma}_{i+1}K_{ii+1} = h; \quad [\hat{\gamma}_i, \hat{\gamma}_j] = 0. \tag{2.30}$$

(The role of Hecke algebras in this context has been stressed in Ref.[7].) This implies that

$$[K_{ij}, \Delta(u)] = 0, \quad \Delta(u) \equiv \prod_i (u - \hat{\gamma}_i), \tag{2.31}$$

and hence that $\Delta(u)$ is a c-number function of u and the $\{\gamma_i, z_i\}$. Since the $\hat{\gamma}_i$ commute with each other and with the spins, a representation of $Y(gl_2)$ is clearly given by

$$\mathbf{T}(u) = \mathbf{T}_1(u - \hat{\gamma}_1) \ldots \mathbf{T}_N(u - \hat{\gamma}_N), \tag{2.32}$$

where $\Delta(u)\mathbf{T}(u)$ is a polynomial of degree N in u.

Now let Π^+ be a projection operator into the subspace that is fully symmetric under simultaneous permutation of spins and the parameters $\{\gamma_i, z_i\}$, so $K_{ij}\Pi^+ = P_{ij}\Pi^+$. The Hecke algebra guarantees the property

$$\Pi^+\mathbf{T}(u)\Pi^+ = \mathbf{T}(u)\Pi^+. \tag{2.33}$$

This ensures that the totally-symmetric subspace is an invariant subspace of $\mathbf{T}(u)$, and hence that the projection of (2.32) into this subspace is *also* a representation of $Y(sl_2)$. The representation can now be evaluated [16] within the totally-symmetric subspace as

$$\mathbf{T}(u) = \mathbf{1} + h \sum_{ij} \mathbf{P}_i((u - L)^{-1})_{ij}, \tag{2.34a}$$

where L_{ij} are elements of a $N \times N$ "quantum Lax matrix"

$$L_{ij} = \gamma_i\delta_{ij} + h(1 - \delta_{ij})(1 - (z_i/z_j))^{-1}P_{ij}. \tag{2.34b}$$

The 2^N states of this representation are all eigenstates of the quantum determinant with the same eigenvalue $\Delta(u+h)/\Delta(u)$ where $\Delta(u)$ is the polynomial of degree N given by $\det|u - L^0|$ and L^0_{ij} is the eigenvalue of L_{ij} acting on the fully-spin-symmetric YHWS with $J^z_0 = N/2$ and $P_{ij} = 1$. This YHWS has the Drinfeld polynomial $P^+(u) = \Delta(u)$.

For generic values of the $\{\gamma_i\}$ and $\{z_i\}$, the representation is irreducible. We can now ask whether for some special choices of these parameters, the representation becomes fully-reducible as a direct sum of smaller representations, which would imply that there is a non-trivial Hermitian operator H that commutes with both $\vec{J}_0$ and $\vec{J}_1$. We can postulate such an operator to be of the form $H = \sum_{i<j} h_{ij}P_{ij}$, and look for solutions of the condition $[H, \vec{J}_1] = 0$. This is satisfied provided

$$\gamma_i - \gamma_j = 0; \quad h_{ij} \propto (w_{ij}^2 + h^2); \quad \sum_{j(\neq i)}{}' w_{ij}h_{ij} = 0. \tag{2.35}$$

We also require that H is Hermitian (h_{ij} real). We find [15] *two* families of translationally-invariant Hamiltonians with quantum symmetry, where $[\exp iK, \vec{J}_1] = [\exp iK, H] = 0$; the Hamiltonian is even parity under spatial reflection, while $\vec{J}_1$ is odd-parity, with $(\vec{J}_1)^\dagger \propto \vec{J}_1$, which ensures that any reducibility is *full* reducibility. The solutions are $\gamma_i = 0$, and $h_{ij} \propto 1/d(i-j)^2$, where in the *hyperbolic models*, defined on an *infinite* chain,

$$w_{ij} = ih \coth(\kappa(i-j)); \quad d(j) = \kappa^{-1}\sinh(\kappa j) \tag{2.36a}$$

with real $z_{i+1}/z_i = \exp 2\kappa$, which can be either positive (κ real) or negative ($\kappa - i\pi/2$ real). In the limit $\kappa \to \infty$, the hyperbolic model becomes the $N = \infty$ Bethe model; in the limit $\kappa \to 0$, and $ih = 2\kappa$, so $h \to 0$, we get a "classical limit " where $Y(sl_2)$ becomes the infinite-dimensional Lie algebra $\widehat{sl}_{2+}$, and we obtain the *rational model*

$$w_{ij} = (i-j)^{-1}; \quad d(j) = j. \tag{2.36b}$$

The other family of solutions are the *trigonometric models*, obtained from the hyperbolic models by the replacement $\kappa \to i\pi/N$, with

$$w_{ij} = h\cot(\pi(i-j)/N); \quad d(j) = (N/\pi)\sin(\pi j/N), \tag{2.36c}$$

defined on a *finite* periodic chain on N sites, with $d(j)$ being the chord distance between lattice sites equally-spaced on a circle. For the choice of real h, the N roots of $\Delta(u)$ form an N-string along the real axis, centered on the origin. The trigonometric model then has $(L_{ij})^\dagger = L_{ji}$, and hence $(T^{\alpha\beta}(u))^\dagger = T^{\beta\alpha}(u^*)$.

In all these models, $H \propto H_2$ is the first member of a family of commuting constants of the motion $\{H_2, H_3, H_4, \ldots\}$ that commute with $\vec{J}_0$, $\vec{J}_1$, and $\exp iK$. The eigenstates of these operators form irreducible representations of $Y(sl_2)$. For example [10],

$$H_3 \propto \sum_{i \neq j \neq k} \frac{z_i z_j z_k}{z_{ij} z_{jk} z_{kl}} \vec{S}_i \cdot \vec{S}_j \times \vec{S}_k, \quad z_{ij} \equiv z_i - z_j, \tag{2.37}$$

but the systematic construction beyond H_4 [15] has not yet been elucidated. So far, no generalization of the spectral-parameter "boost" operator from the Bethe limit to the hyperbolic and trigonometric models has been found. The origin of the Hamiltonian constants of the motion in the models with quantum symmetry is conceptually rather different from that in the finite-N Bethe model. They must commute with all elements of $\mathbf{T}(u)$, but the only elements of $Y(gl_2)$ with this property derive from the quantum determinant, which is a trivial c-number in these representations. Thus the Hamiltonians with quantum symmetry *cannot be expressed as functions of* $\mathbf{T}(u)$, and are independent objects.

The Yangian symmetry of the *hyperbolic* models is essentially similar to that of the Bethe model: it occurs in the thermodynamic limit, and $\vec{J}_1$ is made Hermitian by the usual BAE choice $h = i$. On the other hand, the realization of Yangian symmetry in the *trigonometric* models is essentially new, as it occurs in a *finite periodic* system, which has discrete, highly-degenerate energy levels that can be classified as finite-dimensional irreducible representations of $Y(sl_2)$. In this case $\vec{J}_1$ is made Hermitian by choosing h *real*. The two variants are only connected via the "classical" *rational* model with $h = 0$.

The consequences of Yangian quantum symmetry are very different in the two cases; there is an analogy to the difference between the non-compact Lorentz group in (1+1)-dimensions and $SU(2)$, which have symmetry algebras corresponding to different real forms of the same complex sl_2 Lie algebra. In the hyperbolic case, the representation theory involves the Bethe "rapidity strings" with spacing $h = i$ in the imaginary rapidity direction, which are related to bound states. In the trigonometric case, the string spacing is in the real direction, and is related to the quantization of the momentum of free particles with periodic boundary conditions in units of $2\pi/N$, and to a generalization of the Fock space structure of such a system.

The trigonometric models, and their rational limit, were independently discovered by Haldane [4] and Shastry [5]. The extension to the hyperbolic model, was found by Inozemtsev [10], who also proposed the integrability of an *elliptic* variant with $h_{ij} \propto \wp(i-j)$, where $\wp(u)$ is the doubly-periodic Weierstrass function with periods N and $i\pi/\kappa$. This is a version of the hyperbolic model, made periodic on a finite chain of N sites, and (like the finite-N Bethe model ($\kappa \to \infty$)) does *not* have quantum symmetry. A proof of integrability is missing, but two odd-parity operators that commute with the Hamiltonian are given in [10], making integrability very plausible. One linear combination of these corresponds to H_3 in the trigonometric and hyperbolic limits, the other, which corresponds to $\vec{J}_0 \cdot \vec{J}_1$, is (where $\zeta(u)$ is the elliptic zeta function)

$$\sum_{i \neq j \neq k} \left(\zeta(i-j) + \zeta(j-k) + \zeta(k-i)\right) \vec{S}_i \times \vec{S}_j \cdot \vec{S}_k. \tag{2.38}$$

I have not been able to find a $Y(sl_2)$ generator $\vec{J}_1$ where $\vec{J}_1 \cdot \vec{J}_0$ gives (2.38). I therefore conjecture that the integrability of this model may involve a yet more general "quantum group" algebra with *two* "quantum parameters" $h = 2i\kappa$ and $h' = 2\pi/N$ (which would be scale parameters for two distinct spectral parameters), which only degenerates to the Yangian when either of them vanishes or becomes infinite (the Bethe limit). This would correspond to a "double quantization" of $\widehat{sl}_{2+}$, or a further "quantization" of $Y(sl_2)$. Indeed, Inozemtsev [10] suggests that two spectral parameters play a role in this model. A better understanding of the elliptic model is clearly needed.

To conclude the formal discussion of algebraic aspects of these models, I note that, as in the case of Bethe's model [23], there is a straightforward extension from the $Y(gl_2)$ quantum group with $SU(2)$ symmetry, to the $Y(gl_n)$ quantum group, where $\mathbf{T}(u)$ is an $n \times n$ matrix, with $SU(n)$ symmetry. A further extension is to the graded or *supersymmetric* generalization, $Y(gl_{m|n})$, with $SU(m|n)$ supersymmetry, where a site can be in one of m states with even fermion number or n states with odd fermion number. Let $c^{\dagger}_{i\alpha}$ create on site i a particle of species α, which may be fermionic or bosonic, and impose the constraint

$$\sum_{\alpha} c^{\dagger}_{i\alpha} c_{i\alpha} = 1, \quad \text{all } i. \tag{2.39}$$

The exchange operator becomes

$$P_{ij} = \sum_{\alpha\beta} c^{\dagger}_{i\alpha} c^{\dagger}_{j\beta} c_{i\beta} c_{j\alpha}, \tag{2.40}$$

and the generators of the $sl_{m|n}$ graded Lie algebra and its Yangian extension can be given as the traceless operator-valued matrices

$$J_0^{\alpha\beta} = \sum_i \left(c^{\dagger}_{i\alpha} c_{i\beta} - (m+n)^{-1} \delta^{\alpha\beta} \right), \tag{2.41a}$$

$$J_1^{\alpha\beta} = \tfrac{1}{2} \sum_{i \neq j, \gamma} w_{ij} c^{\dagger}_{i\alpha} c^{\dagger}_{j\gamma} c_{i\gamma} c_{j\beta}. \tag{2.41b}$$

In particular, the graded trigonometric exchange model with $(m|n) = (1|2)$ corresponds to the "supersymmetric t-J model" variant found by Kuramoto [24].

3. Spectrum of the Trigonometric Haldane-Shastry Model, and its Interpretation as a Generalized Ideal Gas

In this section I will focus on the remarkable properties of the trigonometric model, with its realization of Yangian quantum symmetry in a form compatible with periodic boundary conditions. The Hamiltonian, normalized so the spin-wave velocity is v_s in units where the lattice spacing $a = \hbar = 1$, is

$$H = \frac{\pi v_s}{N^2} \sum_{i<j} \frac{P_{ij}}{\sin^2(\pi(i-j)/N)}. \tag{3.1}$$

Let us first consider the case where the symmetry group is $SU(1|1)$, when the model is equivalent to a lattice gas of free *spinless fermions*, with creation operators $a_i^\dagger = c_{i2}^\dagger c_{i1}$, so

$$P_{ij} = 1 - (a_i^\dagger - a_j^\dagger)(a_i - a_j) = -1 + (a_i - a_j)(a_i^\dagger - a_j^\dagger). \tag{3.2}$$

The energy levels of this model are characterized by a sequence of $N-1$ Bloch-orbital occupation numbers $\{n_1, n_2, \ldots, n_{N-1}\}$, taking values 0 or 1, so

$$E = \sum_{m=1}^{N-1} \epsilon_m (n_m - \tfrac{1}{2}); \quad e^{iK} = \prod_{m=1}^{N-1} e^{ik_m n_m} \tag{3.3}$$

where

$$\epsilon_m = \left(\frac{2\pi v_s}{N^2}\right) m(m-N); \quad k_m = 2\pi m/N. \tag{3.4}$$

All eigenstates have a two-fold degeneracy, corresponding to the two possible occupation states of the translationally-invariant Bloch orbital, which has zero energy. This is the supersymmetry: all states are $sl_{1|1}$ doublets, and there is no non-trivial Yangian extension in this case. The ground state of H is the state with maximum occupancy, $\{n_m\} = \{111\ldots11\}$, and the ground state of $-H$ has minimum occupancy $\{000\ldots00\}$; The replacement $1 \leftrightarrow 0$ in any configuration maps E to $-E$ and K to $\pi(N-1) - K$.

It is a remarkable fact that this set of energy-momentum levels is the *complete* set of levels contained in the general $SU(m|n)$ exchange model, but in the general case they have different and *much larger* Yangian multiplicities. Specializing to the $SU(n|0)$ and $SU(0|n)$ cases, some of these multiplicities are zero, and the energy level is absent. The selection rule for allowed multiplets in the case $SU(n|0)$ is that occupation patterns containing a sequence of n or more consecutive 1's are forbidden; in the $SU(0|n)$ case, n or more consecutive 0's are forbidden. Each allowed sequence corresponds to an irreducible Yangian multiplet. In general, the spectrum of H with $SU(m|n)$ symmetry is the same as that of $-H$ with $SU(n|m)$ symmetry.

The above rule was found empirically by examination of numerical diagonalization results [15], but I now specialize to the simplest case, $SU(2|0) \equiv SU(2)$, where the rule is that occupation patterns with consecutive 1's are forbidden, and make contact with the $Y(sl_2)$ representation theory [19].

Recall that each YHWS labeled by ν has an associated Drinfeld polynomial, $P_\nu^+(u)$, and that $\Delta(u)\mathrm{T}(u)$ is a polynomial, with quantum determinant $\Delta(u-h)\Delta(u+h)$. Thus there is a polynomial $f_\nu(u)$ where $\phi_1^+(u)$, (the eigenvalue of $T^{11}(u)$), has the form $f_\nu(u)P_\nu^+(u+h)$, $\phi_2^+(u) = f_\nu(u)P_\nu^+(u)$, and

$$f_\nu(u)f_\nu(u-h)P_\nu^+(u+h)P_\nu^+(u-h) = \Delta(u+h)\Delta(u-h). \tag{3.5}$$

Since the roots of $\Delta(u)$ are an N-string, it is easily seen that $P_\nu^+(u)$ must be a factor of $\Delta(u)$, so $\Delta(u) = P_\nu^+(u)g_\nu(u)$, where $g_\nu(u)$ is a polynomial with roots at the roots of $\Delta(u)$ *not* contained in $P_\nu^+(u)$, and

$$f_\nu(u)f_\nu(u+h) = g_\nu(u-h)g_\nu(u+h). \tag{3.6a}$$

This is a polynomial equation with the elementary solution

$$g(u) = (u-a)(u-a+h), \quad f(u) = (u-a-h)(u-a+h). \tag{3.6b}$$

Any product of such solutions is a solution. This shows that $g_\nu(u)$ must be a product of 2-strings.

We now have the recipe [16] for constructing the possible Drinfeld polynomials (in fact, there is one YHWS in the spectrum of the trigonometric model corresponding to each allowed Drinfeld polynomial [16]). First partition the N-string of roots of $\Delta(u)$ into 1-strings and 2-strings. Between each consecutive pair of roots place a 1 if they belong to the same 2-string, and 0 otherwise. This gives a binary sequence of length $N-1$, *with the constraint that there are no consecutive* 1*'s*, as found empirically. The 1-strings are the roots of the Drinfeld polynomial; if an extra 0 is added at each end of the binary sequence, the 1-strings are located between each consecutive pair of 0's. From the Yangian representation theory, we conclude (in agreement with the empirical findings [15]) that a sequence of $M+1$ consecutive 0's represents an independent $S = M/2$ degree of freedom.

There is a simple physical interpretation: each root of the Drinfeld polynomial represents the presence of a $S = 1/2$ "spinon" excitation. A m-string of such roots represents m spinons "in the same orbital" with a rule that their spin state must be totally symmetric. If the total number of spinons present is N_{sp}, there are $M = (N - N_{sp})/2$ 1's in the occupation pattern, which serve to separate the $N_{\text{orb}} = M+1$ "orbitals" into which the spinons are distributed. The change in the number of available "orbitals" as the spinon number is changed obeys $\Delta N_{\text{orb}}/\Delta N_{sp} = -1/2$, which allows spinons to be identified as excitations with "semionic" fractional statistics, in between Bose and Fermi statistics [25].

If $\{I_i\}$ are the positions of the M non-zero entries in the length-$(N-1)$ binary sequence $\{n_j\}$, "spinon orbital occupations" $\{n_{i\sigma}\}$, $1 \leq i \leq M+1$, $\sigma = \pm 1/2$, are given by $I_{i+1} - I_i = 2 + n_{i\uparrow} + n_{i\downarrow}$, with $I_0 \equiv -1$, and $I_{M+1} \equiv N+1$. We can then treat the $\{n_{i\sigma}\}$ as independent variables subject only to the constraint

$$N_{sp} \equiv \sum_{i\sigma} n_{i\sigma} = N - 2M. \tag{3.7}$$

Each distinct configuration $\{n_{i\sigma}\}$ satisfying this constraint corresponds to an eigenstate of the system. The energy and momentum can now be expressed in terms of the "spinon orbital occupations": it is convenient to relabel $n_{i\sigma}$, with $1 \leq i \leq M+1$, as $n_{k\sigma}$, where k is a crystal momentum in the range $-k_0 \leq k \leq k_0$, $k_0 = \pi M/N$, then if

$$\epsilon(k) = \frac{v_s}{\pi}(k_0^2 - k^2); \quad V(k) = v_s(k_0 - |k|); \tag{3.8}$$

the spectrum is given by

$$E = E_{MN} + \sum_{k\sigma} \epsilon(k) n_{k\sigma} + \frac{1}{2N} \sum_{k\sigma, k'\sigma'} V(k-k') n_{k\sigma} n_{k'\sigma'}, \tag{3.9a}$$

$$e^{iK} = (-1)^M \prod_{k\sigma} e^{ikn_{k\sigma}}; \quad J_0^z = \sum_{k\sigma} \sigma n_{k\sigma}. \tag{3.9b}$$

Here $E_{MN} = \pi v_s(2N(N^2-1) + 4M(M^2+2) - 3MN^2)/6N^2$. From this spectrum, it is straightforward to derive the thermodynamic potentials [11] in the limit $N \to \infty$, including a Zeeman coupling $-hJ_0^z$.

The spinon orbitals can be parameterized by a *rapidity* x in the range $-1 < x < 1$, (where the spinons in that orbital have velocity $v_s x$), and have mean occupation numbers

$$\bar{n}_\sigma(x) = \exp -\beta[\varepsilon(x) - \sigma h(1 + \mu(x))], \tag{3.10}$$

where $\varepsilon(x) = (\pi v_s/4)(1 - x^2)$ and

$$\frac{\sinh(\beta h \mu(x)/2)}{\sinh(\beta h/2)} = \exp(-\beta \varepsilon(x)). \tag{3.11}$$

This gives an easily-solved quadratic equation for $\exp(\beta h\mu(x)/2)$. The free energy per site is given by

$$-\beta f(\beta,h)=\frac{1}{2}\int_{-1}^{1}dx\,\ln\left(\frac{\sinh(\beta h(1+\mu(x))/2)}{\sinh(\beta h/2)}\right). \tag{3.12}$$

In the absence of a magnetic field, the entropy per site is

$$s(\beta,h=0)=k_B\int_{-1}^{1}dx\left(\ln[2\cosh\beta\varepsilon(x)]-\beta\varepsilon(x)\tanh\beta\varepsilon(x)\right), \tag{3.13}$$

which unexpectedly is *even* in β, so is the same for the ferromagnetic and antiferromagnetic models.

The identification of spinons as ***semions*** is supported when the ***wavefunctions*** of the trigonometric $SU(2)$ model are examined. A class of polynomial wavefunctions of the Calogero-Sutherland model of a non-relativistic gas with $1/\sin^2$ interactions was discovered [4, 5] to also give a class of "fully-spin-polarized spinon gas" [11] wavefunctions of the trigonometric model, and these are now identified with the YHWS states. If Z_i (with $(Z_i)^N=1$) are the complex coordinates of the M sites with $\sigma_i=-1/2$, the wavefunctions have the form

$$\Psi_\nu(\{Z_i\})=\Phi_\nu(\{Z_i\})\Psi_0(\{Z_j\}), \tag{3.14}$$

where $\Phi_\nu(\{Z_i\})$ is a symmetric polynomial with degree N_{sp} in each Z_j, and $\Psi_0(\{Z_j\})$ is

$$\Psi_0(\{Z_i\})=\prod_{j<k}(Z_j-Z_k)^2\prod_j Z_j. \tag{3.15}$$

The is essentially the same as the m=2 Laughlin fractional quantum Hall effect state for bosons [26], and for $N_{sp}=0$, this is the unique Yangian singlet state. The symmetric polynomials Φ_ν are solutions of the eigenvalue equation

$$\left(\sum_j x_j^2\frac{\partial^2}{\partial x_j^2}+\lambda\sum_{j\neq k}\left(\frac{x_j^2}{x_j-x_k}\right)\frac{\partial}{\partial x_j}\right)\Phi^{(\lambda)}(\{x_i\})=\mu\Phi^{(\lambda)}(\{x_i\}), \tag{3.16}$$

with $\lambda=2$. (The solutions are known in the Mathematical Literature as the Jack Polynomials [27]). If $\{I_i\}$ are the positions of the non-zero entries in the binary "occupation number" sequence, the Taylor series expansion of the corresponding YHWS wavefunction has the form

$$\Psi=\sum_{\{m_i\}}C(\{m_i\})\sum_P\left(\prod_i(Z_{P(i)})^{m_i}\right),\quad \{m_i\}\leq\{I_i\} \tag{3.17}$$

where $P(i)$ is a permutation, and where $m_i\leq m_{i+1}$, and $\{m_i\}<\{m_i'\}$ means that $\{m_i\}$ can be reached from $\{m_i'\}$ through a sequence of pairwise "squeezing" operations $m_i\to m_i+1$, $m_j\to m_j-1$, with $m_i<m_j-1$, and m_k unchanged for $k\neq i,j$.

The $N_{sp}=0$ Yangian singlet state may also be written in terms of the azimuthal spin variables σ_i: it occurs only for even N, and is the $n=2$ case of the $SU(n)$ singlet wavefunction where the σ_i can take one of n ordered values $\{\alpha\}$:

$$\Psi_0^{(n)}(\{z_i,\sigma_i\})=\prod_{i<j}(z_i-z_j)^{\delta(\sigma_i,\sigma_j)}(i)^{\mathrm{sgn}(\sigma_i,\sigma_j)}\prod_\alpha\delta(N,nN(\alpha)), \tag{3.18}$$

where $N(\alpha)\equiv\sum_i\delta(\alpha,\sigma_i)$. Here the $\{z_i\}$ are a complete set of the N'th roots of unity. Note that this state is essentially the same as the wavefunction for a filled Landau level of $SU(n)$ fermions, and is hence explicitly $SU(n)$-singlet for *arbitrary* $\{z_i\}$.

The $N_{sp} = 1$ states are also particularly simple: they occur only for odd $N = 2M+1$, and are generated by

$$\Psi(z;\{Z_j\}) = \prod_i (z - Z_i) \prod_{i<j} (Z_i - Z_j)^2 \prod_i Z_i. \tag{3.19}$$

Expanding this in powers of z gives a band of Bloch states with crystal momentum $-\pi/2 < K + M\pi < \pi/2$, so the spinon band covers *half* a Brillouin zone. The localized spinon wavefunction $\Psi(z;\{Z_i\})$ is essentially the quasihole excitation of the $m = 2$ bosonic Laughlin state, and from this perspective, clearly describes a *semionic* excitation. If the parameter z is chosen to be a lattice coordinate z_i, the localized spinon wavefunction with $\sigma_i = \sigma$ can be rewritten in terms of spin variables as

$$\delta(\sigma,\sigma_i)\Psi_0^{(2)}(\{z_j,\sigma_j; j \neq i\}). \tag{3.20}$$

This shows that the spinon is *completely* localized on the lattice site, and induces no spin polarization of its local environment. While there are N such fully-localized spinon states, they are fundamentally non-orthogonal, since the expansion in orthogonal Bloch states shows there are only $(N+1)/2$ independent N_{sp}=1 states.

To end this section, I discuss the extension of the state-counting from $SU(2)$ to the general $SU(n|m)$ case. The Yangian counting rules for the $SU(2)$ model are very simple, and allowed the thermodynamic functions to be explicitly obtained [11] in closed-form in the thermodynamic limit, but become much more complicated in the general $SU(m|n)$ case. Recently Sutherland and Shastry [12] showed how to recover the $SU(2)$ results for the thermodynamics, and generalize them to $SU(m|n)$, from a strong-coupling limit of the exchange-generalization of the Calogero-Sutherland model [6, 7, 8, 9].

The spin chain degrees of freedom are essentially those of the spin-1/2 Fermi gas *with the charge degrees of freedom removed.* From this viewpoint, the spinon is the spin-1/2 fermion with the charge degrees of freedom factored out, leaving "half a fermion", and hence a semion. I will interpret Sutherland and Shastry's result [12] as showing how to "put back" charge degrees of freedom into the spin chain to recover a spectrum with the familiar degeneracies of the ideal gas with internal degrees of freedom. This facilitates the computation [12] of the thermodynamics in the general $SU(m|n)$ case.

Let $b_k^\dagger$, $k = 1,\ldots N-1$, be a set of harmonic oscillator creation operators, and add these degrees of freedom to the spin chain as follows:

$$H' = \left(\frac{v\pi}{N^2}\right)\left(\sum_{i<j}\frac{1-P_{ij}}{\sin^2(\pi(i-j)/N)} + \sum_{k=1}^{N-1} 2k(N-k)b_k^\dagger b_k\right), \tag{3.21a}$$

$$e^{iK'} = e^{iK_s}e^{iK_c}; \quad K_c = \frac{2\pi}{N}\left(\sum_{k=1}^{N-1} k b_k^\dagger b_k\right), \tag{3.21b}$$

where $\exp iK_s$ is the spin-chain translation operator, and P_{ij} is the $SU(m|n)$ exchange operator.

Now let $n_{k\alpha} = c_{k\alpha}^\dagger c_{k\alpha}$ be occupation numbers of a set of orbitals for particles of a bosonic or fermionic species α, with $k = 0, \pm1, \ldots \pm\infty$, subject to the constraint that

$$N = \sum_{k\alpha} n_{k\alpha} \tag{3.22}$$

is fixed. From the result of Sutherland and Shastry, the spectrum of H' (with the constraint (3.22)) is identical to that of

$$H'' = \left(\frac{v\pi}{N^2}\right)\sum_{kk'}\sum_{\alpha\beta}|k-k'|n_{k\alpha}n_{k'\beta}, \tag{3.23a}$$

$$K'' = \frac{2\pi}{N}\sum_{k\alpha} k n_{k\alpha}; \quad J_0^{\alpha\beta} = \sum_k (c^\dagger_{k\alpha}c_{k\beta} - (m+n)^{-1}\delta^{\alpha\beta}). \tag{3.23b}$$

The spectrum of H'' is periodic in $K'' \to K'' + 2\pi$, corresponding to a shift $n_{k\alpha} \to n_{k+1\,\alpha}$ of the occupation number pattern. The precise statement is that the spectrum of H'' in one period of K'' coincides with that of H'. In the thermodynamic limit, the free energy of H' can thus be easily calculated, as can the free energy contribution from the extra oscillator modes; the *difference* is the spin-chain free energy [12]. However, this does *not* give a simple method for identifying the Yangian degeneracies of the *discrete* (finite N) spectrum of H *without* the oscillator modes. The explicit expressions for the free energy are in fact only obtained in the limit $N \to \infty$, when the Yangian algebra has degenerated to its "classical" $\widehat{sl_{m|n}}_+$ limit.

4. Dynamical Correlation functions

Let us consider the state $S_i^+|0\rangle$, where $|0\rangle$ is the Yangian singlet ground state of the trigonometric chain. It is a spin-1 state, and is easily seen to be a linear combination of Yangian highest weight states with $N_{sp} = 2$. The action of S_i^+ is to remove a down-spin coordinate at site i. Thus $M = N/2 - 1$ and

$$\Psi(z_i;\{Z_j\}) = \prod_{j=1}^{M}(z_i - Z_j)^2\Psi_0(\{Z_j\}). \tag{4.1}$$

The polynomial prefactor has degree 2, confirming that this state is composed purely of two-spinon eigenstates. Thus if we wish to compute the dynamical correlation function

$$C(i-j, t-t') = \langle 0|S_i^-(t)S_j^+(t')|0\rangle, \tag{4.2}$$

the only intermediate states that contribute to its spectral function are the N_{sp}=2 YHWS.

Recently, it has become possible to compute certain dynamical correlation functions of the Calogero-Sutherland model [2, 3]

$$H = \sum_{i=1}^{N}\frac{p_i^2}{2m} + \frac{\hbar^2}{m}\sum_{i<j}\frac{\lambda(\lambda-1)}{d(x_i-x_j)^2}, \tag{4.3}$$

with $d(x) = (L/\pi)\sin(\pi x/L)$, in the thermodynamic limit at fixed density $\rho = N/L$. The eigenfunctions of the CSM have the form

$$\Phi^{(\lambda)}(\{Z_i\})\Psi_J(\{Z_i\}); \quad \Psi_J = \prod_{i<j}(Z_i - Z_j)^\lambda \prod_i (Z_i)^J \quad (\lambda \geq 0) \tag{4.4}$$

where $\{\Psi_J\}$ is the family of states that includes the ground state and Galilean boosts of it, $Z_i = \exp 2\pi i x_i/L$, and $\Phi^{(\lambda)}(\{Z_i\})$ is a symmetric polynomial solution of (3.16). While the apparent statistics can be modified with a singular gauge transformation, the "natural" statistics of this model are, from (4.4), evidently *fractional*, with statistical parameter $\theta = \pi\lambda$ (*i.e.*, particles carry charge 1 and flux $\pi\lambda$). The elementary excitations are particles with velocities greater than the speed of sound $v_s = \pi\hbar\lambda\rho/m$, and *holes* with velocities less than v_s. The holes carry charge $-1/\lambda$ and flux $-\pi$, and have statistical parameter $\theta_h = \pi/\lambda$.

The calculations can be carried out at one of three special couplings (only two of which are non-trivial), and involve a mapping to the $N \to \infty$ limit of a Gaussian dynamical $N \times N$ matrix-model, with either orthogonal ($\lambda = 1/2$), unitary ($\lambda = 1$), or symplectic ($\lambda = 2$) symmetry. This reduces the problem to a complicated, but tractable Gaussian problem [28].

At integer couplings $\lambda = q$, the natural particle statistics are Bose (even q) or Fermi (odd q). The ground state is Ψ_J with $J = -q(N-1)/2$. The wavefunction of the state $\Psi(x)|J\rangle$ is

$$\prod_i (z - Z_i)^q \Psi_J(\{Z_i\}), \quad z = \exp(2\pi i x/L), \tag{4.5}$$

which is composed only of eigenstates with just q hole excitations, so the spectral function of the retarded single-particle Greens function will only contains contributions from intermediate states of that type. This correlation function will thus have the form

$$\langle 0|\Psi^\dagger(x,t)\Psi(0,0)|0\rangle = \rho \left(\prod_{i=1}^{q} \int_{-v_s}^{v_s} dv_i\right) |f_q(\{v_i\})|^2 e^{i(Px-Et)}, \tag{4.6a}$$

$$P = \sum_i m_h v_i, \quad E = \sum_i \tfrac{1}{2} m_h v_i^2, \quad m_h = -m/q, \tag{4.6b}$$

where $f_q(\{v_i\})$ is a *form factor* that must be calculated.

Using the mapping to the symplectic matrix model, it has been possible [13] to calculate this form factor for $\lambda = q = 2$. The result is remarkably simple:

$$|f_2(v_1, v_2)|^2 = \frac{1}{8v_s}\left(\frac{(v_1 - v_2)^2}{(v_s^2 - v_1^2)(v_s^2 - v_2^2)}\right)^{\frac{1}{2}}. \tag{4.7}$$

The significance of this in the context of the $SU(2)$ trigonometric Haldane-Shastry chain is that the YHWS wavefunctions with $Z_i = \exp(2\pi i x_i/L)$ are also eigenstates of the $\lambda = 2$ CSM, and the matrix elements of $S_i^+(t)$ between two YHWS are equivalent to those of $\Psi(x_i, t)$ between the corresponding CSM states. *The CSM result can immediately be translated to the results for the spin chain in the rational limit, where the Yangian symmetry algebra becomes classical.*

While the calculation [13] is complicated and indirect, the simplicity of the result suggests there should be a simple derivation, not based on the "accident" of a mapping to a Gaussian matrix model. It is very tempting to *conjecture* the extension of the result to the $SU(n)$ chain, for which no such mapping is known. Examination of the action of the operator $c_{i\alpha}^\dagger c_{i\beta}$ on the singlet ground state of the $SU(n)$ chain leads to the conclusion that it produces a two-parameter family of YHWS states with one spinon (of "color" $\bar{\beta}$) moving with rapidity x_1, and a complex of $n-1$ spinons (with net "color" α) moving together (but not as a bound state) with rapidity x_2. In the singlet ground state of the rational $SU(n|0)$ chain state, the correlation function $\langle 0|X_j^{\alpha\beta}(t)X_0^{\gamma\delta}(0)|0\rangle$, where $X_i^{\alpha\beta} \equiv c_{i\alpha}^\dagger c_{i\beta}$, will thus have the form

$$\frac{(-1)^j}{n}\delta^{\alpha\delta}\delta^{\gamma\beta}C(j,t) + \frac{(-1)^j}{n^2}\delta^{\alpha\beta}\delta^{\gamma\delta}(1 - C(j,t)), \tag{4.8}$$

where

$$C(m,t) = \frac{1}{4}\int_{-1}^{1} dx_1 \int_{-1}^{1} dx_2\, |F_n(x_1,x_2)|^2 \left(e^{i(q(x_1)m - \epsilon(x_1)t)}\right)\left(e^{i(q(x_2)m - \epsilon(x_2)t)}\right)^{n-1}, \tag{4.9a}$$

$$q(x) = \frac{\pi x}{n}; \quad \epsilon(x) = \frac{\pi v_s}{2n}(1 - x^2). \tag{4.9b}$$

Here $F_n(x_1, x_2)$ is the form factor that must be found.

The asymptotic form of the correlations are straightforward to compute from bosonization (they are free-fermion correlations with the factor coming from charge degrees of freedom divided out), or from conformal field theory [29], and can be fit to a simple *Ansatz* based on (4.7). I therefore present the *conjecture* for the form factor of the rational $SU(n|0)$ chain (which is a rigorous result for n=2) :

$$|F_n(x_1,x_2)|^2 = A_n \left(\frac{4(x_1-x_2)^2}{(1-x_1^2)(1-x_2^2)}\right)^{1/n}, \quad A_n = \frac{\left(\frac{n-1}{n}\right)\Gamma\left(\frac{n+1}{n}\right)}{\Gamma\left(\frac{n-1}{n}\right)\Gamma\left(\frac{n+2}{n}\right)}. \tag{4.10}$$

It has been verified [30] that this form fits the numerically computed static structure factor of the trigonometric $SU(3)$ chain with $N \leq 18$ extremely well, with very small finite-size corrections, leaving no doubt that this conjecture for $n > 2$ is indeed correct.

As another test of the reliability of generalizing (4.7) "by conjecture", based on its remarkable simplicity, the same type of arguments can be used to obtain the form factor for the retarded Greens function (4.6) of the CSM at a general integer coupling $\lambda = q$. In this case, the states contributing to the spectral function have q hole excitations with independent velocities. The resulting conjecture is

$$|f_q(\{v_i\})|^2 = \frac{1}{2v_s} B_q \left(\prod_{i=1}^{q}(v_s^2 - v_i^2)\right)^{-1+1/q} \left(\prod_{i<j}(v_i - v_j)^2\right)^{1/q}, \tag{4.11a}$$

where the multi-dimensional integral fixing the normalization B_q is hard to do. I have just learned of recent work by Forrester [31], who calculates the equal-time limit of this correlation function using quite different methods based on the celebrated *Selberg trace formula.* His expression *is precisely the equal-time limit of the formula conjectured here!* Moreover, Forrester's result provides the normalization as

$$B_q = \prod_{j=1}^{q} \left(\frac{\Gamma\left(\frac{1+q}{q}\right)}{\left(\Gamma\left(\frac{j}{q}\right)\right)^2}\right). \tag{4.11b}$$

It is striking that the form factors are essentially Laughlin-type wavefunctions for *anyons* with the same statistics as the holes, now with the *rapidities* as coordinates.

The remarkable property of the rational and trigonometric models is that the *local* operators such as the spin on a given site act on the ground state to produce only a very restricted class of excitations. There is a general selection rule, verified empirically [32], that the local spin operator $\vec{S}_i$ acting on any eigenstate cannot change the spinon number N_{sp} by more than ± 2. Such properties are very reminiscent of an ideal gas, and the most natural interpretation of the trigonometric and rational models is as generalizations of the ideal gas Fock-space structure to non-trivial statistics.

5. Conclusion

I have attempted to present, from my perspective, the main results associated with the still-unfolding properties of the Haldane-Shastry spin chain model, stressing the simplest $SU(2)$ or $S = 1/2$ model. I have clearly omitted many aspects, and there is clearly much more yet to emerge.

A direct algebraic treatment in the rational limit, when the "quantum group" becomes classical, would be particularly desirable. It can be no accident that this is the limit in which all the explicit results are obtained, but (disappointingly) so far a direct use of its algebraic properties such as the infinite $\widehat{sl_2}_+$ symmetry not yet been made. An algebraic construction of the highest weight eigenstates (fully spin-polarized spinon gas states) in terms of particle creation operators ("vertex operators") acting on the vacuum is also needed; this would closely parallel a similar treatment needed for the Calogero-Sutherland model.

Other possible lines of investigation are whether a "quantum deformation" $\widehat{sl}_{2+} \to Y(sl_2)$ can be used to calculate the hyperbolic model form factors, and what is the origin of integrability of the elliptic models. There is clearly much more work to be done!

I thank B. S. Shastry for bringing Ref.[31] to my attention, and wish to express my appreciation to the Taniguchi Foundation for its vision in making Symposia such as this one possible.

References

[1] H. A. Bethe, Z. Phys. **71**, 205 (1931).

[2] F. Calogero, J. Math. Phys. **10**, 2191 (1969).

[3] B. Sutherland, J. Math. Phys. **12**, 246 (1971); **12**, 251 (1971).

[4] F. D. M. Haldane, Phys. Rev. Lett. **60**, 635 (1988).

[5] B. S. Shastry, Phys. Rev. Lett. **60**, 639 (1988).

[6] A. P. Polychronakos, Phys. Rev. Lett. **69**, 703 (1992).

[7] I. Cherednik, Adv. Math. (to be published) [preprint RIMS-776, Research Institute for Mathematical Sciences, Kyoto University, (1991)].

[8] Z. N. C. Ha and F. D. M. Haldane, Phys. Rev. B **46**, 9359 (1992).

[9] K. Hikami and M. Wadati, J. Phys. Soc. Japan **62**, 469 (1993).

[10] V. I. Inozemtsev, J. Stat. Phys, **59**, 1143 (1990).

[11] F. D. M. Haldane, Phys. Rev. Lett. **66**, 1529 (1991).

[12] B. Sutherland and B. S. Shastry, Phys. Rev. Lett. **71**, 5 (1993).

[13] F. D. M. Haldane and M. R. Zirnbauer, Phys. Rev. Lett. **71**, 4055 (1993).

[14] V. G. Drinfel'd, Dokl. Acad. Nauk. USSR **283**, 1060 (1985) (Sov. Math. Dokl. **32**, 254 (1985)); "Quantum Groups" (Proc. ICM. Berkeley) p. 798 (1987).

[15] F. D. M. Haldane, Z. N. C. Ha, J. C. Talstra, D. Bernard, and V. Pasquier, Phys. Rev. Lett. **69**, 2021 (1992).

[16] D. Bernard, M. Gaudin, F. D. M. Haldane, and V. Pasquier, J. Phys. A:Math. Gen. **26**, 5219 (1993).

[17] L. D. Faddeev, in "Integrable Models in 1+1-Dimensional Quantum Field Theory", in *Les Houches Lectures 1982*, Elsevier, Amsterdam, 1984.

[18] A. N. Kirillov and N. Yu. Reshetikhin, Lett. Math. Phys. **12**, 199 (1986).

[19] V. Chari and A. Pressley, L'Enseignement Math. **36**, 267 (1990); J. Reine Angew. Math. **417**, 87 (1991).

[20] A. G. Izergin and V. E. Korepin, Sov. Phys. Dokl. **26**, 653 (1981).

[21] M. G. Tetel'man, Sov. Phys. JETP **55**, 306 (1982).

[22] L. Taktajan, Phys. Lett. **87A**, 479 (1982); J. Babudjian, Phys. Lett. **90A**, 479 (1982).

[23] B. Sutherland, Phys. Rev. B **12**, 37 (1975).

[24] Y. Kuramoto and H. Yokoyama, Phys. Rev. Lett. **67**, 1338 (1991).

[25] F. D. M. Haldane, Phys. Rev. Lett. **67**, 937 (1991).

[26] V. Kalmeyer and R. B. Laughlin, Phys. Rev. Lett. **59**, 2095 (1987).

[27] R. P. Stanley, Adv. Math. **77**, 76 (1989).

[28] B. D. Simons, P. A. Lee, and B. L. Altshuler, Phys. Rev. Lett. **70**, 4122 (1993); Phys. Rev. Lett. **72**, 64 (1994).

[29] N. Kawakami, Phys. Rev. B **46**, 3191 (1992).

[30] F. D. M. Haldane and Z. N. C. Ha, unpublished.

[31] P. J. Forrester, Phys. Lett. A **179**, 127 (1993).

[32] J. C. Talstra and F. D. M. Haldane, unpublished.

The $1/r^2$ Integrable System: The Universal Hamiltonian for Quantum Chaos

B.S. Shastry

A.T.&T. Bell Laboratories, 600 Mountain Avenue, Murray Hill, NJ 07974, USA

Abstract. We summarize recent work showing that the $1/r^2$ model of interacting particles in 1-dimension is a universal Hamiltonian for quantum chaotic systems. The problem is analyzed in terms of random matrices and of the evolution of their eigenvalues under changes of parameters. The robustness of bulk space-time correlations of a many particle system to changing boundary conditions is suggested to be at the root of the universality. The explicit density-density correlation functions of the $1/r^2$ model, now available through the above mapping at two values of the coupling constant, are interpreted in the light of Bethe's *Ansatz*, giving a vivid picture of the fractionalization of bare particles or holes into "quark" like Bethe quasi-particles and holes.

1. Introduction

There has been a new and fascinating development recently, spearheaded by results from Altshuler, Simons and coworkers [1], which is of considerable interest to problems of quantum chaos as well as exactly integrable many body systems. A new type of universality has been pointed out to be operative in many quantum chaotic systems, involving energy levels and their variation as some parameter is changed. New and concrete results for a certain correlation function of two variables has been computed by the use of the supersymmetry method of Efetov [2], and it has been conjectured that these correlation functions are the space time correlation functions of density fluctuators in a very well known and well studied many body problem, the $1/r^2$ model introduced by Sutherland [3]and Calogero [4] in the early 70's, and indeed even earlier in a relaxational framework by Dyson during the course of his seminal work on random matrices [5]. This amazing connection between disordered systems (typifying quantum chaotic systems) on the one hand, and exactly integrable systems on the other, is the subject of this article. We have now a route map that shows why such a mapping is appropriate [6, 7] and it is my purpose to explicate the same. We will go over the mapping in two different ways here, one [6] quite elegant following the work of Dyson, and the other more mundane but perhaps more instructive.

Let us consider the following game which typifies the entire problem, let us take two matrices , say real symmetric matrices H_0 and V in an N dimensional space and mix them parametrically

$$H(\phi) = \cos(\Omega\phi)H_0 + \frac{\sin(\Omega\phi)}{\Omega}V. \tag{1}$$

The motivation for this particular form of parametrization will become transparent later, let us just regard Ω as some number of $O(1/\sqrt{N})$ for now, and further H_0 as a diagonal matrix with some given eigenvalues of $O(N)$, and V an off diagonal generic matrix with matrix elements of $O(1)$. The matrix $H(\phi)$ then is characterized by the parameter ϕ and if we should rotate the matrix by any orthogonal transformation, then the typical matrix element $H_{i,j} \sim O(\sqrt{N})$, giving us eigenvalues that are always $O(N)$, and hence their separation of

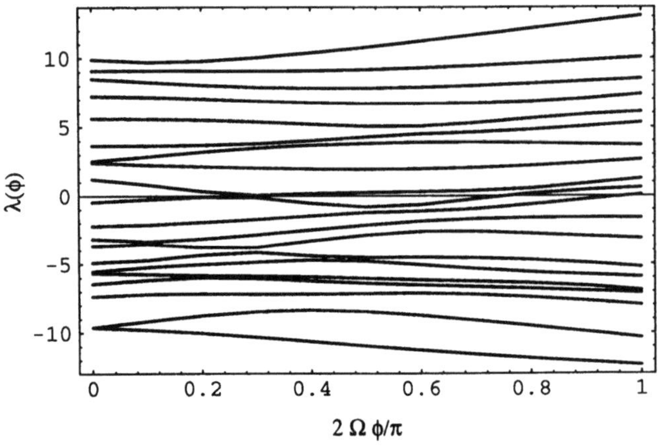

Fig.1 A typical spaghetti of eigenvalues for the mixing problem of Eq(1), where we take a pair of 20×20 matrices and evaluate their eigenvalues as ϕ varies.

$O(1)$. We can ask, how do the eigenvalues vary as we vary the parameter ϕ, Fig.1 shows the result of a typical run with a 20×20 matrix. The eigenvalues never cross: they come close together and then move off; this is the familiar level repulsion at work. What is more, the picture suggests, after rotating by $\pi/2$, that we may view these spaghetti like lines as the world lines of interacting particles, with ϕ a time-like coordinate and λ a space like coordinate. This view happens to be completely correct, and will ultimately be established here. Our choice of the scales was judicious, but if one had a different set of scales for the various parameters, we could simultaneously rescale the total H and the ϕ to get a similar spectrum.

The universality of Altshuler and Simons [1] is obtained in a slight variant of the above obtained by taking the limit $\Omega \to 0$, and that of large N giving a thermodynamic limit and some hope of ergodicity, enabling one to define an average by taking the mean over several energy levels. The next step involves the choice of the matrix V, one can always remove the mean of V, and further it is suggested that the variance of V can be adjusted to be unity by rescaling the parameter ϕ. The independence of the final correlations to the various choices of V and H_0, and indeed the size of the Hilbert space have been checked with various examples [1]. Let us recall that universality of the Random Matrix ensembles is a statement of the universality of the "time independent", or "static" correlations in the above sense [8]. The connection between the static correlations and the ground state correlation functions of the $1/r^2$ many body models are of course well known, and date back to the work of Sutherland [3], who pointed out that the three ensembles, Orthogonal Unitary and Symplectic correspond to just three values of the coupling constant λ namely $\lambda = 1/2, 1, 2$, or equivalently $\beta \equiv 2\lambda = 1, 2, 4$, in the $1/r^2$ model Hamiltonian

$$H_S = -\sum_i \partial^2/\partial x_i^2 + 2\lambda(\lambda - 1) \sum_{i<j} \frac{(\pi/L)^2}{\sin^2(\pi(x_i - x_j)/L)}. \tag{2}$$

The work of Altshuler and Simons [1] then implies that the above model is a Universal Hamiltonian for Quantum chaos in the sense that its time dependent density density correlation functions are also universal!

The plan of this article is as follows, we will recall the random matrix ensembles in Section 2, and focus on the Orthogonal ensemble, where will see the workings of the essentially rigorous mapping [6] between the eigenvalue motion problem and the Sutherland Calogero model in the thermodynamic limit. In Section 3, we will discuss a more elementary proof of the equivalence using simpler tools. This derivation gives a broader equivalence for non-gaussian V's for a specific initial condition providing us with the idea that universality can be viewed as the independence of bulk correlations in a many body system to the boundary effects. In Sections 4.1 and 4.2, we will discuss the results of the calculations for the correlations within the context of the exact solution of the model by Bethe's Ansatz [9], and provide a novel interpretation of the correlation functions as composite objects, made out of Bethe quasiparticles. Section 4.1 summarizes the results of the Bethe's Ansatz for the $1/r^2$ model giving *all* the eigenvalues of the model, and Section 4.2 expresses the calculated correlation functions for the Orthogonal and Symplectic ensembles in terms of the Bethe spectrum. This is a kind of ' quark" spectroscopy in that the bare electrons of the $1/r^2$ system are found to breakup into more basic constituents, "quarks" obeying the exclusion principle of Pauli.

2. The Mapping for Gaussian V

Let us now consider the problem[6] posed by the mixing problem of Eq(1). We are interested in considering the eigenvalues of the matrix H as functionals of the matrix V at a given ϕ. We will consider the matrix V to be a Gaussian, i.e. (suppressing an overall normalization factor everywhere in the article)

$$p(V)\,dV = \exp(-\frac{1}{2}Tr(V^2)) \prod_{i\leq j} dV_{i,j}. \tag{3}$$

Therefore the eigenvalues $\mu \equiv \{\mu_1, \mu_2, \mu_3, \ldots\}$ of H are now also random variables with a certain distribution $P(\mu, \phi|\mu_0)$, and μ_0 the initial eigenvalues, i.e. those of H_0 which may be chosen diagonal without any loss of generality. We are interested in the evolution of P with ϕ, clearly at $\phi = \pi/(2\Omega)$, the matrix $H = \frac{1}{\Omega}V$, and the eigenvalues distribution must equal that of V apart from a simple rescaling. Let us briefly recall that this is easily obtained for any distribution (i.e. Gaussian or otherwise) through the methods of invariant integration in group theory [10]; we summarize the main results below in a small digression. We consider the resolution of identity for an arbitrary real symmetric matrix M,

$$1 = \frac{1}{N!}\int_{-\infty}^{\infty}\cdots\int_{-\infty}^{\infty}\prod_i d\lambda_i \int dO\ \delta_{Weyl}(O\lambda_d O^T - M)\prod_{i<j}|\lambda_i - \lambda_j| \tag{4}$$

where $\lambda_d = \{\lambda_1, \ldots, \lambda_N\}_{diag}$ is a diagonal matrix, and the Weyl delta function [12] for a real symmetric matrix is a product over the number of distinct elements $\nu_G = N + N(N-1)/2$ of the usual δ functions: $\prod_{i\leq j}\delta(M_{i,j})$. The group integration($\int dO \times 1 \equiv 1$) is over the manifold of $N(N-1)/2$ parameters of the Orthogonal group. This identity is easy to establish by realizing that only the proximity of the diagonalizing orthogonal matrix O^* contributes to the group integral, and the λ's are forced to be the eigenvalues. Thus we can write $R_{i,j} \equiv (O^T O^*)_{i,j} = \delta_{i,j} + \epsilon A_{i,j} + O(\epsilon^2)$ whereby the group integration can be shifted to an integration over the antisymmetric matrix $A_{i,j}$, this matrix has precisely the correct number of independent matrix elements. The the delta function can be rotated to $\delta_{Weyl}(\lambda_d - RMR^T)$, which reduces to $\prod_{i<j}\delta(A_{i,j}(\lambda_i - \lambda_j))$ and hence the identity. Note that a similar identity also holds for the Unitary and Symplectic ensembles. In the Unitary case, the matrix M is an arbitrary complex but hermitean matrix and the group integration is over the manifold $[U]$, consisting of the Unitary group $U(N)$ with its N parameter abelian subgroup of diagonal matrices identified with a single element [10], and thus with $N^2 - N$ parameters. Clearly the total number of variables in the integration are N^2, equalling the number of independent

elements of M. For the Symplectic ensemble the integration is over the $2N^2 - N$ parameters of the manifold $[USp(N)]$ consisting of the Unitary Symplectic group of $2N \times 2N$ dimensional matrices, with its $2N$ dimensional abelian subgroup of diagonal matrices identified with a single element. The number $\nu_G = N + \beta N(N-1)/2$ for the various cases with $\beta = 1,2,4$ corresponding to the Orthogonal, Unitary and Symplectic ensembles respectively. With this little digression, we see that given the probability distribution function of any matrix ensemble in the form $p(M) = f(Tr(M^m))$, with an appropriate arbitrary positive function f, we can immediately infer the eigenvalue distribution of M by convoluting with identity and integrating out the Orthogonal group, yielding the familiar result

$$p(\{\lambda_1, \ldots, \lambda_N\}) = \prod_{i<j} |\lambda_i - \lambda_j| f(\sum_i \lambda_i^m). \tag{5}$$

We now make the elementary observation that the gaussian distribution Eq(3) for the matrix V implies for the matrix H of Eq(1) the distribution

$$P(H, \phi | H_0) = (\frac{\Omega}{\sin(\Omega\phi)})^{\nu_G} \exp(-\frac{\Omega^2}{2\sin^2(\Omega\phi)} Tr(H - \cos(\Omega\phi)\, H_0)^2). \tag{6}$$

In writing out the trace explicitly, we extensively use the functions and notation introduced by Dyson [5], $g_{i,j} = 1 + \delta_{i,j}$ and further combine the pair of indices i,j into one single index $\mu \equiv (i,j)$, such that $1 \le \mu \le \nu_G$. The noteworthy feature of the Gaussian is the uncoupling of different $\mu's$, and we can write

$$P(H, \phi | H_0) = \prod_\mu [\frac{\Omega}{\sqrt{1-q^2}} \exp(-\frac{\Omega^2 (H_\mu - q H_{\mu_0})^2}{g_\mu (1-q^2)})], \tag{7}$$

with $q = \cos(\Omega\phi)$, and recognize immediately that this is the solution of a Fokker Planck (F-P) equation for the uncoupled components H_μ

$$\partial P/\partial t = \sum_\mu \frac{\partial}{\partial H_\mu} [\frac{g_\mu}{2} \frac{\partial}{\partial H_\mu} + \Omega^2\, H_\mu] P, \tag{8}$$

with the mapping $t = -\log(\cos(\Omega\phi))/\Omega^2$, subject to the initial condition $P \to \delta_{Weyl}(H - H_0)$ as $t \to 0$. By the usual arguments, this system is equivalent to a Langevin Equation

$$\dot{H}_\mu = -\Omega^2 H_\mu + \zeta_\mu(t), \tag{9}$$

with a delta correlated white noise $< \zeta_\mu(t)\zeta_\nu(0) >= g_\mu \delta_{\mu,\nu} \delta(t)$. The game now is to show that such an equation for the matrix H implies another for the eigenvalue distribution, which contains the physics of level repulsion. This was in fact already proved by Dyson [5], in a very clever and beautiful argument, which emphasizes the underlying group theoretic structure. In essence, Dyson argues that we can establish and exploit the invariance of the above equations under a change of basis to great advantage. The change of basis corresponds to Orthogonal transformations on H induced by an arbitrary O via $H' = OHO^T$. In the component form this transformation gives $H'_\mu = \sum_\nu R(\mu|\nu) H_\nu$, with $\mu \equiv (i,j)$ and $\nu \equiv (k,l)$ so that

$$R(\mu|\nu) = (O_{i,k}O_{j,l} + O_{i,l}O_{j,k})/g_{k,l}. \tag{10}$$

It is easy to show that $\sum_\mu g_\mu R(\rho|\mu)\; R(\nu|\mu) = g_\rho \delta_{\rho,\nu}$, as well as the fact that the inverse of

R is obtainable through $R^{-1}(\mu|\nu) = \frac{1}{g_\nu} R(\nu|\mu) g_\mu$. Armed with these relations we can show the basis independence of the Langevin equations Eq(9) easily since under a change of basis $\zeta'_\mu(t) = \sum_\nu R(\mu|\nu)\zeta_\nu(t)$, and hence the correlator $< \zeta'\zeta' >$ is unchanged. We may also show the invariance of the F-P equation Eq(8) under the change of basis by a direct computation. What is the advantage of showing the basis independence, one may ask. The answer given by Dyson, is that second order perturbation theory gives the exact result for the eigenvalue distribution function, provided one chooses a suitable basis. At time t we have some H with certain eigenalues, and we first perform an orthogonal transformation to diagonalize H. Now if we increase time to $t+\delta t$, the matrix H changes only infinitesimally away from the diagonal with a change $\delta H_{i,j}$ satisfying $< \delta H_{i,j} >= -\delta t\ \delta_{i,j}\ \lambda_i \Omega^2$, and $< (\delta H_{i,j})^2 >= \delta t\ g_{i,j}$. Second order perturbation theory suffices to fix the change of the eigenvalues to $O(\delta t)$, and we take the averages of the equation $\delta\lambda_i = \delta H_{i,i} + \sum_{j\neq i}(\delta H_{i,j})^2/(\lambda_i - \lambda_j)$ and its square to obtain a Langevin equation for eigenvalues or equivalently a F-P equation

$$\frac{\partial}{\partial t} P = \sum_i \frac{\partial}{\partial \lambda_i}[\frac{\partial}{\partial \lambda_i} + \lambda_i \Omega^2 + \sum_{j\neq i} \frac{1}{\lambda_j - \lambda_i}] P. \tag{11}$$

We may perform a similarity transformation followed by a Wick rotation to get a Schroedinger equation of a many body problem. In brief, let us denote the F-P equation by the abstract equation $\partial/\partial t|P >= -\Pi|P >$ with Π the F-P operator having eigenvalues ≥ 0, and with the coordinate space wavefunctions being obtained through the usual projections $P(x) =< x|P >$, with x standing for the collection of $\lambda's$. The special vector $< 0| \equiv \int\ dx < x|$ satisfies $< 0|\Pi = 0$ so that the probability is conserved in time, and hence $\Pi^\dagger|0 >= 0$. We seek to Hermiticize the F-P operator through a similarity transform $S^{-1}\Pi S = H$, where S is hermitean and invertible operator depending only on the coordinates, so we must have the ground state condition $HS|0 >= 0$. We define a wavefunction through $|\psi >= S^{-1}|P >$ giving $H|\psi >= -\partial/\partial t|\psi >$, and hence at equilibrium $|\psi_0 >= S|0 >$ with the notation meant to remind us that this is the ground state wave function of H, hence $|P_{eq} >= S^2|0 >$. The time evolution of the initial condition may be written as $|P(t) >= S\exp(-tH)S^{-1}|P(0) >$, and we can thus make the correspondence between correlation functions of the F-P problem and the Quantum problem by noting that the classical time dependent correlation function $< Q_a(t)Q_b(0) >=< 0|Q_a S\exp(-tH)S^{-1}Q_b|P(0) >$. If we choose $|P(0) >= |P_{eq} >= S|\psi_0 >$ then provided $[Q, S] = 0$, i.e. position dependent variables, we find $< Q_a(t)Q_b(0) >=< \psi_0|Q_a\exp(-tH)Q_b|\psi_0 >$. Similar aguments can be given for multi time correlations, and one has a mapping between the correlations of the F-P problem and the Euclidean Schroedinger eqn. In the present problem, we can carry out the required similarity transformation readily and find

$$H_C = -\sum_i \frac{\partial^2}{\partial \lambda_i^2} - \frac{1}{2}\sum_{i>j} \frac{1}{(\lambda_i - \lambda_j)^2} + \frac{\Omega^4}{4}\sum_i \lambda_i^2. \tag{12}$$

Similar results hold for the Unitary and Symplectic cases with different values of the interaction. This model has a ground state that is known[4, 3] $\psi_0 = \prod_{i<j}|\lambda_i - \lambda_j|\exp(-\frac{\Omega^2}{2}\sum_i \lambda_i^2)$. Regarding $|\psi_0|^2$ as a probability distribution of particles, the density of particles obeys the Wigner semi circle law, $d(\lambda) = d(0)\sqrt{1 - \frac{\Omega^2\lambda^2}{2N}}$ with $d(0) = \frac{\sqrt{2N}}{\pi}\Omega$. We now see that this model has an average interparticle spacing that varies from point to point when λ becomes large enough but there is a central region, where the density is $d(0)$. We expect on thermodynamic grounds that the bulk properties of the system should be identical to those of the Sutherland model Eq(2), a model with periodic boundary conditions and hence no surface effects, provided we choose the two to have the same densities via $d_{Calogero}(0) = d_{Sutherland}$. This is an example of "universality"in the statics, which is recognized to be a statement of the independence of bulk static correlations on the shape of the boundaries that we are

familiar with in statistical mechanics. If we choose a sensible thermodynamical limit then $d_{Calogero}(0) = O(1)$, implies $\Omega = O(1/N^{1/2})$. We infact now set $d(0) = 1$ giving $\Omega = \pi/\sqrt{2N}$ This requires λ to be of $O(1)$, and the surface effects become relevant when λ becomes of $O(N)$, the density remains unity for λ of $O(1)$, and the surface effects become relevant when λ becomes of $O(N)$. Using the above, write finally the density density correlation function for the $1/r^2$ model with r as the spatial separation and time separation $t = -\ln(\cos(\Omega\phi))/\Omega^2$ in the form

$$\begin{aligned} < \rho(r,t)\rho(0,0) > &= \int dH\, dH_0\, P(H,t|H_0)\, \exp(-\frac{\Omega^2}{2}\mathrm{Tr}H_0^2) \\ & < \mathrm{Tr}\Big[\delta(\epsilon - r - H)\Big]\mathrm{Tr}\Big[\delta(\epsilon - H_0)\Big] >_\epsilon \end{aligned} \tag{13}$$

or using Eq(6),

$$\begin{aligned} < \rho(r,t)\rho(0,0) > &= \frac{\Omega^{\nu_G}}{|\sin^{\nu_G}(\Omega\phi)|}\int dH\, dH_0\, < \mathrm{Tr}\Big[\delta(\epsilon - r - H)\Big]\mathrm{Tr}\Big[\delta(\epsilon - H_0)\Big] >_\epsilon \\ & \exp\Big\{ - \frac{\Omega^2}{2\sin^2(\Omega\phi)}\mathrm{Tr}[H_0^2 + H^2 - 2H_0 H\cos(\Omega\phi)]\Big\}, \end{aligned} \tag{14}$$

and taking the limit $\Omega \to$, $N \to \infty$ and averaging over ϵ a simpler expression:

$$< \rho(r,t)\rho(0,0) >= \frac{1}{(2t)^{\frac{\nu_G}{2}}}\int_{-\infty}^{\infty}\frac{da}{2\pi}\int dH\, dH_0\, Tr[e^{-ia(r+H)}]\; Tr[e^{iaH_0}]\, e^{-\frac{1}{4t}Tr(H-H_0)^2}. \tag{15}$$

Generalizations of this are given for all ensembles in [6], and the noteworthy aspect of this formula is the absence of path integration, we have "merely" a two matrix integral! A direct evaluation of this expression is of interest, but the result of [1] is presumably the answer. Similarly we may write down for a three point function by breaking up the time propagation into two intervals:

$$\begin{aligned} < \rho(r_2,t_2)\; \rho(r_1,t_1)\; \rho(0,0) > = \int dH_2\, dH_1\, dH_0\, P(H_2,t_2 - t_1)\; P(H_1,t_1|H_0)\, \exp(-\frac{\Omega^2}{2}\mathrm{Tr}H_0^2) \\ < \mathrm{Tr}\Big[\delta(\epsilon - r_2 - H_2)\Big]\mathrm{Tr}\Big[\delta(\epsilon - r_1 - H_1)\Big]\mathrm{Tr}\Big[\delta(\epsilon - H_0)\Big] >_\epsilon \end{aligned} \tag{16}$$

3. A Direct Method for General distribution of V

We will now consider a generalized mixing problem with real symmetric matrices,

$$H(\phi) = \alpha(\phi)H_0 + \beta(\phi)V, \tag{17}$$

where α and β are some arbitrary functions of the parameter ϕ, and further the probability distribution (P.D.) of V is no longer Gaussian, but say

$$p(V)dV = \exp(-\frac{1}{2}Tr(V^2) - \gamma\, Tr(V^4))\prod_{i\le j} dV_{i,j}, \tag{18}$$

so that we can recover the Gaussian case by letting $\gamma \to 0$. Other non Gaussian distributions may be treated in a similar fashion. The eigenvalues λ_n of Eq(17) are functionals of V and we are interested in the P. D. function

$$P(\{\mu_n\}|\phi) = \int\prod_{i=1}^{N}\delta(\mu_i - \lambda_i[V])\, p(V)dV, \tag{19}$$

and begin by writing its ϕ derivative

$$\partial P/\partial\phi = -\sum_n \int \prod_{i\neq n} \delta(\mu_i - \lambda_i)\, \delta'(\mu_n - \lambda_n)\, [\frac{\partial\lambda_n}{\partial\phi}]\, p(V)\, dV \tag{20}$$

with the derivative $\frac{\partial\lambda_n}{\partial\phi} =< n|H'(\phi)|n >$ in the running eigenbasis of H, i.e. $H(\phi)|n >= \lambda_n(\phi)|n >$. We now introduce the orthogonal transformation which diagonalizes H, i.e. $\{\lambda_i\}_{diag} = OHO^T$. The eigenvalue equation for H in component form reads $H_{i,j}\psi_j^n = \lambda_n\psi_i^n$ and hence the orthogonal matrix has elements $O_{i,j} = \psi_j^i$. Thus

$$\frac{\partial\lambda_n}{\partial\phi} = 2\sum_{i\leq j} O_{n,i}O_{n,j}\frac{H'_{i,j}}{g_{i,j}}, \tag{21}$$

and further we rewrite $H' = a(\phi)H + b(\phi)V$, with $a(\phi) = \frac{\alpha'}{\alpha}$ and $b(\phi) = \frac{\alpha\beta'-\alpha'\beta}{\alpha}$. Hence we find

$$\partial P/\partial\phi = -a(\phi)\sum_n \frac{\partial}{\partial\mu_n}(\mu_n P) - 2b(\phi)\sum_n \int \prod_{i\neq n} \delta(\mu_i - \lambda_i)\, \delta'(\mu_n - \lambda_n) \sum_{k\leq l} O_{n,k}O_{n,l}\frac{V_{k,l}}{g_{k,l}} p(V)\, dV. \tag{22}$$

Let us now define differentiation w.r.t. the matrix element $V_{k,l}$ of any functional of a symmetric matrix, such as $Tr(V^m)$ through $\frac{\partial}{\partial V_{k,l}} Tr(V^m) = 2m(V^{m-1})_{k,l}/g_{k,l}$, i.e. the two components $V_{k,l}$ and $V_{l,k}$ are *not* regarded as independent variables, but identical ones. Using this definition, and integration by parts, we verify the identity for any functional $f[V]$,

$$\int f[V]p(V)\, dV\, \frac{V_{k,l}}{g_{k,l}} = \frac{1}{2}\int p(V)\, dV \frac{\partial}{\partial V_{k,l}} f[V] - 4\,\gamma \int p(V) dV f[V]\frac{(V^3)_{k,l}}{g_{k,l}}. \tag{23}$$

Hence

$$\begin{aligned}\partial P/\partial\phi = &-a(\phi)\sum_n \frac{\partial}{\partial\mu_n}(\mu_n P) - b(\phi)\sum_n\sum_{k\leq l} \int p(V)\, dV \frac{\partial}{\partial V_{k,l}}[\prod_{i\neq n} \delta(\mu_i - \lambda_i)\, \delta'(\mu_n - \lambda_n) O_{n,k}O_{n,l}] \\ &+4\gamma b(\phi)\sum_n \int \prod_{i\neq n} \delta(\mu_i - \lambda_i)\, \delta'(\mu_n - \lambda_n)\sum_{k,l} O_{n,k}O_{n,l}(V^3)_{k,l}.\end{aligned} \tag{24}$$

We are therefore obliged to take the derivatives of the eigenvalues and the orthogonal matrix O w.r.t. $V_{k,l}$, these follow from perturbative arguments and we summarize the very useful results

$$\frac{\partial\lambda_m}{\partial V_{k,l}} = 2\frac{\beta(\phi)}{g_{k,l}} O_{m,k}O_{m,l} \tag{25}$$

$$\frac{\partial O_{n,p}}{\partial V_{k,l}} = -\frac{\beta(\phi)}{g_{k,l}} \sum_{m\neq n} \frac{1}{\lambda_m - \lambda_n} O_{m,p}(O_{m,k}O_{n,l} + O_{m,l}O_{n,k}). \tag{26}$$

An immediate consequence of these is the simple result

$$\sum_{k\leq l} \frac{\partial O_{n,k}O_{n,l}}{\partial V_{k,l}} = -\frac{\beta(\phi)}{g_{k,l}} \sum_{m\neq n} \frac{1}{\lambda_m - \lambda_n}. \tag{27}$$

We therefore find Eq(24) takes the form

$$\partial P/\partial\phi = -a(\phi)\sum_n \frac{\partial}{\partial\mu_n}(\mu_n P) + \beta(\phi)b(\phi)\sum_n \frac{\partial}{\partial\mu_n}[\frac{\partial}{\partial\mu_n} + \sum_{m\neq n}\frac{1}{\mu_m - \mu_n}]P$$
$$+4\gamma b(\phi)\sum_n \frac{\partial}{\partial\mu_n}\int\prod_i \delta(\mu_i - \lambda_i)\sum_{k,l} O_{n,k}O_{n,l}(V^3)_{k,l}\,p(V)dV. \qquad (28)$$

We can now define a time variable $t = \int_0^\phi \beta(\phi)b(\phi)d\phi$ and $\xi(t) = -a/b\beta$ as well as $\eta(t) = 1/\beta^4$, in terms of which Eq(28) becomes

$$\partial P/\partial t = \sum_n \frac{\partial}{\partial\mu_n}[\frac{\partial}{\partial\mu_n} + \xi(t)\mu_n + \sum_{m\neq n}\frac{1}{\mu_m - \mu_n}]P$$
$$+4\gamma\,\eta(t)\sum_n \frac{\partial}{\partial\mu_n}\int\prod_i \delta(\mu_i - \lambda_i)[O(\mu_{diag} - \alpha(t)H_0)^3 O^T]_{n,n}\,p(V)dV. \qquad (29)$$

The last term in Eq(29) prevents one from obtaining a closed form F-P equation involving only the eigenvalues $\mu's$ unless $\gamma = 0$, showing that the Gaussian is a very special distribution. We summarize the various functions for the original parametrization in Eq(1): $\alpha(\phi) = \cos(\phi\Omega)$, $\beta(\phi) = \sin(\phi\Omega)/\Omega$, $a(\phi) = -\Omega\tan(\phi\Omega)$, $b(\phi) = 1/\cos(\phi\Omega)$ and $\xi(\phi) = \Omega^2$, $\eta(\phi) = \Omega^4/\sin^4(\phi\Omega)$, with of course $t = -\ln\cos(\phi\Omega)/\Omega^2$. If $\gamma \neq 0$, however, there is one set of initial conditions, namely $H_0 = c\mathbf{1}$ the totally degenerate initial conditions, where the last term in becomes simple; $4\gamma\eta(t)\sum_n \frac{\partial}{\partial\mu_n}(\mu_n - c)^3 P$. Although the initial conditions are not "generic", this case does have the advantage of showing that the added term is translated into a quartic potential well, and corresponds to a different kind of a box for the many body system. In fact, dimensionally, the last term for the parametrization of Eq(1) may be regarded as arising from an energy functional of $O(\frac{\mu^4}{N^4})$ and hence *both* the box boundary terms, namely the quadratic well and the 'quartic well' become visible to the eigenvalues μ when they become of $O(N)$. Therefore provided the initial conditions correspond to a uniform distribution of $\mu's$ (as in the center of the Wigner semicircle), one expects a large window of space and hence of time wherein the boundaries are irrelevent. We reach "equilibrium", provided we have $\alpha = 0$ and $\dot{\xi}, \dot{\eta} \to 0$. The last term does simplify for an arbitrary initial H_0, we get the distribution to be that of a logarithmic gas of Wigner- Dyson type with quartic and quadratic confinement. It appears from the above that the robustness of static *and* dynamic correlations to the boundary effects is at the root of the universality of Altshuler and Simons.

4.1 Asymptotic Bethe's Ansatz and Quasiparticle Energies

We now summarize the results of the asymptotic Bethe's Ansatz [3, 13], which gives an explicit expression for the "particle-hole" like excitations underlying the system. The *complete* excitation spectrum for the $1/r^2$ model Eq(2.) with $\beta = 2\lambda$ can be described in remarkably simple terms as follows. The total energy of a state of the system is expressible as

$$E = \sum_n p_n^2\,, \qquad (30)$$

with the "pseudo-momenta" p_n satisfying the equation,

$$p_n = k_n + \frac{\pi(\beta - 2)}{2L}\sum_{n\neq m}\text{sign}(k_n - k_m)\,. \qquad (31)$$

The total momentum of the state is

$$P = \sum_n p_n = \sum_n k_n \ . \tag{32}$$

The bare momenta are given by $k_n = 2\pi J_n/L$, where the J_n's are fermionic quantum numbers $J_1 < J_2 < J_3 \ldots < J_N$. Note that at $\beta = 2$ the interaction is turned off and we recover the free-fermion results. The important point is that the totality of states for the N particle sector is obtained by allowing the integers J_n to take on all values consistent with Fermi statistics, not only for $\beta = 2$, but for *all* $\beta \in [1, +\infty]$. The summation in Eq. (31) is trivial to carry out and we find

$$p_n = k_n + \frac{(\beta-2)\pi}{L}\left(n - \frac{N+1}{2}\right) \ . \tag{33}$$

We can now select an arbitrary state of the system by specifying that states $\{k_1, k_2, \ldots\}$ are occupied, i.e. by introducing the fermionic occupation numbers $n(k_j) = 0, 1$, such that

$$E = \sum_n \varepsilon(k_n) n(k_n) + \sum_{n \neq m} v(k_n - k_m) n(k_n) n(k_m) + \left[\frac{\pi(\beta-2)}{2}\right]^2 , \tag{34}$$

with $\varepsilon(k) = k^2$ and $v(k) = \pi(\beta - 2)|k|/2L$. For future reference, the ground state is represented by $n_0(k_n) = 1$ for $|k_n| < k_F$ and $n_0(k_n) = 0$ otherwise, where $k_F = \pi d$.

We now consider the excitation spectrum near the ground state, wherein we excite a particle-hole pair in the free Fermi system and ask what the energy of the interacting system is by including the Hartree-Fock back flow term. From this point onwards we measure all momenta in units of k_F and energies in units of the Fermi energy. Let us suppose that one of the particles described by Eq. (34) has initially a momentum k, with $|k| < 1$; we promote it to some state labeled by $k + q$, with $|k + q| > 1$. The energy cost in units of the Fermi energy is equal to

$$\begin{aligned} \Delta(q,k) &= \varepsilon(k+q) - \varepsilon(k) + 2\sum_{|k'|<1} [v(k+q-k') - v(k-k')] \\ &= q^2 + 2kq + \frac{(\beta-2)}{2}(2|k+q| - k^2 - 1) , \end{aligned} \tag{35}$$

and the momentum of this state is simply q. This implies that we can associate a generalized energy corresponding to a particle $\varepsilon_>(k)$ (i.e. $|k| > 1$) and a hole $\varepsilon_<(k)$ (i.e. $|k| < 1$):

$$\begin{aligned} \varepsilon_>(k) &= k^2 + (\beta-2)|k| \\ \varepsilon_<(k) &= \frac{\beta}{2}k^2 + \left(\frac{\beta}{2} - 1\right) , \end{aligned} \tag{36}$$

such that

$$\Delta(q,k) = \varepsilon_>(k+q) - \varepsilon_<(k) \ . \tag{37}$$

Note that $\varepsilon_>(k)$ and $\varepsilon_<(k)$ are continuous and have continuous derivatives across the Fermi surface.

We will introduce in the usual way, particle operators $A^\dagger(k)$ and hole operators $B^\dagger(k)$ with the convention that the momenta corresponding to these are constrained by $|k| > k_F$ for particles and $|k| \leq k_F$ for holes, with excitation energies

$$\begin{aligned} E_A(k) &= \varepsilon_>(k) - \mu \\ E_B(k) &= \mu - \varepsilon_<(-k) , \end{aligned} \tag{38}$$

where $\mu \equiv \varepsilon_>(k_F)$ is the "chemical potential". The quasi-particle quasi-hole excitation created by the operator $A^\dagger(k+q)B^\dagger(-k)$ then has energy $E_A(k+q) + E_B(-k)$, which of course is equal to $\Delta(k,q)$. Having introduced the underlying fermionic quasi-particles quasi-holes through Eqs. (36), we would like to see if the excitations generated by the bare density fluctuation operator ρ_q can be expressed in terms of the latter. One of our objectives then, is to express the excitations of the system probed by the bare density fluctuation operator ρ_q in terms of the quasi-particle quasi-hole operators. Recall that in Landau's Fermi Liquid Theory [14] one expresses the bare particles $c(k)$ in a series involving quasi-particles and quasi-holes of the form

$$c(k) = \sqrt{z_k}\; B^\dagger(-k) + \sum_{(p,l)} M[k,p,l]\; B^\dagger(p)B^\dagger(l)A^\dagger(-k-p-l) + \ldots , \tag{39}$$

where $|k| \leq k_F$, and a similar expansion for particles, where z_k is the quasi-particle residue. The density fluctuation operator $\rho_q = \sum_k c^\dagger(k+q)c(k)$ then has a development in terms of $1,2,3,\ldots$ pairs of (quasi) particle-hole excitations. In one dimension, we expect z_k to vanish for arbitrary non-zero interactions, and hence the particle-hole series is expected to be such that the single pair should not appear. The expansions are somewhat non-unique, in view of the fact that we can add an arbitrary number of "zero energy" and "zero-momentum" particle-hole excitations to any given scheme.

4.2 Dynamical Correlation Functions and their Interpretation

We will now recall the results of [1] for the correlations of the two non trivial ensembles, the Orthogonal and the Symplectic cases, and indicate their interpretation following Ref.[9]. We begin by writing the structure function

$$S(q,\omega) = \frac{1}{2\pi d}\int dr \int dt\; k(r,t)\; e^{-i(qr-\omega t)} .$$

where the correlation function

$$k(r,t) = \langle \rho(\bar{r}-r, \bar{t}+t)\rho(\bar{r},\bar{t}\,)\rangle - d^2 , \tag{40}$$

with $\rho(r,t) = \sum_{i=1}^N \delta(r - r_i(t))$. $S(q,\omega)$ has a representation in terms of the excited states of the system:

$$S(q,\omega) \equiv \frac{1}{N}\sum_{\nu\neq 0} |\langle\nu|\rho_q|0\rangle|^2 \delta(\omega - E_\nu + E_0) , \tag{41}$$

where $H|\nu\rangle = E_\nu|\nu\rangle$, and

$$\rho_q = \int dr\ \rho(r)\ e^{-iqr} = \sum_{i=1}^{N} e^{-iqr_i}\ . \tag{42}$$

For the Orthogonal ensemble the result is [1]

$$\begin{aligned} k^o(r,t) &= d^2 \int_{-1}^{1} d\lambda \int_{1}^{\infty} d\lambda_1 \int_{1}^{\infty} d\lambda_2\ \frac{(1-\lambda^2)(\lambda_1\lambda_2-\lambda)^2}{(\lambda_1^2+\lambda_2^2+\lambda^2-2\lambda\lambda_1\lambda_2-1)^2} \\ &\times\ \exp[-ik_F^2 t(2\lambda_1^2\lambda_2^2-\lambda_1^2-\lambda_2^2-\lambda^2+1)/2]\cos[k_F r(\lambda_1\lambda_2-\lambda)]\ . \end{aligned} \tag{43}$$

Taking the Fourier transform of $k^o(r,t)$ in space and time yields

$$\begin{aligned} S^o(q,\omega) &= \frac{2q^2}{k_F^4}\int_{1}^{\infty} d\lambda_1 \int_{1}^{\infty} d\lambda_2\ \frac{[1-(\lambda_1\lambda_2-|q|/k_F)^2]}{(\lambda_1^2+\lambda_2^2+q^2/k_F^2-\lambda_1^2\lambda_2^2-1)^2} \\ &\times\ \delta(\lambda_1^2+\lambda_2^2+q^2/k_F^2-1-\lambda_1^2\lambda_2^2-2\lambda_1\lambda_2|q|/k_F+2\omega/k_F^2) \\ &\times\ \theta(\lambda_1\lambda_2-|q|/k_F+1)\ \theta(1-\lambda_1\lambda_2+|q|/k_F)\ . \end{aligned} \tag{44}$$

Also note that in the Symplectic ensemble the results are

$$\begin{aligned} k^s(r,t) &= \frac{d^2}{2}\int_{1}^{\infty} d\lambda \int_{-1}^{1} d\lambda_1 \int_{-1}^{1} d\lambda_2\ \frac{(\lambda^2-1)(\lambda-\lambda_1\lambda_2)^2}{(\lambda_1^2+\lambda_2^2+\lambda^2-2\lambda\lambda_1\lambda_2-1)^2} \\ &\times\ \exp[-4ik_F^2 t(\lambda_1^2+\lambda_2^2+\lambda^2-2\lambda_1^2\lambda_2^2-1)]\cos[2k_F r(\lambda-\lambda_1\lambda_2)]\ . \end{aligned} \tag{45}$$

We take the Fourier transform of $k^s(r,t)$ to get

$$\begin{aligned} S^s(q,\omega) &= \frac{q^2}{64k_F^4}\int_{-1}^{1} d\lambda_1 \int_{-1}^{1} d\lambda_2\ \frac{[(\lambda_1\lambda_2+|q|/2k_F)^2-1]}{(\lambda_1^2+\lambda_2^2+q^2/4k_F^2-\lambda_1^2\lambda_2^2-1)^2} \\ &\times\ \delta(\lambda_1^2+\lambda_2^2+q^2/4k_F^2-1-\lambda_1^2\lambda_2^2+\lambda_1\lambda_2|q|/k_F-\omega/4k_F^2) \\ &\times\ \theta(\lambda_1\lambda_2+|q|/2k_F-1)\ . \end{aligned} \tag{46}$$

We now turn to an interpretation of these results in the light of the Bethe excitations. A useful change of variables is

$$\begin{aligned} u &= \lambda_1\lambda_2 \\ z &= \lambda_1+\lambda_2\ . \end{aligned} \tag{47}$$

Further defining $k = u - q$ we find the expression for the Orthogonal ensemble:

$$S^o(q,\omega) = \frac{q^2}{4}\int_{\substack{|k|\le 1\\ k+q>1}} dk\ \frac{[-\varepsilon_<(k)]}{[\omega-q(k+q)]^2}\ \frac{\theta(\omega-\Delta(k,q)+(k+q-1)^2/2)\ \theta(\Delta(k,q)-\omega)}{\sqrt{\Delta(k,q)-\omega}\ \sqrt{\Delta(k,q)+2(k+q)-\omega}}\ . \tag{48}$$

An examination of the energy conserving delta function shows that we can view the excited

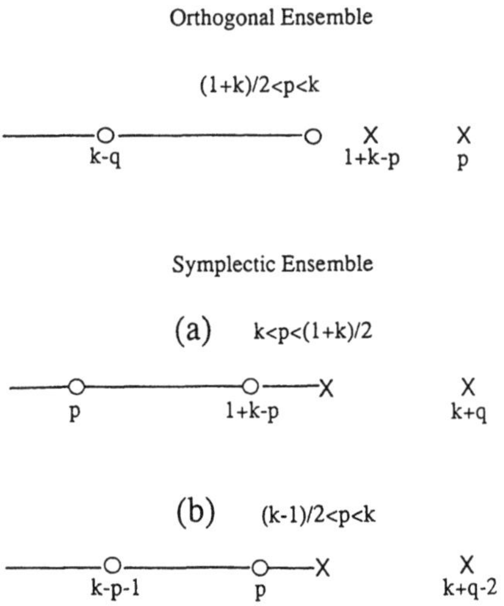

Fig.2 Two quasi-particle and quasi-hole scheme for the Orthogonal and Symplectic ensembles. The x's denote quasi-particles and o's the quasi-holes. The solid line is the fermi surface with $1 \geq k \geq -1$. The Orthogonal ensemble is interpreted as having a fractionalization of the inserted bare particle but not of the bare hole. The Symplectic case on the other hand corresponds to the bare hole being fractionalized. The two pieces S_a and S_b correspond to the two figures (a) and (b).

particle state as a combination of a particle and particle-hole pair, as follows. We write a schematic development for $k \geq 1$

$$c^{\dagger}(k) \sim \sum_{\frac{1+k}{2} \leq p \leq k} A^{\dagger}(p) A^{\dagger}(k-p+1) B^{\dagger}(-1) . \tag{49}$$

The excitation energy of this complex is readily seen from Eqs. (36,38) to be $E_A(p) + E_A(k-p+1) + E_B(-1)$, with $\frac{1+k}{2} \leq p \leq k$. The density fluctuation ρ_q is then seen to be formally a two quasi particle-hole object: writing $c(k) \sim B^{\dagger}(-k)$, we have

$$c^{\dagger}(k) c(k-q) \sim \sum_{\frac{1+k}{2} \leq p \leq k} A^{\dagger}(p) A^{\dagger}(k-p+1) B^{\dagger}(-1) B^{\dagger}(q-k) . \tag{50}$$

The above scheme for the density fluctuation operator $c^{\dagger}(k)c(k-q)$ is indicated in Fig. 2. We may therefore regard the density fluctuation as being built up from a particular set of (non-interacting) pair states consisting of annihilating two particles at momenta $k-q$ and 1, and creating a pair with total momentum $k+1$, distributed over all possible relative momenta with appropriate form factors.

We can perform a similar analysis for the Symplectic ensemble, we recall for the Symplectic case, the Bethe energies $\varepsilon_{>}(p) = p^2 + 2|p|$ and $\varepsilon_{<}(p) = 2p^2 + 1$ (Eq. 36)). We now rewrite Eq. (46) using the same variables as in the previous case (Eq. (47)). We find the result breaks up naturally into two pieces S_a and S_b, with the second piece S_b only arising for $q > 2$:

$$S^{s}(q,\omega) = S_a(q,\omega) + \theta(q-2)\; S_b(q,\omega) . \tag{51}$$

We write $u = (1+k)/2$ in S_a and $u = (l-1)/2$ in S_b, in terms of which the result may be written as

$$S_a(q,\omega) = \frac{q^2}{4}\int_{\substack{|k|\le 1\\ k+q>1}} dk \frac{[\varepsilon_>(k+q)-3]}{[\omega-\Delta(k,q)-(k-1)^2+q^2]^2} \times \frac{\theta(\omega-\Delta(k,q))\,\theta(\Delta(k,q)+(k-1)^2-\omega)}{\sqrt{\omega-\Delta(k,q)}\sqrt{\omega-\Delta(k,q)+8(1+k)}}, \tag{52}$$

and

$$S_b(q,\omega) = \frac{q^2}{4}\int_{\substack{|l|\le 1\\ l+\alpha>1}} dl \frac{[\varepsilon_>(l+\alpha)-3]}{[\omega-\Delta(l,\alpha)-(l+1)^2+q^2]^2} \times \frac{\theta(\omega-\Delta(l,\alpha))\,\theta(\Delta(l,\alpha)+(l+1)^2-\omega)}{\sqrt{\omega-\Delta(l,\alpha)}\,\sqrt{\omega-\Delta(l,\alpha)+8(1-l)}}. \tag{53}$$

An examination of the energy conserving delta function shows that in this expression, the excited particle is "normal", whereas the hole is apparently further fractionalized into two "quasi-holes" and a "drone quasiparticle" living at the fermi surface, i.e. a schematic decomposition

$$c(k) \sim \sum_{k\le p\le \frac{k+1}{2}} B^\dagger(p-k-1)B^\dagger(-p)A^\dagger(1)\,. \tag{54}$$

The restriction on the range of p is such that we avoid double counting the pair and have a natural ordering of the two quasi-holes. The energy of the effective hole is then $E_B(-p) + E_B(p-k-1) + E_A(1)$, with the constraint $k \le p \le \frac{k+1}{2}$. The operator ρ_q is seen to be formally a two quasi particle-hole object,

$$c^\dagger(k+q)c(k) \sim \sum_{k\le p\le \frac{k+1}{2}} B^\dagger(p-k-1)B^\dagger(-p)A^\dagger(1)A^\dagger(k+q)\,. \tag{55}$$

This scheme for the density fluctuation is illustrated in Fig. 2.

In the second piece of S^s, we have schematically

$$c(l) \sim \sum_{\frac{l-1}{2}\le p\le l} B^\dagger(-p)A^\dagger(-1)B^\dagger(1+p-l)\,. \tag{56}$$

Using the quasi-energies Eqs. (36,38) this complex has energy $E_B(p)+E_B(1+p-l)+E_A(-1)$, with the physical constraint $\frac{l-1}{2} \le p \le l$.. Owing to momentum conservation, we must regard the creation operator $c^\dagger(l+q)$ as $A^\dagger(l+q-2)$ times a particle-hole pair with energy zero and momentum 2, i.e. $A^\dagger(1)B^\dagger(1)$. We may eliminate a 'zero pair' $A^\dagger(-1)B^\dagger(1)$ and thus obtain the scheme for the density fluctuation operator ρ_q,

$$c^\dagger(l+q)c(l) \sim \sum_{\frac{l-1}{2}\le p\le l} A^\dagger(l+q-2)A^\dagger(1)B^\dagger(-p)B^\dagger(1+p-l)\,. \tag{57}$$

The term S_b then evidently may be regarded as a two quasi particle-hole object, and is illustrated in Fig. 2. Summarizing, in process (a), we may regard the density fluctuation as being built up from a particular set of (non-interacting) pair states consisting of creating two particles at momenta 1 and $k+q$, and destroying a pair with total momentum $k+1$, distributed over all possible relative momenta with appropriate form factors. Likewise, in process (b), we may regard the density fluctuation as being built up from a particular set of (non-interacting) pair states consisting of creating two particles at momenta 1 and $l+q-2$, and destroying a pair with total momentum $l-1$, distributed over all possible relative momenta with appropriate form factors.

5. Conclusions

In this article, we have reviewed the equivalence between the problem of eigenvalue "motion" with a parameter, and the dynamics of the $1/r^2$ many body problem, within a random matrix framework. The special features of the Gaussian distribution lead to a closed Fokker Planck equation that is equivalent to the $1/r^2$ model Hamiltonian, and non Gaussian distributions are shown to lead to extra terms involving the "angular variables" of the diagonalizers. These however, dimensionally correspond to different kinds of confining boxes for the system, suggesting that the universality of the dynamics found by Altshuler and Simons has the thermodynamic robustness of bulk correlations to the boundaries at its root. We have further discussed the results of the dynamical density correlations of the $1/r^2$ model within the context of Bethe's ansatz excitations. This phenomenological spectroscopy reveals an interesting decomposition of the bare particles into constituents, which is different for the different ensembles.

Acknowledgements

It is a pleasure to thank Boris Altshuler and Onuttom Narayan as well as Eduardo Mucciolo, Ben Simons and Bill Sutherland for lively and stimulating discussions.

References

[1] B.D. Simons and B.L. Altshuler, Phys. Rev. Letts. **70**, 4063 (1993); B.D. Simons, P.A. Lee and B.L. Altshuler, Phys. Rev. Letts. **70**, 4122 (1993); B. D. Simons, A. Szafer, and B. L. Altshuler, *JETP Lett.* **57**, 276 (1993).

[2] K. B. Efetov, Adv. in Phys. **32**, 53 (1983).

[3] B. Sutherland, J. Math. Phys. **12**, 246 (1971); ibid **12**, 251 (1971); Phys. Rev. A**4**, 2019 (1971); ibid **5**, 1372 (1972).

[4] F. Calogero, J. Math. Phys. **10**, 2191, 2197 (1969).

[5] F. Dyson, J. Math. Phys. **3** 1191 (1962), J. Math. Phys. **3**, 140, 157, 166 (1962).

[6] O. Narayan and B. S. Shastry, Phys. Rev. Letts. **71**, 2106 (1993).

[7] B. Simons, P. A. Lee and B. L. Altshuler, "*Matrix Models, One Dimensional Fermions and Quantum Chaos*", Preprint (1993). see also, C. W. J. Beenaker, Phys. Rev. Letts. **70**, 4126 (1993).

[8] P. Pechukas, Phys. Rev. Lett. **51**, 943 (1983); S. Wojciechowski, Phys. Lett. A**111**, 101 (1985); T. Yukawa, Phys. Rev. Lett. **54**, 1883 (1985); K. Nakamura and M. Lakshmanan, Phys. Rev. Lett. **57**, 1661 (1986).

[9] E. Mucciolo, B. S. Shastry, B. Simons and B. L. Altshuler, "*Exact Dynamical Correlations of the* $1/r^2$ *Model*", Preprint (1993).

[10] H. Weyl, *Classical Groups*, Princeton University Press (1946).

[11] M. L. Mehta, *Random Matrices*, 2nd. ed. (Academic Press, San Diego, 1991).

[12] There are several places in recent literature where a variant of the resolution of unity Eq(4) occurs for the Unitary case, however, it seems to us that the counting of parameters is not usually done carefully enough, unlike in the careful treatments of [10, 11]. The $\delta_{Weyl}(M)$ for a real symmetric has a useful representation displaying its invariances: $\delta_{Weyl}(A) = \lim_{\alpha\to 0} 1/(\alpha^{\nu_G}\pi^{\nu_G/2})\exp(-\frac{1}{\alpha^2}\,Tr(A^2))$. The same expression holds for other ensembles after the number ν_G is redefined as indicated in the text.

[13] B. Sutherland, in *Exactly Solvable Problems in Condensed Matter and Relativistic Field Theory*, edited by B. S. Shastry *et al.*, Lecture Notes in Physics (Springer-Verlag, Berlin, 1985), Vol. 242.

[14] L. D. Landau, Sov. Phys. JETP **30**, 1058 (1956).

Role of Supersymmetry in Exactly Soluble Multi-Component Fermion Models

Y. Kuramoto

Department of Physics, Tohoku University, Sendai 980, Japan

This paper discusses two examples where the presence of supersymmetry makes models for strongly correlated fermions exactly soluble: the long-range SU(ν) t-J model in one dimension and a continuum SU(ν,1) model in arbitrary dimensions. The symmetry property of the models are clarified. It is shown for the first model with spin-charge separation that all eigenstates are mapped into those of SU($\nu-1,1$) generalization of the Sutherland model in continuum space. The second model realizes, after renormalization, free SU(ν) fermions without spin-charge separation. A remarkable feature common to these models is that eigenstates display the fixed-point behavior for all energies.

1. Introduction

Supersymmetry is associated with transformation of a fermion into a boson and vice versa. It has been investigated extensively in field theory partly because certain divergences cancel each other in models with supersymmetry. In this paper we show that supersymmetry brings drastic simplification also in strongly correlated fermion models for condensed-matter physics. Namely there is a group of models where the presence of supersymmetry makes the models exactly soluble. Moreover, the excitation spectra of these models display the fixed-point behavior for all energies. This is in strong contrast with more familiar models such as the Hubbard model where the presence of marginally irrelevant operators causes complication in dynamics and in asymptotic behavior of correlation functions.

We discuss in this paper two types of fermion models where supersymmetry permits us to obtain the exact solution. The first type is the one-dimensional t-J model with long-range interaction. This model realizes a free Luttinger liquid with spin-charge separation for all energies [1-6]. If one takes the high-density limit and freezes the charge degrees of freedom, the model is reduced to the long-range exchange model proposed by Haldane and Shastry [7,8]. Another type is a newly found supersymmetric continuum model [9] which constitutes a fixed point of free non-relativistic SU(ν) fermions. The unique feature of the latter model is that it is solvable in arbitrary dimensions. The model has a similarity to the Wess-Zumino-Witten Lagrangian [10] (or the non-linear sigma model with a topological term) and should in fact become equivalent to it in the (1+1)-dimensional relativistic limit.

If one replaces the SU(ν,1) symmetry in the second model by the SU($\nu+1$) symmetry, the resultant model represents a continuum limit of the ferromagnetic Heisenberg model, being generalized from the usual SU(2) case. It is shown that the eigenstates consist of free magnons out of completely polarized vacuum. In contrast to the Heisenberg lattice or the classical continuum Heisenberg model, there are no interactions among magnons in the present quantum model. In order to make this paper complementary to previously published ones [1-3], we discuss here mainly on symmetry properties of the models.

Springer Series in Solid-State Sciences, Vol. 118
Correlation Effects in Low-Dimensional Electron Systems
Editors: A. Okiji and N. Kawakami

2. Supersymmetry in Lattice Models

Let us first introduce a scheme to use supersymmetry in lattice models. Similar ideas have also been used in the literature [11]. We consider a model on a lattice and assume that each site can accommodate at most one electron because of infinitely strong on-site repulsion. Then it is convenient to represent the hard-core electron at site i with an internal degree of freedom $\alpha = 1, \ldots, \nu$ as a composite particle of a fermion with a creation operator $c_i^\dagger$ and a boson with $b_{\alpha i}^\dagger$. Here the fermion represents the charge degrees of freedom and the boson does the spin degrees of freedom in the case of $\nu = 2$, or its SU(ν) generalization which we call color. We represent the creation operator of a hard-core electron as

$$X_i^{\alpha 0} \equiv b_{\alpha i}^\dagger c_i^\dagger. \tag{1}$$

The hard-core constraint $X_i^{\alpha 0} X_i^{\beta 0} \equiv 0$ is automatically satisfied by this representation. The color flipping operator from β to α is defined by $X_i^{\alpha\beta} = b_{\alpha i}^\dagger b_{\beta i}$. We further define $X_i^{00} = c_i c_i^\dagger$, which is the projection operator onto the fermionic vacuum. A general X-operator is represented by X^{ab} where a and b denote either 0 or one of α.

The commutation rules of these X-operators at a site obeys those of a Lie superalgebra associated with the supergroup SU(ν,1) [12]. Hence the X-operators generate, like the spin operators in the case of SU(2), "superrotation" which mixes $c_i^\dagger$ and $b_{\alpha i}$. Supersymmetry in this case refers to invariance of the quantity

$$I = X_i^{00} + \sum_\alpha X_i^{\alpha\alpha}$$

under the superrotation, i.e. $[X^{ab}, I] = 0$.

The supersymmetry at the single-site level is thus merely a representation of the hard-core constraint of fermions. However supersymmetry imposes a nontrivial constraint when one introduces intersite interactions. To see this in a concise way we introduce a $(\nu + 1)$-dimensional vector operator Ψ_i whose conjugate $\Psi_i^\dagger$ has components

$$\Psi_i^\dagger = (b_{1i}^\dagger, \cdots, b_{\nu i}^\dagger, h_i^\dagger), \tag{2}$$

where $h_i^\dagger = c_i$ creates the vacant state. Our Fock space consists of states constructed by application of $X_i^{\alpha 0}$ to the vacuum, and each site has $\nu + 1$ kinds of states. Thus we have the constraint $I = \Psi_i^\dagger \Psi_i = 1$. The important fact is that an intersite operator $\Psi_i^\dagger \Psi_j$ is also invariant under the global $(\nu + 1)$-dimensional superrotation which is common to all sites. Thus if the intersite interaction in a model is expressible in terms of $\Psi_i^\dagger \Psi_j$, there remains a global supersymmetry.

We introduce a graded permutation operator defined by

$$\tilde{P}_{ij} = :(\Psi_i^\dagger \Psi_j)(\Psi_j^\dagger \Psi_i): = -\sum_{a,b} X_i^{ab} X_j^{ba} \theta_b, \tag{3}$$

where $: \cdots :$ indicates the normal ordering of component operators. If the ordering makes exchange of two fermion operators, the sign is reversed. This operation leads to the second equality where $\theta_b = 1$ if $b = 0$, and $\theta_b = -1$ otherwise. The presence of the factor θ_b is characteristic of the graded permutation. We note that $\tilde{P}_{ij}$ is invariant under the superrotation as apparent from the first equality.

3. Long-Range t-J Model in One Dimension

The t-J model is a popular model for studying strong correlation of electrons in copper oxides and other related systems. It is defined by

$$\mathcal{H} = \mathcal{P}\sum_{i\neq j}[-t_{ij}\sum_{\sigma} f_{i\sigma}^{\dagger} f_{j\sigma} + J_{ij}(\boldsymbol{S}_i \cdot \boldsymbol{S}_j - \frac{1}{4}n_i n_j)]\mathcal{P}, \tag{4}$$

where $f_{j\sigma}$ represents the annihilation operator of an electron at site j and spin σ, and $\mathcal{P}$ is a projection operator to exclude double occupation of any site. The hopping is represented by $-t_{ij}$, and J_{ij} denotes exchange interaction between spins $\boldsymbol{S}_i$ and $\boldsymbol{S}_j$. The presence of the number operator $n_i = \sum_\sigma f_{i\sigma}^{\dagger} f_{i\sigma}$ causes the J_{ij} term to vanish if the pair $\boldsymbol{S}_i$ and $\boldsymbol{S}_j$ make a triplet.

As the one-dimensional lattice we consider a ring which consists of N (even) lattice points x_i with unit spacing and impose the condition $t_{ij} = J_{ij}$. Then the t-J model in this case turns out to be the SU(2,1) specialization of the following model:

$$\mathcal{H} = \sum_{i\neq j} J_{ij}(\tilde{P}_{ij} - 1 + X_i^{00}). \tag{5}$$

From the result of Section 2, it is clear that the model without the X_i^{00} term has supersymmetry. This term breaks the SU(ν,1) supersymmetry just as the magnetic field breaks the SU(2) symmetry of the Heisenberg model. Once the model without the X_i^{00} term is solved, inclusion of the term is easily done.

In the case where J_{ij} is nonzero only for nearest neighbors, the model has been solved exactly by the use of the Bethe ansatz [13,14]. In another case where the interaction has the long-range form: $t_{ij} = J_{ij} = tD(x_i - x_j)^{-2}$ with $D(x_i - x_j) = (N/\pi)\sin[\pi(x_i - x_j)/N]$, the model can also be solved exactly [1-6]. We concentrate on the latter case and assume here $t > 0$.

If there were no two-body interactions, the single-particle spectrum would be given by Fourier transform of J_{ij} or t_{ij}. This is calculated as

$$J(q) = \sum_{i} J_{ij}\exp(-iq\theta_{ij}) = \frac{1}{2}(q - \frac{N}{2})^2 - \frac{1}{24}(N^2 + 2), \tag{6}$$

where $\theta_{ij} = \theta_i - \theta_j = 2\pi(x_i - x_j)/N$ and q is an integer with $0 \leq q < N$. The unit of energy is so chosen that $J = t = (N/\pi)^2$.

A remarkable feature of this spectrum is that $J(q)$ is a quadratic function of q. If one shifts the origin of q to the edge $N/2$ of the Brillouin zone, the spectrum is nearly the same as that of free particles in continuum space, except for the presence of the cut off in q. Using this similarity [15] we can also work conveniently with the first-quantized representation with the completely polarized state in the direction of $\alpha = \nu$ as the reference state. Then for each site the vacant state with respect to the original vacuum is represented as a hole, and a state with color $\alpha\,(\neq \nu)$ is represented by ($\nu - 1$)-kinds of hard-core bosons. The complete set for the system consists of the product of one-body states inside the Brillouin zone with proper symmetrization. In terms of $z_i = \exp(i\theta_i)$, where θ_i denotes a coordinate with either a color other than ν or a hole, the one-body states are spanned by monomials z_i^k with $0 \leq k < N$. We now shift the origin of the wave number to the edge $N/2$ of the Brillouin zone. Within the many-body Hilbert space defined in this way, we may replace $q - N/2$ in eq.(6) by $-i\partial/\partial\theta$. Then the Hamiltonian is given by

$$\mathcal{H}=\sum_{i=1}^{M}p_i^2+\frac{1}{4}\sum_{i\neq j}\sin^{-2}(\frac{1}{2}\theta_{ij})(1+\hat{P}_{ij})+E_M, \tag{7}$$

where $p_i=-i\partial/\partial\theta_i$ and M is the number of holes plus bosons with colors different from ν. The graded permutation operator $\hat{P}_{ij}$ acts now in the space of $(\nu-1)$ colors and holes. The constant E_M appears as a result of transformation.

By taking the hole picture in the first quantization, the minus sign in the hopping combines nicely with the plus sign in J_{ij} in leading to eq.(7), where p_i^2 is common to holes and bosons. If all bosons have the same color α, then we have $\hat{P}_{ij}=1$. The resultant Hamiltonian, written as $\mathcal{H}_S$, looks the same as the Sutherland model [16] in the continuum space. It is known in the continuum case [17] that $\mathcal{H}_S$ is related to the Laplace operator in a fictitious Riemannian space, and the eigenfunctions are described by generalized plane waves. By using the fact that the Laplace operator is a Casimir invariant in the Lie algebra, we can derive all the eigenvalue $E(\boldsymbol{l})$ of $\mathcal{H}_S$ in terms of the M-dimensional vector $\boldsymbol{l}$ which is called the highest weight. Then the result given by the asymptotic Bethe ansatz [18-20] is reproduced [3].

On the other hand if all particles are holes, we get $\hat{P}_{ij}=-1$. Then eq.(7) has no interaction term and each eigenfunction is given by a Slater determinant of M free fermions. In a general case, eq.(7) can be regarded as the SU$(\nu-1,1)$ generalization of the Sutherland model in continuum space. Thus all eigenfunctions of the SU$(\nu$,1) t-J model can be mapped into those of the SU$(\nu-1,1)$ Sutherland model. However the reverse is not true because of restriction of wave functions to the Brillouin zone in the t-J model.

4. SU(ν,1) Continuum Model in Arbitrary Dimensions

In the momentum space the t-J model is written as

$$\mathcal{H}=-\frac{1}{2}\sum_{k=1}^{N}(k-\frac{N}{2})^2\sum_{a,b}X^{ab}(k)X^{ba}(-k)\theta_b+E_0, \tag{8}$$

where $X^{ab}(k)$ is the Fourier transform of X_i^{ab} and E_0 is a constant dependent on the number of vacant sites. As we have considered polynomial wave functions of $z_i=\exp(i\theta_i)$ in the first quantization, we can also restrict operators to polynomials of X_i^{ab}. Under this restriction $(q-N/2)$ in eq.(6) is replaced by $-i\partial/\partial\theta$ which acts on X-operators. Then the model can also be written as

$$\mathcal{H}=-N\int_0^{2\pi}\frac{d\theta}{4\pi}\sum_{a,b}\frac{\partial X_\theta^{ab}}{\partial\theta}\cdot\frac{\partial X_\theta^{ba}}{\partial\theta}\theta_b+E_0. \tag{9}$$

Inspired by this form, we have recently introduced a new supersymmetric model defined in a d-dimensional continuum space [9]. The model is given by

$$\mathcal{H}=\frac{1}{2}g\int d\boldsymbol{r}\sum_{a,b}\nabla X^{ab}(\boldsymbol{r})\cdot\nabla X^{ba}(\boldsymbol{r})\theta_b+C, \tag{10}$$

where $\nabla=\partial/\partial\boldsymbol{r}$. The constants g and C are determined so as to give zero energy for the vacuum and finite energy for a finite number of hard-core fermions. Note that the overall sign in eq.(10) is reversed from that of eq.(9).

In the continuum space, products of X-operators at the same space point $\boldsymbol{r}$ become highly singular and renormalization is required. We have to give meaning to the short-range limit of the δ-function. A convenient regularization is to replace $\delta(x)$ by $\delta_\varepsilon(x) := (\varepsilon/\pi)(\varepsilon^2 + x^2)^{-1}$. Then in d dimensions we define $\delta_\varepsilon(\boldsymbol{r} = 0) \equiv \delta_\varepsilon(x = 0)^d \equiv g^{-1}$, and $\varepsilon \to 0$ is performed always in the end. The Fock space in which we work is characterized by the vacuum $|0\rangle$ such that $b_\alpha(\boldsymbol{r})|0\rangle = c(\boldsymbol{r})|0\rangle = 0$, and by n-particle states $|n\rangle$ spanned by linear combinations of $X^{\alpha_1 0}(\boldsymbol{r}_1)X^{\alpha_2 0}(\boldsymbol{r}_2)\cdots X^{\alpha_n 0}(\boldsymbol{r}_n)|0\rangle$.

We can construct normalized wave functions also in the first quantization. A state vector $|\Psi\rangle$ with n particles is expanded as

$$|\Psi\rangle = \int d\boldsymbol{r}_1 \sum_{\alpha_1} \cdots \int d\boldsymbol{r}_n \sum_{\alpha_n} \Psi(\boldsymbol{r}_1\alpha_1, \ldots \boldsymbol{r}_n\alpha_n) g^{n/2} X^{\alpha_1 0}(\boldsymbol{r}_1) \ldots X^{\alpha_n 0}(\boldsymbol{r}_n)|0\rangle. \tag{11}$$

With the wave-function renormalization factor $g^{n/2}$ in eq.(11), the first-quantized wave function is properly normalized in the d-dimensional box, i.e.

$$\int d\boldsymbol{r}_1 \sum_{\alpha_1} \cdots \int d\boldsymbol{r}_n \sum_{\alpha_n} |\Psi(\boldsymbol{r}_1\alpha_1, \ldots \boldsymbol{r}_n\alpha_n)|^2 = 1.$$

In ref.[9] we have shown that all wave functions $\Psi(\boldsymbol{r}_1\alpha_1, \ldots \boldsymbol{r}_n\alpha_n)$ and eigenvalues are the same as those of ν-component free fermions. The presence of supersymmetry is essential in this surprising cancellation of the hard-core constraint against the exchange-type interaction.

5. Relation to Some Bosonic Models

Let us now consider a case where the X-operators are constructed by replacing $b^\dagger_{\alpha i}$ in eq.(1) by a fermion creation operator $d^\dagger_{\alpha i}$. Then $X_i^{\alpha 0} = d^\dagger_{\alpha i} c^\dagger_i$ creates a hard-core boson. There emerges the SU$(\nu + 1)$ symmetry by this replacement. Instead of the graded permutation operator $\tilde{P}_{ij}$ the usual permutation operator P_{ij} should be used in constructing the model. P_{ij} is simply obtained by putting $\theta_b = -1$ for all b in eq.(3). Then we obtain a bosonic Hamiltonian where $\tilde{P}_{ij}$ in eq.(5) is replaced by P_{ij}. Again the complete SU$(\nu+1)$ symmetry is broken by the X_i^{00} term. This is interpreted in such a way that the external magnetic field of critical strength makes complete polarization of the ground state into the direction of the 0-th component. Then a spin-flip (creation of a boson α) needs no threshold energy. The bosonic model so obtained is in fact the same as the "supersymmetric bosonic model" of Ha and Haldane [6]. It is also equivalent to the SU$(\nu+1)$ generalization of the Haldane-Shastry model. Thus the charge excitation in the model of Ha and Haldane is a disguise of a color excitation under finite polarization.

In the continuum case we can also obtain a bosonic model by putting $\theta_b = 1$ in eq.(10) for all b. Note that the resultant model is ferromagnetic and the vacuum corresponds to completely polarized state. It can be shown that the plane wave states of hard-core bosons constitute exact eigenstates of the model in arbitrary dimensions. The bosons represent free magnons out of the completely polarized ground state.

The absence of interaction among magnons is in striking contrast with case of the Heisenberg chain and the classical Heisenberg model with finite spin density. The infinite density of polarization in the present quantum model seems to be responsible for the free nature of magnons.

6. Discussions

In this paper we have discussed the role of supersymmetry in exactly soluble fermion models. Recently a new type of symmetry called the Hopf algebra has been used to explain the enormous degeneracy of the spectrum in the Haldane-Shastry model [21]. It seems possible to generalize the new symmetry to a supersymmetric one. Then we will have better understanding of the degeneracy found in the t-J model [5].

In ref.[18] it is remarked that "free spinons" in the Haldane-Shastry model are related to excitations in the SU(2) Wess-Zumino-Witten (WZW) Lagrangian with the level $k = 1$. On the other hand, if the WZW model is supplemented by a bosonic part describing the charge degrees of freedom, the excitations are just those of free Dirac fermions [22]. In the case of the t-J model, the spin-charge separation [1] does not permit interpretation in terms of Dirac fermions. On the other hand, the continuum SU(ν,1) model does not show the spin-charge separation and the Hamiltonian given by eq.(10) is very similar to the main part of the SU(ν) WZW model.

In order to make connection to the WZW model we consider the relativistic limit of our continuum model in the one-dimensional case. Suppose one keeps temperature T of the system constant and increases the Fermi energy ε_{F} of each color equally. In the limit of $\varepsilon_{\mathrm{F}}/T \gg 1$, the curvature of the single-particle spectrum becomes negligible in the relevant energy range. Then with the common Fermi velocity identified as the light velocity, the excitation spectrum can be approximated by relativistic chiral fermions with SU(ν)×SU(ν) symmetry. Thus the spectrum is seen to be equivalent to that described by the WZW Lagrangian supplemented by a charge boson part [22]. It should also be possible to make a direct proof that the present supersymmetric model in one-dimension is reduced to the WZW Lagrangian in the relativistic limit of hard-core fermions. This route of proof can make direct comparison of wave functions, and may supplement the standard proof [10] which makes use of equivalent current algebras.

The results of this paper are mostly obtained by cooperation with H. Yokoyama and J. Zittartz. The author thanks them for stimulating discussions. The author would also acknowledge useful discussions with F.D.M. Haldane and N. Kawakami particularly on the long-range t-J model.

References

[1] Y. Kuramoto and H. Yokoyama, Phys.Rev.Lett. **67**, 1338 (1991); Y. Kuramoto and H. Yokoyama, Physica **C185-189**, 1557 (1991).

[2] H. Yokoyama and Y. Kuramoto, J.Phys.Soc.Jpn. **61**, 3046 (1992).

[3] Y. Kuramoto, Physica **B186-188**, 831 (1993).

[4] N. Kawakami, Phys.Rev. B**45**, 7525 (1992).

[5] D.F. Wang, J.T. Liu and P. Coleman, Phys.Rev. B**46**, 6639 (1992).

[6] Z.N.C. Ha and F.D.M. Haldane, Phys.Rev. B**46**, 9359 (1992).

[7] F.D.M. Haldane, Phys.Rev.Lett. **60**, 635 (1988).

[8] B.S. Shastry, Phys.Rev.Lett. **60**, 639 (1988).

[9] Y. Kuramoto and J. Zittartz, to be published.

[10] E. Witten: Commun. Math. Phys. **92**, 455 (1984).

[11] P.B. Wiegmann, Phys. Rev. Lett. **60**, 821 (1988); D. Förster, Phys. Rev. Lett. **63**, 2140 (1989).

[12] J.F. Cornwell, *Group Theory in Physics III, Supersymmetries and Infinite Dimensional Algebras* (Academic Press, 1989).

[13] B. Sutherland, Phys. Rev. B**12**, 3795 (1975).

[14] P. Bares, G. Blatter and M. Ogata, Phys. Rev. B**44**, 130 (1991).

[15] B. Sutherland, Phys.Rev. B**38**, 5589 (1988).

[16] B. Sutherland, Phys. Rev. A**4**, 2019 (1971).

[17] M.A. Olshanetsky and A.M. Perelomov, Phys. Rep. **94**, 313 (1983).

[18] F.D.M. Haldane, Phys. Rev. Lett. **66**, 1529 (1991).

[19] N. Kawakami, Phys. Rev. B**46**, 1005 (1992).

[20] B. Sutherland and S. Shastry, Phys. Rev. B**71**, 5 (1993).

[21] F.D.M. Haldane et al., Phys. Rev. Lett. **69**, 2021 (1992).

[22] V.G. Knizhnik and A.B. Zamolodchikov, Nucl.Phys. B**247**, 83 (1984).

Hierarchical Extension of Quantum Models with Inverse-Square Interaction

N. Kawakami

Yukawa Institute for Theoretical Physics, Kyoto University,
Kyoto 606, Japan

Abstract. Hierarchical electron models with inverse-square interaction are solved in one dimension. The construction of the models is related to that for the fractional quantum Hall effect (FQHE). Under the chiral constraint the model describes characteristic properties of edge states for the FQHE. In particular the excitation spectrum is classified in terms of the topological-order matrix of the FQHE.

1. Introduction

One-dimensional (1D) exactly solvable quantum models with $1/r^2$ interaction have close relationship with fundamental ideas in condensed matter physics [1–18]. Various applications of the models have been proposed so far. In this paper we solve a hierarchy of multicomponent electron models with $1/r^2$ interaction by the asymptotic Bethe ansatz method [18]. The idea to construct the models is essentially same as that for the hierarchical FQHE with the filling factor $\nu = 1/(p_1 - 1/(p_2 - \cdots - 1/p_N)\cdots)$ [19–21]. Imposing the chiral constraint, the relationship to the edge states of the FQHE is clarified [22]. In particular the matrix deduced from the energy spectrum is shown to coincide with the topological-order matrix which characterizes the internal structure of the FQHE state [23].

2. Hierarchy of Models

Let us introduce a family of N-component electron models with $1/r^2$ interaction in the periodic ring of length L [18],

$$H = -\frac{1}{2}\sum_i \frac{\partial^2}{\partial x_i^2} + \sum_{\alpha \leq \beta}\sum_{i<j} d_{ij}^{-2}\eta_\beta(\eta_\beta + P_{ij}^{\alpha\beta}), \tag{1}$$

with chord distance $d_{ij} = (L/\pi)\sin[\pi(x_i - x_j)/L]$, where $P_{ij}^{\alpha\beta}$ is the spin exchange operator with spin indices $\alpha, \beta = 1, 2, \cdots, N$, and $\eta_\beta = \sum_{\gamma=1}^{\beta} \tau_\gamma$ with $\tau_\gamma \geq 0$. Starting from noninteracting SU(N) electrons

Springer Series in Solid-State Sciences, Vol. 118
Correlation Effects in Low-Dimensional Electron Systems
Editors: A. Okiji and N. Kawakami

(zero-th stage), let us specify each stage of the hierarchy. The first stage is defined by introducing interaction τ_1 to noninteracting electrons, and the resultant model coincides with the SU(N) Sutherland model [7]. The m-th family of the model, for which $\tau_j > 0$ (=0) for $j \leq m$ ($j > m$), is introduced by turning on the interaction τ_m to the $(m-1)$-th stage of the model. Note that the interaction τ_m acts on particles with spin indices $\alpha = m, m+1, \cdots, N$. We shall see that the construction of the present Hamiltonians is essentially same as that for the FQHE [20,21].

3. Asymptotic Bethe Ansatz

A powerful and systematic way to deal with the family of $1/r^2$ models is provided by the asymptotic Bethe ansatz (ABA) [1,8]. The essence of the idea is that the many-body S-matrix in this family can be decomposed into two-body matrices in spite of long-range nature of interaction [1,8]. This conjecture works for all the known integrable $1/r^2$ systems. For the present hierarchical models, the two-body S-matrix is given as [18]

$$S_{ij} = \lim_{\epsilon \to 0} \frac{k_i - k_j - i\epsilon P_{ij}^{\alpha\beta}}{k_i - k_j - i\epsilon} e^{-i\phi_{\alpha\beta}(k_i - k_j)}, \tag{2}$$

where the first factor arises from non-interacting SU(N) electrons, while the phase shift in the second factor, $\phi_{\alpha\beta}(k) = \eta_\beta \pi \mathrm{sgn}(k)$ for $\alpha \leq \beta$, reflects the $1/r^2$ interaction. Under periodic boundary conditions, many body S-matrix out of two-body S-matrices can be diagonalized by the standard nested Bethe ansatz. We then get ABA equations for N-kinds of rapidities $k_j^{(\alpha)}$ [18],

$$k_j^{(1)} L = 2\pi I_j^{(1)} + \sum_m \Phi(k_m^{(2)} - k_j^{(1)}) + \tau_1 \sum_l \Phi(k_j^{(1)} - k_l^{(1)}), \tag{3}$$

$$(\tau_\alpha + 1) \sum_l \Phi(k_m^{(\alpha)} - k_l^{(\alpha)}) + 2\pi I_m^{(\alpha)} = \sum_{q=\pm 1} \sum_j \Phi(k_m^{(\alpha)} - k_j^{(\alpha+q)}), \tag{4}$$

for $2 \leq \alpha \leq N$, where $\Phi(k) = \pi \mathrm{sgn}(k)$ and $I_j^{(\alpha)}$ is an integer or a half integer which classifies the charge and spin excitations. The total energy is given by $E = (1/2) \sum_j (k_j^{(1)})^2$. Following a standard technique, we can compute bulk quantities. Remarkably enough, all the static quantities are scaled by only one parameter [18]

$$\nu = 1/[p_1 - 1/(p_2 - \cdots - 1/p_N) \cdots], \tag{5}$$

where we have introduced the new parameters $p_i = \tau_i + 2 - \delta_{i1}$. For example, the compressibility is given as, $\kappa_c = (\nu/\pi)^2 n^{-1}$, where n is the total electron density. Also, the specific heat coefficient is given by $\gamma = (\pi/3)\sum_{\alpha=1}^{N}(1/v_\alpha)$ with the velocity $v_\alpha = \pi n/(2\nu)$. The compressibility and the charge velocity satisfy the Luttinger-liquid relation, $\pi\kappa_c v_1 = \nu/2$ [24]. Note that ν coincides with the filling factor of the FQHE when the model is applied to the edge states by imposing the chiral constraint (see below).

Let us now glance at the excitation spectrum. The excitation energy is specified as,

$$\epsilon = \frac{2\pi v}{L}[\frac{1}{4}\vec{M}^t \mathbf{T}\vec{M} + \vec{J}^t(\mathbf{T})^{-1}\vec{J}], \tag{6}$$

where the $N \times N$ matrix $\mathbf{T}$ is defined by

$$T_{ij} = p_i\delta_{ij} - \delta_{i(i\pm1)}. \tag{7}$$

Here the N-component vector $\vec{M}$ consists of quantum numbers which classify the charge and spin excitations, whereas the vector $\vec{J}$ is out of quantum numbers which carry the current $2J_\alpha k_F$ with the Fermi momentum k_F [18]. It should be noted that the matrix $\mathbf{T}$ deduced from the energy spectrum coincides with the topological-order matrix for the FQHE of the filling ν [23].

4. Chiral Liquids and Edge States of FQHE

Let us see what is modified if we consider only right (or left)-moving electrons in the 1D system. This restriction is called chiral constraint which is important for edge states of the FQHE in disk geometry. We will see that low-energy excitations of the present model exhibit characteristic properties of edge states of the FQHE [22]. In order to get chiral theory, we employ a simple trick which is valid for the $1/r^2$ systems. To suppress the current $\vec{J}$ which does not appear in the chiral case, we first add two electrons at the left and right Fermi points. This gives rise to the energy increase both for the right and left branches. By splitting the energy increment into two parts and discarding the left-moving part, we get the excitation spectrum for right-going electrons as,

$$\epsilon = \frac{\pi v}{L}\vec{M}^t \mathbf{T}\vec{M}. \tag{8}$$

Based on conformal field theory, we read the critical exponents for the correlation functions. For example, the critical exponent θ_α for the momentum distribution function is obtained as $\sum_{j=1}^{\alpha}(p_j - 2 + \delta_{j1})$. Note that all the results deduced from (8) agree with those of effective field theory (chiral Luttinger liquids) for the edge states of FQHE [22]. In particular the key matrix $\mathbf{T}$ in (6) and (8) coincides exactly with the topological order matrix which characterizes the internal structure of the FQHE with the filling fraction ν of (5) [18].

5. Lattice Effects

We now wish to observe what happens for the lattice case. We introduce a family of lattice models with $1/r^2$ interaction,

$$\mathcal{H} = \sum_{\alpha, i\neq j} d_{ij}^{-2} c_{i\alpha}^{\dagger} c_{j\alpha} + \sum_{\alpha\leq\beta, i<j} d_{ij}^{-2} \eta_{\beta}(\eta_{\beta} + P_{ij}^{\alpha\beta}), \tag{9}$$

with interaction parameters η_α (or p_α) defined in (1) and (7), where configurations with more than one electron at each site are assumed to be prohibited. At the first level the model reduces to the SU(N) t-J model discussed in [7]. We can solve the above family of lattice models and find ABA equations to be given in the same formula as in (3) and (4). For the lattice case, however, we encounter two crucial constraints [18]: (i) parameter p_α should be an even positive integer, and (ii) the rapidity ranges in $[-\pi, \pi]$. According to (i), the parameter ν in (5) turns out to be a fraction with odd denominator, from which one can see the analogy to the FQHE more explicitly. More remarkably the constraint (ii) brings about a singular property of the system at the density $n = \nu$ where Luttinger liquid state breaks down for $n > \nu$ [18]. This implies that the band-edge singularity for non-interacting SU(N) lattice electrons at the density $n = N$ is modified into the singularity at $n = \nu$ in the presence of the $1/r^2$ interaction. We note that this phenomenon is quite similar to the evolution of the FQHE starting from the integer QHE with the filling $n = N$ [21]. We briefly mention this point below.

6. Relationship to Composite Fermion Theory

We have seen that the present hierarchical models describe characteristic properties for the edge states of the FQHE with filling fraction ν. The correspondence is clearly seen in the role played by the phase shift function $r_\alpha \pi \mathrm{sgn}(k)$ produced by the $1/r^2$ interaction. It is known that in the composite fermion theory for the FQHE [21], the even number of flux quanta are attached to electrons in order to evolve the hierarchy starting from noninteracting electrons [21]. An important point is that attaching r_α flux quanta in the FQHE corresponds to introducing the phase shift function $r_\alpha \pi \mathrm{sgn}(k)$ in the present 1D system, which can be appropriately supplied by the $1/r^2$ interaction. Therefore one can clearly see from this correspondence why the family of models with $1/r^2$ interaction describes characteristic properties of a hierarchy of the FQHE state remarkably.

7. Holon-Spinon Basis and Electron Basis

We conclude this paper by mentioning another instructive relationship to the FQHE. We have classified the excitation spectrum (6) and (8) in terms of the matrix $\mathbf{T}$ and the corresponding quantum numbers out of charge and spin excitations. This basis, which may be referred to as *holon-spinon* basis, shows up naturally when one uses the Bethe ansatz method. There is another important basis in 1D, i.e. *electron* basis. Though the electron basis does not specify the excitation spectrum appropriately in case of different velocities, it still describes the critical behavior of correlation functions correctly [25]. Quantum numbers in the electron basis are obtained from those of holon-spinon basis via a linear transformation, $\vec{N} = \mathbf{U}\vec{M}$, and $\vec{D} = (\mathbf{U}^t)^{-1}\vec{J}$, with $U_{\alpha\beta} = \delta_{\alpha\beta} - \delta_{\alpha(\beta-1)}$. After the transformation, the matrix $\mathbf{T}$ is modified into the $N \times N$ symmetric matrix $\tilde{T}_{\alpha\beta}$ with elements ($\alpha \leq \beta$)

$$\tilde{T}_{1\beta} = \tau_1 + \delta_{\beta 1}, \tilde{T}_{2\beta} = \tau_1 + \tau_2 + \delta_{\beta 2}, \cdots, \tilde{T}_{N\beta} = \sum \tau_j + \delta_{\beta N}. \qquad (10)$$

Remarkably enough, we find the exactly same matrix in the 2D FQHE with the filling fraction ν [23]. The above two kinds of bases are called the *symmetric* basis and the *hierarchical* basis in the FQHE, both of which characterize the same topological order of the Abelian hierarchical FQHE [23,26].

This work is partly supported by Monbusho International Scientific Research Program and also by Grant-in-Aid from the Ministry of Education, Science and Culture.

REFERENCES

[1] B. Sutherland, Phys. Rev. A**4**, 2019 (1971); A**5**, 1372 (1971)
[2] F. D. M. Haldane, Phys. Rev. Lett. **60**, 635 (1988); **66**, 1529 (1991); **67**, 937 (1991)
[3] B. S. Shastry, Phys. Rev. Lett. **60**, 639 (1988); **69**, 164 (1992)
[4] Y. Kuramoto and H. Yokoyama, Phys. Rev. Lett. **67**, 1338 (1991)
[5] N. Kawakami and S.-K. Yang, Phys. Rev. Lett. **67**, 2493 (1991)
[6] F. D. M. Haldane, Z. N. C. Ha, J. C. Telstra, D. Bernard, and V. Pasquier, Phys. Rev. Lett. **69** 2021 (1992)

[7] Z. N. C. Ha and F. D. M. Haldane, Phys. Rev. **B46**, 9359 (1992)
[8] N. Kawakami, Phys. Rev. **B46**, 1005, 3192 (1992)
[9] A. P. Polychronakos, Phys. Rev. Lett. **69**, 703 (1992)
[10] H. Kiwata and Y. Akutsu, J. Phys. Soc. Jpn. **61**, 2161 (1992)
[11] K. Hikami and M. Wadati, J. Phys. Soc. Jpn. **62**, 469 (1993)
[12] P. J. Forrester, Nucl. Phys. **B388**, 671 (1992)
[13] B. S. Shastry and B. Sutherland, Phys. Rev. Lett. **70**, 4029 (1993)
[14] B. Sutherland and B. S. Shastry, Phys. Rev. Lett. **71**, 5 (1993)
[15] B. D. Simons, P. A. Lee and B. L. Altshuler, Phys. Rev. Lett. **70**, 4122 (1993)
[16] M. Fowler and J. A. Minahan, Phys. Rev. Lett. **70**, 2325 (1993)
[17] H. Frahm, J. Phys. **A26**, L437 (1993)
[18] N. Kawakami, Phys. Rev. Lett. **71**, 275 (1993)
[19] R. B. Laughlin, Phys. Rev. Lett. **50**, 1395 (1983)
[20] F. D. M. Haldane, Phys. Rev. Lett. **51**, 605 (1983); B. I. Halperin, Phys. Rev. Lett. **52**, 1583 (1984)
[21] J. K. Jain, Phys. Rev. Lett. **63**, 199 (1989); Phys. Rev. **B41**, 7653 (1990)
[22] X. G. Wen, Phys. Rev. Lett. **64**, 2206 (1990); Phys. Rev. **B41**, 12838 (1990); Mod. Phys. Lett. **B5**, 39 (1991)
[23] B. Blok and X. G. Wen, Phys. Rev. **B42**, 8133(1990); N. Read, Phys. Rev. Lett. **65**, 1502 (1990)
[24] F. D. M. Haldane, J. Phys. **C14**, 2585 (1981)
[25] N. Kawakami, J. Phys. Soc. Jpn. **62**, 2270, 2419 (1993)
[26] X. G. Wen and A. Zee, Phys, Rev. **B46**, 2290 (1992); J. Frölich and A. Zee, Nucl. Phys. **B364**, 517 (1991); M. Greiter and F. Wilczek, Mod. Phys. Lett. **B4**, 1063 (1990)

Confined System with $1/r^2$ Exchange Interaction and Its Application to Resonances in Narrow Channels

K. Vacek[1], *A. Okiji*[1], *and N. Kawakami*[2]

[1]Department of Applied Physics, Osaka University, Suita, Osaka 565, Japan
[2]Yukawa Institute for Theoretical Physics, Kyoto University, Kyoto 606, Japan

Abstract. An integrable model for SU(ν) electrons with inverse-square interaction is studied for the system with harmonic confinement. We obtain exactly the ground-state wavefunction of Jastrow type and the excited-state wavefunctions out of Hermite polynomials. The corresponding energy spectrum is computed exactly. The results are then applied to single-electron charging phenomena in narrow channels of semiconductor nanostructures.

1. Introduction

One-dimensional quantum systems with the inverse-square ($1/r^2$) interaction have attracted considerable attention [1-4]. There are several classes of integrable Hamiltonians related to this issue. In particular, a family of models with periodic boundary conditions has been studied intensively with the use of the multicomponent generalization of the $1/r^2$ models [5-9].

The inverse-square models with *harmonic confining potential* are also integrable [1,2]. This class of confined models may be quite interesting because of its realistic applications to phenomena in mesoscopic systems, such as conductance oscillations [10-12]. Unfortunately, multicomponent models (including electron models) with confinement have not been studied in detail so far [13], although the integrability for a class of the SU(ν) models was proven by Minahan and Polychronakos [14]. They also attempted to construct the ground-state wavefunction, and obtained its explicit formula up to $N = 6$ particles. The general expression for the ground-state wavefunction for many particles, however, has not been obtained yet. Moreover, the systematic construction of the excited-state wavefunctions seems to be much more complicated, which remains as a challenging problem.

Here, we solve the eigenvalue problem for the confined SU(ν) electron systems with $1/r^2$ interaction, and explicitly construct the eigenfunctions both for the ground-state as well as the excited states. We will show our strategy using the ordinary SU(2) electron model, from which the SU(ν) generalization follows straightforwardly. The model Hamiltonian is introduced in Sec. 2. In Sec. 3, the ground-state eigenfunction of Jastrow-product form and the ground-state energy are obtained exactly. The excited-state eigenfunctions are then constructed explicitly in Sec. 4, and the corresponding energy spectrum is computed. In Sec. 5, the results are applied to experiments on the conductance of narrow channels in semiconductor nanostructures.

2. Confined Model with Exchange Interaction

In the Calogero-Sutherland model [1,2], a one-dimensional electron system is considered to be confined by the harmonic potential $\frac{1}{2}m\omega_0^2x^2$ where m is the mass of a particle. We consider the following generalization [6,14] of the original model: In addition to the

Springer Series in Solid-State Sciences, Vol. 118
Correlation Effects in Low-Dimensional Electron Systems
Editors: A. Okiji and N. Kawakami

inverse-square pair interaction term, which is proportional to $\sum_{j>i}(x_j - x_i)^{-2}$, the spin-exchange term is introduced explicitly [6,8]. This leads to the following model Hamiltonian in units of $\hbar^2/m$ for the system of N electrons [14],

$$H = -\frac{1}{2}\sum_{i=1}^{N}\frac{\partial^2}{\partial x_i^2} + \frac{m^2\omega_0^2}{2\hbar^2}\sum_{i=1}^{N}x_i^2 + \sum_{j>i}\frac{\lambda(\lambda + P_{ij}^{\sigma})}{(x_j - x_i)^2}, \tag{1}$$

where $\lambda \geq 0$ is the dimensionless interaction parameter and P_{ij}^{σ} is the spin-exchange operator of two particles. This Hamiltonian with harmonic confinement is a variant of the periodic model introduced for SU(ν) systems in Refs. [6,8]. As it will be shown later, the spin-exchange is chosen here so as to reproduce the noninteracting limit correctly. Note that this interaction can remove a pathological degeneracy of the ground state for the electron model [12] arising from peculiar short-range properties of $1/r^2$ interaction in one-dimension. This point is crucial to confront the results with experiments, for example with conductance-oscillation phenomena (see Sec. 5).

3. Ground-State Properties

In construction of an eigenfunction, we follow a general conjecture [3,4,6] that the eigenfunction for integrable models with the inverse-square interaction is a product of the Jastrow factors $|x_j - x_i|^{\lambda}$ multiplied by the eigenfunction for the noninteracting case,

$$\Psi(x_1\sigma_1,\ldots,x_N\sigma_N) = \left\{\prod_{j>i}|x_j - x_i|^{\lambda}(x_j - x_i)^{\delta_{\sigma_j\sigma_i}}\exp\left[i\frac{\pi}{2}\mathrm{sgn}(\sigma_j - \sigma_i)\right]\right\}\prod_{i=1}^{N}\exp(-\frac{m\omega_0}{2\hbar}x_i^2), \tag{2}$$

where the particle spin is denoted by σ_i. Let us briefly show that this function is indeed the eigenfunction of the Hamiltonian (1). The application of the kinetic and potential terms results in the expression:

$$\frac{1}{\Psi}\left[\frac{1}{2}\sum_{i=1}^{N}\left(-\frac{\partial^2}{\partial x_i^2} + \frac{m^2\omega_0^2}{\hbar^2}x_i^2\right)\right]\Psi = \frac{m\omega_0}{2\hbar}\left[\lambda N(N-1) + N_{\uparrow}^2 + N_{\downarrow}^2\right] - A, \tag{3}$$

where the number of particles with the spin up (down) is $N_{\uparrow}$ ($N_{\downarrow}$), $N_{\uparrow} + N_{\downarrow} = N$. The term A can be rearranged as

$$A = A^{(\mathrm{I})} + A^{(\mathrm{II})} + A^{(\mathrm{III})} = \sum_{k<\ell}\frac{\lambda(\lambda-1)}{(x_k - x_\ell)^2} + \sum_{k<\ell}\frac{2\lambda\delta_{\sigma_k\sigma_\ell}}{(x_k - x_\ell)^2} + \sum_{i\neq k\neq \ell}\frac{\lambda\delta_{\sigma_i\sigma_k}}{(x_i - x_k)(x_i - x_\ell)}. \tag{4}$$

The action of the interaction term on the ansatz function (2) results in

$$B = \frac{1}{\Psi}\left[\sum_{j>i}\frac{\lambda(\lambda + P_{ij}^{\sigma})}{(x_j - x_i)^2}\right]\Psi = B^{(\mathrm{I})} + B^{(\mathrm{II})} + B^{(\mathrm{III})}$$

$$= \sum_{k<\ell}\frac{\lambda(\lambda-1)}{(x_k - x_\ell)^2} + \sum_{k<\ell}\frac{2\lambda\delta_{\sigma_k\sigma_\ell}}{(x_k - x_\ell)^2} + \sum_{k<\ell}\frac{\lambda}{(x_k - x_\ell)^2}\left[1 - \prod_{i\neq k\ell}\left(\frac{x_i - x_\ell}{x_i - x_k}\right)^{\delta_{\sigma_i\sigma_k} - \delta_{\sigma_i\sigma_\ell}}\right](1 - \delta_{\sigma_k\sigma_\ell}). \tag{5}$$

It can be easily seen that $A^{(\mathrm{I})} = B^{(\mathrm{I})}$ and $A^{(\mathrm{II})} = B^{(\mathrm{II})}$. We have proven elsewhere [15] that $A^{(\mathrm{III})} = B^{(\mathrm{III})}$ holds as well.

We thus find from Eq. (3) that the eigenenergy $E(N_{\uparrow}, N_{\downarrow})$ is expressed in the simple form as follows

$$E(N_{\uparrow}, N_{\downarrow}) = \frac{1}{2}\hbar\omega_0\left[\lambda(N_{\uparrow} + N_{\downarrow})(N_{\uparrow} + N_{\downarrow} - 1) + N_{\uparrow}^2 + N_{\downarrow}^2\right]. \tag{6}$$

Although (2) and (6) are the exact expressions for the many-particle eigenfunction and the eigenenergy of the Hamiltonian (1), it is not straightforward to show that the eigenfunction (2) and the eigenenergy (6) correspond to those for the ground state. Nevertheless, one can see the evidence for the ground-state in several limiting cases. In the limit of $\omega_0 \to 0$, the Hamiltonian (1) becomes a positive definite operator with the lowest eigenvalue of zero, as was shown earlier [8]. One readily sees that the formula (6) indeed produces the value of zero in such limiting case. Furthermore, in case of electrons with a single spin component, the eigenfunction (2) with the energy (6) reduces to the well-known exact ground-state wavefunction [1,2]. The ground-state wavefunction also agrees with the result for $N \leq 6$ obtained through a different method by Minahan and Polychronakos [14]. From the above observations we are confident that the eigenfunction (2) and the energy (6) generally describe the exact N-electron ground state of Hamiltonian (1).

The ground state with the eigenenergy $E_0(N_\uparrow, N_\downarrow)$ is the state with the minimal spin, $N_\uparrow = N_\downarrow$ (for N even) or $N_\uparrow = N_\downarrow \pm 1$ (for N odd). The formulas for the ground-state eigenfunction (2) and eigenenergy (6) are valid for *any* value λ of the interaction strength. Note that in the limit of weak interactions ($\lambda \to 0$), the formula (6) reproduces the eigenenergy expected for non-interacting electrons with the internal spin degrees of freedom, $E(N_\uparrow, N_\downarrow, \lambda \to 0) = \frac{1}{2}\hbar\omega_0(N_\uparrow^2 + N_\downarrow^2)$. We emphasize that the correct noninteracting energy could not be reproduced if we omit the exchange term from the Hamiltonian. This pathological behavior comes from a peculiar property of $1/r^2$ interaction (short-range divergence property) in one dimension, which gives rise to a large number of degeneracies in the ground state. Therefore, the exchange term is crucial to correctly describe interacting electrons in terms of $1/r^2$ interaction *continuously* from the *weakly* to the *strongly* correlated regimes [6].

It is easy to generalize the SU(2) model for the SU(ν) case where ν is the number of internal spin degrees of freedom. The expression of the ground-state eigenfunction is unchanged, but the spin indices σ_i previously denoted by $\uparrow$ and $\downarrow$ are now replaced by integers $1, 2, \ldots, \nu$. The expression for the SU(ν) ground-state eigenergy is computed straightforwardly,

$$E(N_1, N_2, \ldots, N_\nu) = \frac{1}{2}\hbar\omega_0 \left[\lambda N(N-1) + \sum_{\alpha=1}^{\nu} N_\alpha^2\right], \tag{7}$$

where $N = \sum_{\alpha=1}^{\nu} N_\alpha$. The manipulation for the SU(ν) case can be done similarly to the analysis for the SU(2) case [15]. The ground-state energy (7) has been conjectured from the field-theoretical approach [16].

4. Excited-State Wavefunctions and Energy Spectrum

Let us remark first that it is not straightforward to construct the eigenfunctions of excited states even for the single-component Calogero-Sutherland model with harmonic confinement. To our knowledge, it was treated by several authors, but the systematic construction of the eigenfunctions still remains as an open question. For example, Calogero [1] obtained the eigenfunctions for $N = 3$ and $N = 4$ for the single-component case. Later, the wavefunctions for $N = 5$ were constructed by Gambardella [17]. More recently, several authors have obtained the formal solution to the single-component [8,18] and multicomponent [14] Calogero-Sutherland problem by constructing anihilation and creation operators. Even if one employs such algebraic approach, however, the eigenfunctions quickly become cumbersome because of the sums included in the definition of operators [18].

Refraining from the formal algebraic operator approach, we have successfully constructed the excited-state wavefunctions even for multicomponent particles in systems with parabolic confinement [19]. Here, we would like to present the explicit wavefunctions of excited states for the SU(2) form of the Hamiltonian (1). We find the excited-state eigenfunctions as,

$$\Psi(x_1\sigma_1,\ldots,x_N\sigma_N) = \left\{\prod_{j>i} |x_j - x_i|^{\lambda}(x_j - x_i)^{\delta_{\sigma_j\sigma_i}} \exp\left[i\frac{\pi}{2}\operatorname{sgn}(\sigma_j - \sigma_i)\right]\right\}$$

$$\times \left[\sum_{m_1+m_2+\ldots+m_N=I} \prod_{i=1}^{N} \frac{1}{m_i!} H_{m_i}\left(\sqrt{\frac{m\omega_0}{\hbar}}x_i\right)\right] \prod_{i=1}^{N} \exp\left(-\frac{m\omega_0}{2\hbar}x_i^2\right), \tag{8}$$

where H_{m_i} are Hermite polynomials and $m_i = 0, 1, 2, \ldots$ are quantum numbers describing the excitations. The sum in the above formula is taken over all possible combinations of quantum numbers m_i, $i = 1, 2, \ldots, N$ with the indicated constraint. The energy spectrum is computed using the eigenfunction (8),

$$E(N_\uparrow, N_\downarrow) = \frac{1}{2}\hbar\omega_0 \left[\lambda(N_\uparrow + N_\downarrow)(N_\uparrow + N_\downarrow - 1) + N_\uparrow^2 + N_\downarrow^2\right] + \hbar\omega_0 I. \tag{9}$$

The level of the excitation is therefore determined by the quantum number $I = \sum_{i=1}^{N} m_i$. The ground state is obtained by $I = 0$, which is consistent with the results of the previous section. The lowest lying excitation is obtained for $I = 1$, the next one for $I = 2$, etc. Note that the spectrum consists of the correlated ground-state energy and excitations of free particles in a harmonic well. The correlations via $1/r^2$ interaction between particles in the harmonic well show up only through the ground-state energy [16,18]. The excitations do not include any effects of interactions provided that the number of electrons N is kept fixed [16,18]. The details and the proof of the formulas (8) and (9) will be published elsewhere [19].

It is straightforward to generalize our results for the SU(ν) model. The excited-states spectrum can be generally described as the renormalized harmonic oscillator spectrum proposed by the modified-nested Bethe ansatz [16]. The explicit construction of the wavefunctions (9) therefore establishes the microscopic foundation for the renormalized-harmonic oscillator description.

5. Single-Electron Charging Phenomena

The Hamiltonian (1) may have some interesting applications to the transport phenomena in semiconductor nanostructures. We apply the results to single-electron charging phenomena in narrow channels of Si metal-oxide-semiconductor field-effect transistors (MOSFET's) and GaAs-Al$_x$Ga$_{1-x}$As heterostructures [10]. The single-electron tunneling (SET) reveals that the correlation effects are important for the conductance of narrow channels [10]. Although the $1/r^2$ systems were already considered as models of SET [11,12], the results did not take into account the internal spin degrees of freedom correctly. For example, the main difficulty of the electron model in [12] arises from peculiar properties of $1/r^2$ interactions which prevent any exchange of particles in one-dimensional Calogero-Sutherland Hamiltonian [1,2]. As we have pointed out [15,20,21], the exchange-correlation effects may be studied in SET models described by Hamiltonian (1).

A simple model of the SET device can be presented as follows. The system of N electrons described by Hamiltonian (1) is weakly coupled to two black-body electrodes of adjustable potential $\mu_1 \approx \mu_2$. The weak coupling does not change the spacing of energy levels (9), although each level is slightly broadened into resonance. When the potential of one reservoir is slightly increased, another electron can enter the system via resonant tunneling. Such resonance occurs when the potential of the reservoir is equal to the lowest unoccupied level of the system. The tunneling of the electron into the system causes a peak in the conductance. For simplicity, we limit further discussion to vicinity of zero temperature. The position of the conductance peak $N \to N+1$ is then given by the chemical potential $\mu(N) = E_0(N+1) - E_0(N)$. Experimentally, only the spacing δ of adjacent peaks, *e.g.* $N \to N+1$ and $N+1 \to N+2$, can be observed. The spacing $\delta(N) = \mu(N+1) - \mu(N)$ depends on exchange-correlation effects in our model. Particularly, *two independent periods* δ_1 and δ_2 of conductance oscillations appear,

$$\delta_1 = \hbar\omega_0\lambda, \quad \delta_2 = \hbar\omega_0(\lambda+1),$$

which reflect the exchange effect caused by the spin degrees of freedom. Although such exchange effect is important in real systems, it was not taken into account in the previous models [11,12]. For example, the model of Johnson and Payne concludes that a single period of conductance oscillations $\delta = \hbar\omega_0(\lambda+1)$ appears for any interaction strength [12]. Using the empirical parameters of Johnson and Payne $\delta \approx 8.5\hbar\omega_0$ [12], we predict that two periods should appear as $\delta_1 = 7.5\hbar\omega_0$ and $\delta_2 = 8.5\hbar\omega_0$. For weaker interactions (smaller λ), the above exchange effect becomes relatively larger with the two distinct periods δ_1 and δ_2. For finite temperatures, the resonant effects will be affected by excited states calculated in the previous section. The further discussion of the behavior for the above SET model at finite temperatures will be presented elsewhere.

The presented SET model can be related to the experiments on periodic conductance oscillations in narrow channels [10]. The period of conductance resonances was observed to be not only a single one [10]. The main period in the Fourier spectrum is accompanied by several peaks at anharmonic frequencies. The origin of multiple periods has been explained by the possible existence of several segments with comparable pinning energies [10]. It can be interesting to investigate experimentally whether one may observe two periods in the conductance oscillations due to the exchange effect for a single pinned segment. This may provide further knowledge of the exchange-correlation effects in one-dimensional mesoscopic systems.

Acknowledgments. We gratefully acknowledge fruitful discussions with H. Kasai, Y. Kuramoto, J. Mašek, H. Nakanishi, S. Suga and S.-K. Yang. This work was partly supported by a Grant-in-Aid from the Ministry of Education, Science and Culture. K. V. also acknowledges support from the Japanese Government (Monbusho) Scholarship Program, and N. K. acknowledges support from the Monbusho International Scientific Research Program.

References

[1] F. Calogero, J. Math. Phys. **10**, 2197 (1969); *ibid* **12**, 419 (1971).
[2] B. Sutherland, J. Math. Phys. **12**, 246 (1971); Phys. Rev. A **4**, 2019 (1971); *ibid.* A **5**, 1372 (1971).
[3] F. D. M. Haldane, Phys. Rev. Lett. **60**, 635 (1988); *ibid.* **66**, 1529 (1991).
[4] B. S. Shastry, Phys. Rev. Lett. **60**, 639 (1988); *ibid.* **69**, 164 (1992).
[5] Y. Kuramoto and H. Yokoyama, Phys. Rev. Lett. **67**, 1338 (1991).
[6] Z. N. C. Ha and F. D. M. Haldane, Phys. Rev. B **46**, 9359 (1992).
[7] N. Kawakami, Phys. Rev. B **46**, 1005, 3192 (1992); J. Phys. Soc. Jpn. **62**, 2270 (1993).
[8] A. P. Polychronakos, Phys. Rev. Lett. **69**, 703 (1992); K. Hikami and M. Wadati, Phys. Lett. A **173**, 263 (1993).
[9] B. Sutherland and B. S. Shastry, Phys. Rev. Lett. **71**, 5 (1993).
[10] S. B. Field, M. A. Kastner, U. Meirav, J. H. F. Scott-Thomas, D. A. Antoniadis, H. I. Smith, and S. J. Wind, Phys. Rev. B **42**, 3523 (1990).
[11] S. Tewari, Phys. Rev. B **46**, 7782 (1992).
[12] N. F. Johnson and M. C. Payne, Phys. Rev. Lett. **70**, 1513 (1993); *ibid.* **70**, 3523 (1993).
[13] Note that the spectrum for the SU(ν) spin chain has been studied numerically by H. Frahm, J. Phys. A **26**, L473 (1993), under a specific condition for the positions of particles.
[14] J. A. Minahan and A. P. Polychronakos, Phys. Lett. B **302**, 265 (1993).
[15] K. Vacek, A. Okiji, and N. Kawakami, Phys. Rev. B **49**, 15 February (1994), in press.
[16] N. Kawakami, J. Phys. Soc. Jpn. **62**, 4163 (1993).
[17] P. J. Gambardella, J. Math. Phys. **16**, 1172 (1975).
[18] L. Brink, T. H. Hansson, and M. A. Vasiliev, Phys. Lett. B **286**, 109 (1992).
[19] K. Vacek, A. Okiji, and N. Kawakami, to be published.
[20] N. Kawakami, Kyoto preprint YITP/K-1013 (1993).
[21] K. Vacek, A. Okiji, and N. Kawakami, J. Phys. Soc. Jpn. **62**, 3818 (1993).

Part II

Symmetries and Excitations for Integrable Models

The Role of Symmetries in Systems of Strongly Correlated Electrons

V.E. Korepin and F.H.L. Essler

Institute for Theoretical Physics, State University of New York at Stony Brook, Stony Brook, NY 11794-3840, USA

Abstract. We discuss the role of the $SO(4)$ symmetry for the dynamics of the Hubbard model. The elementary excitations in the half-filled band (in the one-dimensional case) form the fundamental representation of $SU(2) \times SU(2)$. The exact scattering matrix has the same symmetry. By making use of this strong influence of symmetries on the dynamics, it is possible to extend the Hubbard model such that it becomes superconducting. One should add additional nearest neighbour interactions to the hamiltonian, which enlarge the symmetry of the model from $SO(4)$ to the superalgebra $SU(2|2)$.

1. The One-Dimensional Hubbard Model

We first consider the Hubbard model in one space and one time dimension. Electrons on a lattice are described by operators $c_{j,\sigma}$, $j = 1,\ldots,L$, $\sigma = \pm 1$, where $L = 2 \times \text{odd}$ is the (even) total number of lattice sites. These are canonical Fermi operators with anti-commutation relations given by $\{c^\dagger_{i,\sigma}, c_{j,\tau}\} = \delta_{i,j}\delta_{\sigma,\tau}$. The state $|0\rangle$ (the Fock vacuum) satisfies $c_{i,\sigma}|0\rangle = 0$. The Hubbard hamiltonian has the form

$$H(U) = -\sum_{j=1}^{L}\sum_{\sigma=\uparrow,\downarrow}\left(c^\dagger_{j,\sigma}c_{j+1,\sigma} + c^\dagger_{j+1,\sigma}c_{j,\sigma}\right) + 4U\sum_{j=1}^{L}(n_{j,\uparrow} - \frac{1}{2})(n_{j,\downarrow} - \frac{1}{2}) \tag{1.1}$$

where $n_{i,\sigma} = c^\dagger_{i,\sigma}c_{i,\sigma}$ denotes the number operator for electrons with spin σ on site i. Both kinetic and potential energy terms in (1.1) are separately invariant under rotations in spin space. The corresponding generators are

$$S = \sum_{j=1}^{L} c^\dagger_{j,\uparrow}c_{j,\downarrow}\,, \quad S^\dagger = \sum_{j=1}^{L} c^\dagger_{j,\downarrow}c_{j,\uparrow}\,, \quad S^z = \sum_{j=1}^{L}\frac{1}{2}(n_{j,\downarrow} - n_{j,\uparrow})\,. \tag{1.2}$$

The hamiltonian (1.1) is also known to exhibit various discrete symmetries[1,2]. Under the particle-hole transformation

$$c_{j,\sigma} \leftrightarrow (-1)^j c^\dagger_{j,\sigma}\,, \tag{1.3}$$

Springer Series in Solid-State Sciences, Vol. 118
Correlation Effects in Low-Dimensional Electron Systems
Editors: A. Okiji and N. Kawakami

(1.1) transforms into itself. If we perform the particle-hole transformation for spin-up only

$$c_{j,\uparrow} \leftrightarrow (-1)^j c_{j,\uparrow}^\dagger , \tag{1.4}$$

it transforms according to $H(U) \to H(-U)$. Under the same transformation the spin-SU(2) (1.2) will turn into another SU(2) algebra with generators

$$\eta = \sum_{j=1}^{L} (-1)^j c_{j,\uparrow} c_{j,\downarrow} , \quad \eta^\dagger = \sum_{j=1}^{L} (-1)^j c_{j,\downarrow}^\dagger c_{j,\uparrow}^\dagger , \quad \eta^z = \frac{1}{2} \sum_{j=1}^{L} (n_{j,\uparrow} + n_{j,\downarrow} - 1) . \tag{1.5}$$

This algebra is called η-SU(2) algebra[3]. The generators (1.5) commute with the generators (1.2), and with the hamiltonian (1.1) as well. As the lattice length L is even we observe that

$$\eta^z + S^z = \text{integer}, \tag{1.6}$$

which implies that the complete continuous symmetry algebra of the Hubbard model is $SO(4) = SU(2) \times SU(2)/Z_2$[4–7]. This symmetry has essential consequences for the dynamics of the model. Let us consider the half-filled band, where the number of electrons is equal to the number of sites of the lattice. The half-filled ground state of the model was determined by E.H. Lieb and F.Y. Wu in [1]. The excitation spectrum in the repulsive case $U > 0$ was studied by various authors[8–14]. Here we give a complete classification of the excitation spectrum by means of the $SO(4)$ symmetry of the model. The spectrum is described in terms of the *scattering states* of four elementary excitations only. These four elementary particles form the fundamental representation of $SU(2) \times SU(2)$. In the case of repulsion ($U > 0$) the situation is as follows:

(i) Two particles carry spin but no charge. Thus they transform in the spin-$\frac{1}{2}$ representation of the spin-SU(2) and are singlets with respect to the η-SU(2). The corresponding $SU(2) \times SU(2)$ representation is denoted by $(\frac{1}{2}, 0)$. These "spinons" are similar to the two elementary excitations in the spin-$\frac{1}{2}$ Heisenberg XXX antiferromagnet at zero magnetic field[15–17]. Energy and momentum of the spinons are given by[8,9]

$$\begin{aligned} p_s(\lambda) &= \frac{\pi}{2} - \int_0^\infty \frac{d\omega}{\omega} \frac{J_0(\omega)\sin(\omega\lambda)}{\cosh(\omega U)} \qquad 0 < p_s(\lambda) < \pi \\ \epsilon_s(\lambda) &= 2 \int_0^\infty \frac{d\omega}{\omega} \frac{J_1(\omega)\cos(\omega\lambda)}{\cosh(\omega U)} . \end{aligned} \tag{1.7}$$

(ii) The other two elementary excitations, the "holon" and "antiholon", carry charge $\mp e$ but no spin. They form an irreducible spin-$\frac{1}{2}$ representation of the η-SU(2) (1.5), and thus the $(0, \frac{1}{2})$ representation of $SU(2) \times SU(2)$. The momentum of holon and antiholon differs by π as $\eta^\dagger$ does not commute with the momentum operator but changes the momentum by π. Energy and momentum of the holon are given by[10]

$$
\begin{aligned}
p_h(k) &= \frac{\pi}{2} - k - \int_0^\infty \frac{d\omega}{\omega} \frac{J_0(\omega)\sin(\omega\,\sin(k))e^{-\omega U}}{\cosh(\omega U)} = p_{ah}(k) + \pi \\
\epsilon_h(k) &= 2U + 2\,\cos(k) + 2\int_0^\infty \frac{d\omega}{\omega} \frac{J_1(\omega)\cos(\omega\,\sin(k))e^{-\omega U}}{\cosh(\omega U)} = \epsilon_{ah}(k)\,.
\end{aligned}
\tag{1.8}
$$

To illustrate the interpretation of the excitation spectrum in terms of scattering states of elementary particles let us now consider the two particle sector. There are a total of 12 two-particle states forming the following irreducible $SO(4)$ representations

$$
(1,0)\oplus(0,0)\oplus(0,1)\oplus(0,0)\oplus(\frac{1}{2},\frac{1}{2})\,. \tag{1.9}
$$

Here $(1,0)\oplus(0,0)$ corresponds to the spin-SU(2) triplet and singlet scattering states of two spinons, $(0,1)\oplus(0,0)$ corresponds to the η-SU(2) triplet and singlet scattering states of two holons/antiholons, and $(\frac{1}{2},\frac{1}{2})$ describes the scattering of one spinon on one holon/antiholon. The lowest weight states af all these $SO(4)$-multiplets can be constructed by means of the Bethe Ansatz[18]. The derivation of the two-particle spectrum from the Bethe Ansatz can be found in [9,10].

In the case of *attraction* ($U < 0$) the situation is as follows: again there are four elementary excitations, forming the fundamental representation of $SO(4)$. The chargeless spin-carriers in the $(\frac{1}{2},0)$ representation of $SO(4)$ have the dispersion

$$
\begin{aligned}
p_s(k) &= \pi - k - \int_0^\infty \frac{d\omega}{\omega} \frac{J_0(\omega)\sin(\omega\,\sin(k))e^{-\omega|U|}}{\cosh(\omega U)} \\
\epsilon_s(k) &= 2|U| + 2\,\cos(k) + 2\int_0^\infty \frac{d\omega}{\omega} \frac{J_1(\omega)\cos(\omega\,\sin(k))e^{-\omega|U|}}{\cosh(\omega U)}\,,
\end{aligned}
\tag{1.10}
$$

which up to the shift of momentum by $\frac{\pi}{2}$ is the same as for the holon in the repulsive case. The spinless charge-carriers (one particle, one hole in the $(0,\frac{1}{2})$ representation of $SO(4)$) have the dispersions[19]

$$
\begin{aligned}
p_c^p(\lambda) &= \pi - \int_0^\infty \frac{d\omega}{\omega} \frac{J_0(\omega)\sin(\omega\lambda)}{\cosh(\omega U)} = \pi - p_c^h(\lambda)\,, \\
\epsilon_c^p(\lambda) &= 2\int_0^\infty \frac{d\omega}{\omega} \frac{J_1(\omega)\cos(\omega\lambda)}{\cosh(\omega U)} = \epsilon_c^h(\lambda)\,,
\end{aligned}
\tag{1.11}
$$

which are similar to the spinon dispersions in the repulsive case.

This shows that repulsive and attractive Hubbard model are literally related by an interchange of charge and spin degrees of freedom. There is a simple argument, which gives a first hint at this fact (but does of course not prove it): The particle-hole transformation for spin-up (1.4) relates repulsive and attractive Hubbard model as the hamiltonian transforms as $H(U) \to H(-U)$. As we have seen above, the transformation (1.4) interchanges the spin-SU(2) (1.2)

and the η-SU(2) (1.5). This corresponds to an interchange of charge and spin degrees of freedom.
The $SO(4)$-symmetry does not only define the excitation spectrum as discussed above, but also the exact scattering matrix. The S-matrix can be constructed from the Bethe Ansatz by the method of [20], and is found to be $SO(4)$ invariant. It obeys the Yang-Baxter equation, which implies the two-particle reducibility of the N-body problem. The two-particle S-matrix is a 16×16 dimensional and blockdiagonal

$$S = \begin{pmatrix} S_{ss}(\mu_1) & 0 & 0 & 0 \\ 0 & S_{sc}(\mu_2) & 0 & 0 \\ 0 & 0 & S_{cs}(\mu_3) & 0 \\ 0 & 0 & 0 & S_{cc}(\mu_4) \end{pmatrix} , \tag{1.12}$$

where in the repulsive case

$$\begin{aligned}
S_{ss}(\mu_1) &= -\frac{\Gamma(\frac{1+i\mu_1}{2})\Gamma(1-\frac{i\mu_1}{2})}{\Gamma(\frac{1-i\mu_1}{2})\Gamma(1+\frac{i\mu_1}{2})}\left(\frac{\mu_1}{\mu_1+i}I + \frac{i}{\mu_1+i}\Pi\right) , \quad \mu_1 = \frac{\lambda_1^h - \lambda_2^h}{2U}, \\
S_{sc}(\mu_2) &= -i\,\frac{1+i\,\exp(\pi\mu_2)}{1-i\,\exp(\pi\mu_2)}I , \quad \mu_2 = \frac{\lambda^h - \sin(k^h)}{2U} > 0 , \\
S_{cs}(\mu_3) &= -i\,\frac{1+i\,\exp(\pi\mu_3)}{1-i\,\exp(\pi\mu_3)}I , \quad \mu_3 = \frac{\sin(k^h) - \lambda^h}{2U} > 0 , \\
S_{cc}(\mu_4) &= \frac{\Gamma(\frac{1-i\mu_4}{2})\Gamma(1+\frac{i\mu_4}{2})}{\Gamma(\frac{1+i\mu_4}{2})\Gamma(1-\frac{i\mu_4}{2})}\left(\frac{\mu_4}{\mu_4-i}I - \frac{i}{\mu_4-i}\Pi\right), \mu_4 = \frac{\sin(k_1^h) - \sin(k_2^h)}{2U}.
\end{aligned} \tag{1.13}$$

Here I and Π are the 4×4 identity and permutation matrices

$$I = \begin{pmatrix} 1 & 0 & 0 & 0 \\ 0 & 1 & 0 & 0 \\ 0 & 0 & 1 & 0 \\ 0 & 0 & 0 & 1 \end{pmatrix} , \Pi = \begin{pmatrix} 1 & 0 & 0 & 0 \\ 0 & 0 & 1 & 0 \\ 0 & 1 & 0 & 0 \\ 0 & 0 & 0 & 1 \end{pmatrix} . \tag{1.14}$$

Here S_{ss} describes the scattering of two spinons with rapidities λ_1^h and λ_2^h. As a function of spectral parameters it is identical to the S-matrix for the spin-$\frac{1}{2}$ Heisenberg antiferromagnet[15,16]. The blocks S_{sc} and S_{cs} describe the scattering on one spinon on one holon/antiholon. In S_{sc} the spinon is the active scatterer, in S_{cs} the holon/antiholon. The S-matrix for scattering of two holons/antiholons is given by S_{cc}. Clearly the four blocks in the complete S-matrix individually obey the Yang-Baxter equation, and so does S. The S-matrix in the attractive case is identical to (1.12) as a function of *spectral parameter* up to the interchange of spin-and charge degrees of freedom.

To further elucidate the meaning of the $SO(4)$ symmetry in the Hubbard model let us consider the electronic space of states over *one* site of the lattice (with index j). This space is four-dimensional. There are two bosonic basis vectors: the empty vacuum $|0\rangle_j$, and the doubly occupied state $|\uparrow\downarrow\rangle_j = c_{j,\uparrow}^\dagger c_{j,\downarrow}^\dagger|0\rangle_j$. The other two basis vectors are fermionic: $|\uparrow\rangle_j = c_{j,\uparrow}^\dagger|0\rangle_j$ and $|\downarrow\rangle_j = c_{j,\downarrow}^\dagger|0\rangle_j$. $SO(4)$ invariance means that the dynamics is invariant under independent rotation of the bases in the bosonic and fermionic subspaces. It is possible to extend this symmetry by modifying the hamiltonian (1.1). The *maximal* possible symmetry corresponds to invariance under rotations between any of the four basis vectors. The resulting symmetry algebra contains both bosonic and fermionic generators and is called $SU(2|2)$.

2. The $SU(2|2)$ Model of Superconductivity

In what follows we consider a d-dimensional lattice with L sites. Links between the nearest neighbour sites j and k are denoted by $< jk >$. The superalgebra $SU(2|2)$ has 16 generators, 8 of which are bosonic and fermionic respectively. The eight bosonic operators fall into two $su(2)$ and two $u(1)$ subalgebras. The two $su(2)$ algebras are generated by the spin-operators (1.2) and by the η-like operators (note that they are different from (1.5))

$$\eta = \sum_{j=1}^{L} c_{j,\uparrow} c_{j,\downarrow} , \quad \eta^\dagger = \sum_{j=1}^{L} c_{j,\downarrow}^\dagger c_{j,\uparrow}^\dagger , \quad \eta^z = \sum_{j=1}^{L} -\frac{1}{2} n_j + \frac{1}{2} . \tag{2.1}$$

The two $u(1)$ charges are given by the identity operator and the Hubbard interaction

$$X = \sum_{j=1}^{L} (n_{j,\uparrow} - \frac{1}{2})(n_{j,\downarrow} - \frac{1}{2}) . \tag{2.2}$$

The eight fermionic generators are given by

$$Q_\sigma = \sum_{j=1}^{L} (1 - n_{j,-\sigma}) c_{j,\sigma} , \quad \tilde{Q}_\sigma = \sum_{j=1}^{L} n_{j,-\sigma} c_{j,\sigma} , \ \sigma = \uparrow, \downarrow \tag{2.3}$$

and their hermitean conjugates. The electronic hamiltonian that commutes with all generators (1.2),(2.1),(2.2),(2.3) is $H^0 = -\sum_{<jk>} H_{j,k}^0$, where

$$\begin{aligned} H_{j,k}^0 = {} & c_{k,\uparrow}^\dagger c_{j,\uparrow}(1 - n_{j,\downarrow} - n_{k,\downarrow}) + c_{j,\uparrow}^\dagger c_{k,\uparrow}(1 - n_{j,\downarrow} - n_{k,\downarrow}) \\ & + c_{k,\downarrow}^\dagger c_{j,\downarrow}(1 - n_{j,\uparrow} - n_{k,\uparrow}) + c_{j,\downarrow}^\dagger c_{k,\downarrow}(1 - n_{j,\uparrow} - n_{k,\uparrow}) \\ & + \frac{1}{2}(n_j - 1)(n_k - 1) + c_{j,\uparrow}^\dagger c_{j,\downarrow}^\dagger c_{k,\downarrow} c_{k,\uparrow} + c_{j,\downarrow} c_{j,\uparrow} c_{k,\uparrow}^\dagger c_{k,\downarrow}^\dagger \\ & - \frac{1}{2}(n_{j,\uparrow} - n_{j,\downarrow})(n_{k,\uparrow} - n_{k,\downarrow}) \\ & - c_{j,\downarrow}^\dagger c_{j,\uparrow} c_{k,\uparrow}^\dagger c_{k,\downarrow} - c_{j,\uparrow}^\dagger c_{j,\downarrow} c_{k,\downarrow}^\dagger c_{k,\uparrow} \\ & + (n_{j,\uparrow} - \frac{1}{2})(n_{j,\downarrow} - \frac{1}{2}) + (n_{k,\uparrow} - \frac{1}{2})(n_{k,\downarrow} - \frac{1}{2}) . \end{aligned} \tag{2.4}$$

This does obviously not contain the Hubbard coupling constant. However, the hamiltonian H^0 can be simultaneously diagonalized with the three generators of the Cartan-subalgebra of $SU(2|2)$, which are X, S^z, and η^z. Thus it is possible to study the more general hamiltonian[21]

$$H(\mu,U,h) = H^0 + U\sum_{j=1}^{L}(n_{j,\uparrow}-\frac{1}{2})(n_{j,\downarrow}-\frac{1}{2}) - \mu\sum_{j=1}^{L} n_j + h\sum_{j=1}^{L}(n_{j,\uparrow}-n_{j,\downarrow}), \quad (2.5)$$

where U is the Hubbard coupling constant, h is an external magnetic field, and μ is the chemical potential. As compared to the Hubbard hamiltonian (1.1), the new model (2.5) features a number of additional interactions. These fall into three categories:

(i) Nearest neighbour spin-spin and charge-charge interactions, which are similar to the ones of the supersymmetric t-J model[22,23]

(ii) So-called "bond-charge repulsion" terms, which were proposed by Hirsch in his model of hole superconductivity[24](see also [25])

(iii) A pair-hopping term (allowing local electron pairs to move coherently), which was previously studied in the Penson-Kolb-Hubbard models[26–28].

The couplings of the additional interactions (i)-(iii) are fixed by the requirement of $SU(2|2)$-supersymmetry of the hamiltonian (2.4). Like the Hubbard model the $SU(2|2)$ model has a discrete symmetry under the particle-hole transformation $c^\dagger_{j,\sigma} \leftrightarrow c_{j,\sigma}$, where $H(\mu,U,h)$ transforms (up to a constant) into $H(-\mu,U,h)$. Despite the seemingly complicated form of the hamiltonian (2.5) the mathematical structure of the model is quite simple. This is because $H^0_{j,k}$ acts like a *graded permutation operator* between sites j and k. Let us illustrate this point in more detail: there are four allowed configurations $|0\rangle_j$, $|\uparrow\rangle_j$, $|\downarrow\rangle_j$, $|\uparrow\downarrow\rangle_j$ at any site j. $H^0_{j,k}$ acts by permuting the configurations of sites j and k, picking up a minus sign if *both* of the configurations are fermionic, *i.e.*

$$\begin{aligned} H^0_{j,k}|0\rangle_j \times |0\rangle_k &= |0\rangle_j \times |0\rangle_k \\ H^0_{j,k}|\downarrow\uparrow\rangle_j \times |\downarrow\uparrow\rangle_k &= |\downarrow\uparrow\rangle_j \times |\downarrow\uparrow\rangle_k \\ H^0_{j,k}|0\rangle_j \times |\sigma\rangle_k &= |\sigma\rangle_j \times |0\rangle_k \\ H^0_{j,k}|\tau\rangle_j \times |\sigma\rangle_k &= -|\sigma\rangle_j \times |\tau\rangle_k \quad , \qquad \sigma,\tau = \uparrow,\downarrow \\ &\textit{etc.} \end{aligned} \quad (2.6)$$

Models of the permutation type were first studied by B. Sutherland in [29]. From the above it is clear that the individual numbers of electrons with spin up and spin down, the number of empty and doubly occupied sites are conserved quantities. Furthermore, in the sector with no doubly occupied sites the model is identical to the supersymmetric t-J model. This is because the t-J hamiltonian at the supersymmetric point $J = \pm 2t$ is a graded permutation operator on the three permitted configurations $|0\rangle_j$, $|\uparrow\rangle_j$, $|\downarrow\rangle_j$ (in the t-J model there is a constraint projecting out doubly occupied sites).

The zero-temperature phase diagram of the $SU(2|2)$ model was determined for arbitrary dimension in [30]. The ground state in the attractive regime ($U < 0$, $\mu = 0$) is unique and given by

$$|\Omega\rangle = (\eta^\dagger)^N |0\rangle \ ,$$

where N is fixed by the imposed density. This state exhibits off-diagonal long-range order (ODLRO)[31,32] and is thus *superconducting*. This fact is established by considering the following off-diagonal matrix element of the reduced density matrix ρ_2 for the state $|\Omega\rangle = (\eta^\dagger)^N|0\rangle$:

$$(\rho_2)_{kl} = \langle(k,\downarrow)(k,\uparrow)|\rho_2|(l,\uparrow)(l,\downarrow)\rangle = \frac{\langle\Omega|c^\dagger_{k,\downarrow}c^\dagger_{k,\uparrow}c_{l,\uparrow}c_{l,\downarrow}|\Omega\rangle}{\langle\Omega|\Omega\rangle}. \qquad (2.7)$$

In the thermodynamic limit $L \to \infty$, $D = \frac{N}{L}$ fixed, the matrix element (2.7) approaches a *nonzero* value A for large distances $|k - l| \gg 1$

$$A = \lim_{L\to\infty} \frac{1}{L^2} \sum_{k,l} (\rho_2)_{kl} = D(1-D) \ . \qquad (2.8)$$

This estabilishes the property of ODLRO. The mechanism of superconductivity is condensation of zero momentum local electron pairs. It is an extremely interesting feature of the model, that superconductivity also exists for on-site repulsion $U_c > U > 0$ at $\mu = 0$. Here U_c is the critical value of coupling constant, for which superconductivity is destroyed. In one dimension we find $U_c = 4\ln(2)$. The ground state is this region is given by the expression

$$(\eta^\dagger)^n |st - J\rangle \ , \qquad (2.9)$$

where $|st - J\rangle$ is the ground state of the supersymmetric t-J model at induced value of chemical potential $\mu_{tJ} = \frac{U}{2}$, and n is fixed by the imposed density. The ground state (2.9) exhibits off-diagonal long-range order as well and is thus superconducting. The other phases at zero temperature are a "metallic" phase, where the ground state is the same as for the supersymmetric t-J model below half-filling ($0 < \mu + \frac{U}{2} < U_c$), and an insulating phase where the ground state is equal to the half-filled ground state of the supersymmetric t-J model (for $\mu + \frac{U}{2} \geq U_c$). The ground state for $\mu + \frac{U}{2} \leq 0, \mu < 0$ is the completely empty state.

So far our discussion of the $SU(2|2)$ model has been general as far as dimensionality is concerned. Let us now consider the case of one spatial dimension in more detail. In one dimension the $SU(2|2)$ model is solvable by means of the Bethe Ansatz[33,34]. There exist 6 different representations for the Bethe equations. This is very similar to the situation in the supersymmetric t-J model, where there exist 3 different representations, which were derived by Sutherland[29], Lai[35], and Eßler and Korepin[36] respectively. All Bethe states are lowest weight states with respect to the $SU(2|2)$-symmetry algebra of the model. The ground state is in general a singlet with respect to the bosonic $SO(4)$ subalgebra of $SU(2|2)$, but not with respect to the action of all fermionic symmetry generators. It belongs to a $SU(2|2)$ multiplet of dimension 16. This is drastically different from the Hubbard model, where the ground

state was a singlet with respect to the $SO(4)$ symmetry. The spectrum of low-lying excitations can be determined exactly in one dimension. We will consider the case of less then half-filling explicitly. At half-filling there are additional superselection rules, and the excitation spectrum is somewhat different.
The Bethe Ansatz excitations for *repulsive* on-site interaction are described in terms of the following quasiparticles: There are two elementary excitations, which carry spin but no charge. They transform in the spin-$\frac{1}{2}$ representation of the spin-SU(2) and are very similar to the *spinons* in the Heisenberg XXX antiferromagnet and the Hubbard model. They are gapless and have a linear dispersion around their Fermi surface. Like in the Hubbard model they appear only in *pairs*.

There is one gapless particle-hole excitation which carries charge but no spin. The hole is called holon, the particle antiholon. Holon and antiholon always appear together, which is different from the Hubbard model. The dispersion is gapless and linear around the Fermi surface.

The excitations described above are already present in the supersymmetric t-J model[37], which we recall is a submodel of the $SU(2|2)$model. In addition there exist electronic excitations with quadratic dispersions and a gap, and excitations involving local electron pairs. These also have a quadratic dispersion and a gap for $\mu < 0$, but become gapless for $\mu = 0$. For *attractive* on-site interaction we find an infinite number of excitations with cosine-like dispersions. Amongst these are electrons and local pairs and their bound states. The quadratic behaviour of the gapless dispersion of the localons in the superconducting repulsive sector clearly indicates non-conformal properties of the model. The methods of conformal field theory[38–43] can therefore not be applied directly to evaluate the critical exponents. However, by using the algebra of the $SU(2|2)$ generators it is possible to express the asymptotics of correlators in the $SU(2|2)$ model in terms of asymptotics of correlators in the supersymmetric t-J model. Thus we are able to determine the critical exponents of the $SU(2|2)$ by using the conformal field theory results for the t-J model[44,45], which were derived by the general method of [39,40]. Let us illustrate this on a few examples. We recall that the ground state in the sector under consideration ($4\ln(2) = U_c > U > 0$, $\mu = 0$) is of the form $|\psi\rangle = \eta^{\dagger n}|\Psi\rangle$, where $|\Psi\rangle$ is the ground state of the supersymmetric t-J model for a value of chemical potential equal to $\frac{U}{2}$. In particular $|\Psi\rangle$ contains no doubly occupied sites and thus $\eta|\Psi\rangle = 0$. Furthermore $\eta^\dagger|\Psi\rangle \neq 0$ as $U < U_c$, which corresponds to less than half-filling of the t-J ground state $|\Psi\rangle$. The number of single electrons in Ψ divided by the lattice lenght L is equal to the normal state density $D_n < 1$. The superconducting density is $D_s = \frac{n}{L}$. In other words

$$D_n = \frac{\langle\psi|\frac{1}{2} - \frac{2X}{L}|\psi\rangle}{\langle\psi|\psi\rangle}\,,\ D_s = \frac{\langle\psi|\frac{1}{4} + \frac{X-\eta^z}{L}|\psi\rangle}{\langle\psi|\psi\rangle}\,. \tag{2.10}$$

The asymptotics of the pair correlator can be evaluated directly

$$\langle c^{\dagger}_{k,\downarrow} c^{\dagger}_{k,\uparrow} c_{l,\uparrow} c_{l,\downarrow} \rangle = \frac{\langle \psi | c^{\dagger}_{k,\downarrow} c^{\dagger}_{k,\uparrow} c_{l,\uparrow} c_{l,\downarrow} | \psi \rangle}{\langle \psi | \psi \rangle} \underset{|k-l|>1}{\longrightarrow} D_s(1 - D_s - D_n) \,. \tag{2.11}$$

This holds in any dimension and establishes the property of ODLRO. The asymptotics of the electron correlator can be reduced by using the algebra of the generators (2.1)

$$\begin{aligned}
\langle c^{\dagger}_{k,\sigma} c_{l,\sigma} \rangle &= \frac{\langle \psi | c^{\dagger}_{k,\sigma} c_{l,\sigma} | \psi \rangle}{\langle \psi | \psi \rangle} = \frac{\langle \Psi | \eta^n c^{\dagger}_{k,\sigma} c_{l,\sigma} \eta^{\dagger^n} | \Psi \rangle}{\langle \Psi | \eta^n \eta^{\dagger^n} | \Psi \rangle} \\
&= \frac{D_n + D_s - 1}{D_n - 1} \frac{\langle \Psi | c^{\dagger}_{k,\sigma} c_{l,\sigma} | \Psi \rangle}{\langle \Psi | \Psi \rangle} + \frac{D_s}{D_n - 1} \frac{\langle \Psi | c^{\dagger}_{l,-\sigma} c_{k,-\sigma} | \Psi \rangle}{\langle \Psi | \Psi \rangle} \\
&= \frac{D_n + D_s - 1}{D_n - 1} \langle c^{\dagger}_{k,\sigma} c_{l,\sigma} \rangle_{stJ} + \frac{D_s}{D_n - 1} \langle c^{\dagger}_{l,-\sigma} c_{k,-\sigma} \rangle_{stJ} \,.
\end{aligned} \tag{2.12}$$

Here $\langle\ \rangle_{stJ}$ denotes correlators in the supersymmetric t-J model, the asymptotics of which were evaluated by means of the Conformal Field Theory approach in [44,45]. The spin correlation function is not changed at all by the presence of local electron pairs and we find

$$\begin{aligned}
\langle S^z_k S^z_l \rangle &= \frac{\langle \Psi | \eta^n \frac{1}{4}(n_{k,\downarrow} - n_{k,\uparrow})(n_{l,\downarrow} - n_{l,\uparrow}) \eta^{\dagger^n} | \Psi \rangle}{\langle \Psi | \eta^n \eta^{\dagger^n} | \Psi \rangle} \\
&= \frac{\langle \Psi | \frac{1}{4}(n_{k,\downarrow} - n_{k,\uparrow})(n_{l,\downarrow} - n_{l,\uparrow}) | \Psi \rangle}{\langle \Psi | \Psi \rangle} = \langle S^z_k S^z_l \rangle_{stJ} \,.
\end{aligned} \tag{2.13}$$

The same relation holds for all other spin-spin correlators as well. The density-density correlation function is reduced to the corresponding t-J correlator as follows

$$\begin{aligned}
\langle n_k n_l \rangle = \langle n_k n_l \rangle_{stJ} &\left(1 + \frac{4D_s}{D_n - 1} + \frac{4D_s^2}{(D_n - 1)^2}\right) \\
+ \ &\frac{2D_s}{1 - D_n}\left(2D_s + 2D_n - \frac{2D_n D_s}{1 - D_n}\right) \,.
\end{aligned} \tag{2.14}$$

All other correlators in the $SU(2|2)$ model can be reduced to linear combinations of correlators in the t-J model by similar computations.

We would now like to illustrate the superconducting properties of the model. One characteristic property of superconductors is that in response to external magnetic fields they develop superconducting currents. This then leads to the Meissner effect and to the phenomenon of flux quantization. For the model (2.5) (with $U < 0$, $\mu = 0$) on a one-dimensional lattice, the wavefunctions that carry the superconducting current can be constructed explicitly as follows.

We first recall that the ground state in the attractive regime (in the sector with a total number of N electrons) is the state $(\eta^{\dagger})^{\frac{N}{2}}|0\rangle$. States in the same sector that contain single electrons (*i.e.* broken pairs) have an energy which is higher than the ground state energy by an amount of at least $-U$. The response of the system to small perturbations (involving energies that are small in comparison with the mass gap $-U$), can thus be studied in a submodel that only contains localon states.

In the localon submodel the hamiltonian H^0 acts once more as a permutation. This shows that the submodel is actually equivalent to the Heisenberg spin-$\frac{1}{2}$ XXX model, which is exactly solvable in one dimension. By using the XXX Bethe Ansatz, we can therefore write down exact eigenfunctions of H^0 in the sector with only localons. These eigenfunctions are parametrized by rapidities λ_j, whose allowed values follow, through the Bethe equations, from the boundary conditions on the wavefunction.

Imposing an external magnetic flux on the system has the effect of twisting the boundary conditions on the wave function. This changes the allowed values of the rapidities, and thereby the form and the energy of the eigenfunctions (compare with [46,47]). We can study the behaviour of a particular wavefunction as a function of the external magnetic flux. In the absence of a magnetic flux the $N/2$ rapidities that correspond to the ground state $(\eta^\dagger)^{N/2}|0\rangle$ are all equal to infinity, and the ground state therefore corresponds to zero momentum and zero electric current. We have found that in the presence of a magnetic flux the corresponding rapidities form an $N/2$-string (bound state) in the λ-plane. This implies that in the twisted ground state the localons move coherently at non-zero total momentum (in one pulse) and that they thus carry a non-vanishing supercurrent.

ACKNOWLEDGEMENTS

V.E.K. thanks the organizers of the Taniguchi conference for their hospitality and support.

REFERENCES

1. E.H. Lieb, F.Y. Wu, *Phys. Rev. Lett.* **20** (1968) 1445.
2. B.Sutherland, *Lecture notes in Physics, Vol 242* (Springer Verlag Berlin, 1985) p.1.
3. C.N.Yang, *Phys. Rev. Lett.* **63** (1989) 2144.
4. O.J. Heilmann, E.H. Lieb, *Ann. New York Acad. Sci.* **172** (1971) 583.
5. C.N.Yang and S.Zhang, *Mod. Phys. Lett.* **B4** (1990) 759.
6. M.Pernici, *Europhys. Lett.* **12** (1990) 75.
7. I.Affleck in talk given at the Nato Advanced Study Institute on *Physics, Geometry and Topology*, Banff, August 1989.
8. A.A. Ovchinnikov, *Sov. Phys. JETP* **30** (1970) 1160.
9. F. Woynarovich, *J. Physics* **C16** (1983) 5293.
10. F. Woynarovich, *J. Physics* **C15** (1982) 85.

11. F.Woynarovich, *J. Physics* **C15** (1982) 97.
12. A. Klümper, A. Schadschneider, J. Zittartz, *Z.Phys.* **B78** (1990) 99.
13. T.C. Choy, W. Young, *J. Physics* **C15** (1982) 521.
14. N Kawakami, A. Okiji, *Phys. Rev.* **B40** (1989) 7066.
15. L.D. Faddeev, L. Takhtajan, *J. Soviet Math.* **24** (1984) 241.
16. L.D. Faddeev, L. Takhtajan, *Phys. Lett.* **85A** (1981) 375.
17. L. Takhtajan, *Phys. Lett.* **87A** (1982) 479.
18. F.H.L. Eßler, V.E. Korepin, K. Schoutens, *Nucl. Phys.* **B372** (1992) 559, *Nucl. Phys.* **B384** (1992) 431, *Phys. Rev. Lett.* **67** (1991) 3848.
19. F. Woynarovich, *J. Physics* **C16** (1983) 6593.
20. V.E. Korepin, *Theor. Mat. Phys.* **76** (1980) 165.
21. F.H.L.Eßler, V.E.Korepin, K.Schoutens, *Phys. Rev. Lett.* **68** (1992) 2960.
22. F.C.Zhang, T.M.Rice, *Phys. Rev.* **B37** (1988) 3759.
23. P.Schlottmann, *Phys. Rev.* **B36** (1987) 5177.
24. J.E. Hirsch, *Physica* **C158** (1990) 326.
25. R.Z. Bariev, A. Klümper, A. Schadschneider, J. Zittartz, *J. Physics* **A26** (1993) 1249.
26. K.A. Penson, M. Kolb, *Phys. Rev.* **B33** (1986) 1663.
27. M. Kolb, K.A. Penson, *J. Stat. Phys.* **44** (1986) 129.
28. I. Affleck, J.B. Marston, *J. Physics* **C21** (1988) 2511.
29. Bill Sutherland, *Phys. Rev. Lett.* **20** (1968) 98.
30. F.H.L.Eßler, V.E.Korepin, K.Schoutens, *Phys. Rev. Lett.* **70** (1993) 73.
31. C.N. Yang, *Rev. Mod. Phys.* **34** (1962) 694.
32. G. Sewell, *J. Stat. Phys.* **61** (1990) 415.
33. F.H.L.Eßler, V.E.Korepin, K.Schoutens, *preprint* ITP-92-57
34. F.H.L.Eßler, V.E.Korepin, *preprint* ITP-93-15
35. C.K.Lai, *J. Math. Phys.* **15** (1974) 1675.
36. F.H.L.Eßler, V.E.Korepin, *Phys. Rev.* **B46** (1992) 9147.
37. P.A.Bares, G.Blatter, M.Ogata, *Phys. Rev.* **B44** (1991) 130.
38. A.A. Belavin, A.A. Polyakov, A.B. Zamolodchikov, *Nucl. Phys.* **B241** (1984) 333.
39. H. Frahm, V.E. Korepin, *Phys. Rev.* **B42** (1990) 10533.
40. H. Frahm, V.E. Korepin, *Phys. Rev.* **B43** (1991) 5653.
41. I. Affleck, *Phys. Rev. Lett.* **56** (1986) 746.
42. H.W.J. Blöte, J.L. Cardy, M.P. Nightingale, *Phys. Rev. Lett.* **56** (1986) 742.
43. J.L. Cardy, *Nucl. Phys.* **B270** (1986) 186.
44. N.Kawakami, S.K.Yang, *J. Physics* **C3** (1991) 5983.
45. N.Kawakami, S.K.Yang, *Phys. Rev. Lett.* **67** (1990) 2309.
46. M. Fowler, N. Yu, *Phys. Rev.* **B45** (1992) 11795 .
47. B.S. Shastry, B. Sutherland, *Phys. Rev. Lett.* **65** (1990) 243.

Symmetry Approach to Solvable Lattice Models

T. Miwa

Research Institute for Mathematical Sciences, Kyoto University,
Kyoto 606, Japan

Abstract. We give an introduction to the symmetry approach to solvable lattice models. The correlation functions of the six-vertex model is shown to be given by a trace formula in terms of the quantized vertex operators.

1. Introduction

The reason that many mathematicians got interested in Quantum Field Theory and Statistical Mechanics is that the problems therein are those kinds of infinite degrees of freedom. Such problems are usually not exactly solvable. But if the system in question has infinite symmetries, it is often solvable:

$$\frac{\text{Physics } \infty}{\text{Symmetry } \infty} = \text{Mathematics}.$$

In this talk I will explain how to compute the correlation functions for the XXZ model:

$$H = -\sum(\sigma^x_{k+1}\sigma^x_k + \sigma^y_{k+1}\sigma^y_k + \Delta\sigma^z_{k+1}\sigma^z_k)$$

in the anti-ferromagnetic regime $\Delta < -1$ by using the quantum affine symmetry of $U_q(\widehat{sl}(2))$.

2. Conformal Field Theory

Before starting our subject let us briefly review the success in Conformal Field Theory. Suppose that a massless one dimensional Hamiltonian H is given. We are to compute the partition function

$$Z = \text{tr } e^{-\beta H}.$$

Since the Hamiltonian is massless, the low-lying spectrums are behaving

$$h - h_0 = O\left(\frac{1}{M}\right),$$

where M is the size. The CFT limit is [1]

$$\lim_{\substack{T\to 0\\ M\to\infty\\ MT:finite}} \sum_h e^{-\beta(h-h_0)}.$$

Springer Series in Solid-State Sciences, Vol. 118
Correlation Effects in Low-Dimensional Electron Systems
Editors: A. Okiji and N. Kawakami

In this limit we throw out most of the states except for the low-lying ones, and get the character of the physical Hilbert space of states

$$\mathcal{H} = \bigoplus_i (V_i \otimes \overline{V}_i)$$

on which the infinite symmetries of $Vir \otimes \overline{Vir}$ act. In this mathematical language the Hamiltonian is nothing but

$$H_{CFT} = L_0 + \overline{L_0},$$

and the partition function is

$$Z_{CFT} = \mathrm{tr}_{\mathcal{H}} q^{L_0} \overline{q}^{\overline{L_0}}$$

where $q = e^{-\beta/M}$. Furthermore, the CFT limit of the correlation functions is also given by a mathematical formula in terms of representation theory:

$$\mathrm{tr}\ q^{L_0} \overline{q}^{\overline{L_0}} \Phi(z_1, \overline{z}_1) \cdots \Phi(z_n, \overline{z}_n),$$

where $\Phi(z, \overline{z})$ is called the vertex operator such that

$$\Phi(z, \overline{z}) : \mathcal{H} \to \mathcal{H}$$

is characterized by certain property in representation theory. A remarkable fact is that the correlation function satisfies a system of linear differential equations and admits integral representations. This was the great success of CFT.

3. Massive Integrable Field Theory

Now let us turn to massive field theory. For example, the anti-ferromagnetic regime $\Delta < -1$ in the XXZ model is massive. To obtain a field theory limit we put $T = 0$ and consider

$$\lim_{\substack{m \to \infty \\ \Delta \to -1}} \langle vac | \sigma_1^z \sigma_m^z | vac \rangle.$$

This will give two-point function in the cont. limit.

$$\langle vac | \sigma(0) \sigma(x) | vac \rangle.$$

One of the principal question to be asked is what is the structure of the Hilbert space of the states

$$\mathcal{H}(\Delta < -1; T = 0).$$

We will answer to this question by using the representation theory of $U_q(\widehat{sl}(2))$. Before going to the lattice problem, let us recall the *bootstrap approach* [2] to massive integrable field theory. In that approach, the Hilbert space is assumed to be spanned by states

$$|\beta_1, \cdots, \beta_n; \varepsilon_1, \cdots, \varepsilon_n\rangle$$

parametrized by the rapidities β_j and the isospins ε_j. The exchange of (β_j, ε_j) with $(\beta_{j+1}, \varepsilon_{j+1})$ is governed by the S-matrix:

$$\sum_{\varepsilon'_j,\varepsilon'_{j+1}} S(\beta_j - \beta_{j+1})_{\varepsilon_j,\varepsilon_{j+1}}^{\varepsilon'_j,\varepsilon'_{j+1}} |\cdots,\beta_j,\beta_{j+1},\cdots;\cdots\varepsilon'_j,\varepsilon'_{j+1}\cdots\rangle$$
$$= |\cdots\beta_{j+1}\beta_j\cdots;\cdots\varepsilon_{j+1}\varepsilon_j\cdots\rangle.$$

The consistency condition for the S-matrix is called the Yang-Baxter equation.
For the correlation functions, we want the form factors:

$$G(\beta_1,\cdots,\beta_n)_{\varepsilon_1\cdots\varepsilon_n} = \langle vac|\sigma(0)|\beta_1\cdots;\varepsilon_1\cdots\rangle.$$

They obviously obey the exchange relation:

$$S_{jj+1}(\beta_j - \beta_{j+1})G(\cdots\beta_j\beta_{j+1}\cdots) = G(\cdots\beta_{j+1}\beta_j\cdots).$$

For the sake of the locality of the field $\sigma(x)$, we further impose a kind of cyclicity:

$$G(\beta_2\cdots\beta_n\beta_1 + 2\pi i)_{\varepsilon_2\cdots\varepsilon_n\varepsilon_1} = G(\beta_1\beta_2\cdots\beta_n)_{\varepsilon_1\varepsilon_2\cdots\varepsilon_n}.$$

We call these conditions Smirnov's difference equation because Smirnov solved these equation to get the form factors for the massive field theory corresponding to the said continuum limit at $\Delta \to -1$.

4. The Anisotropic Limit $\Delta \to -\infty$

We start solving the XXZ model for $\Delta < -1$ at $T = 0$ in the thermodynamic limit $M \to \infty$. Consider first the extreme anisotropic limit $\Delta = -\infty$, where the Hamiltonian reduces to a diagonal one

$$H(\Delta = -\infty) \sim \sum_k \sigma^z_{k+1}\sigma^z_k.$$

In this limit the ground states are either

$$\cdots v_+ \otimes v_- \otimes v_+ \otimes v_- \cdots$$

or

$$\cdots v_- \otimes v_+ \otimes v_- \otimes v_+ \cdots$$

Let us explain the notation. We set

$$V = \mathbf{C}v_+ \oplus \mathbf{C}v_-.$$

This is the space on which the Pauli matrices act. the Hamiltonian H formally acts on the infinite tensor product

$$\cdots V \otimes V \otimes V \otimes V \cdots.$$

This is too big to handle with, and we will later replace it with some mathematical space tractable by representation theory.
Returning to the trivial case $\Delta = -\infty$, we distinguish four types of the boundary conditions.

$$\begin{aligned}
\mathcal{H} &= \bigoplus_{i,j=0,1} \mathcal{H}_{i,j} \\
\mathcal{H}_{i,j} &= \bigoplus_{p} \mathbf{C} \bigotimes_{k} v_{p(k)} \\
\text{s.t. } p(k) &= (-1)^{k+i}, k \gg 0 \\
&= (-1)^{k+j}, k \ll 0
\end{aligned}$$

i.e., one of the following;

$$\begin{array}{lll}
\cdots v_+ \otimes v_- \otimes & \cdots\cdots\cdots & \otimes v_+ \otimes v_- \cdots \\
\cdots v_+ \otimes v_- \otimes & \cdots\cdots\cdots & \otimes v_- \otimes v_+ \cdots \\
\cdots v_- \otimes v_+ \otimes & \cdots\cdots\cdots & \otimes v_+ \otimes v_- \cdots \\
\cdots v_- \otimes v_+ \otimes & \cdots\cdots\cdots & \otimes v_- \otimes v_+ \cdots
\end{array}$$

Let us write $|p\rangle$ for the vector $\bigotimes_k v_{p(k)}$. If p satisfy one of the boundary conditions. At $\Delta = -\infty$, $|p\rangle$ diagonalizes the Hamiltonian. For $\Delta < -1$, we expect to have eigenvectors in the form

$$\sum_p c(p)|p\rangle$$

where the coefficients $c(p)$ are expanded in, say Δ^{-1}. They are *very complicated*. Nevertheless, we will give a mathematical machinery which in principle computes these coefficients.

5. The Six-Vertex Model

Let $|vac\rangle_i$ be the lowest eigenvector in $\mathcal{H}_{i,i}$ $(i, = 0, 1)$.

We want to compute the matrix element

$${}_i\langle vac|\mathcal{O}|vac\rangle_i$$

of a local operator $\mathcal{O}$ e.g. $\mathcal{O} = \sigma_0^z$.
Actually, Baxter computed this case [3]:

$${}_0\langle vac|\sigma_0^z|vac\rangle_0 = \prod_{n=1}^{\infty} \left(\frac{1-q^{2n}}{1+q^{2n}}\right)^2$$

where

$$\Delta = \frac{q+q^{-1}}{2}.$$

The first idea is to *extend* the problem. Namely we consider the correlation functions for the related $2D$ lattice model, i.e., the 6-vertex model. The Boltzmann weights of the $6V$ model are dependent on the said variable q and an extra variable ζ. Explicitly they are given by

$$\begin{array}{ccc} & + & \\ + & \!+\! & + \\ & + & \end{array} \;=\; \begin{array}{ccc} & - & \\ - & \!+\! & - \\ & - & \end{array} \;=\; 1$$

$$\begin{array}{ccc} & + & \\ - & + & - \\ & + & \end{array} \;=\; \begin{array}{ccc} & - & \\ + & + & + \\ & - & \end{array} \;=\; \frac{(1-\zeta^2)q}{1-q^2\zeta^2}$$

$$\begin{array}{ccc} & + & \\ + & + & - \\ & - & \end{array} \;=\; \begin{array}{ccc} & - & \\ - & + & + \\ & + & \end{array} \;=\; \frac{(1-q^2)\zeta}{1-q^2\zeta^2}$$

We identify the local variable $\pm$ with $v_\pm \in V$, and define the R-matrix

$$\overline{R}(\zeta)_{\varepsilon_1'\varepsilon_2'}^{\varepsilon_1\varepsilon_2} = \begin{array}{ccc} & \varepsilon_1 & \\ \varepsilon_2' & + & \varepsilon_2 \\ & \varepsilon_1' & \end{array}, \quad \overline{R}(\zeta) : V \otimes V \to V \otimes V.$$

We normalize the R-matrix by the partition function per site κ

$$R(\zeta) = \frac{1}{\kappa(\zeta)}\overline{R}(\zeta).$$

The R-matrix satisfies the Yang-Baxter equation. As a consequence, the row-to-row transfer matrix commutes

$$[T(\zeta), T(\zeta')] = 0.$$

The XXZ Hamiltonian is contained in this commuting family

$$H = \zeta\frac{\partial}{\partial\zeta}\log T(\zeta)\Big|_{\zeta=1}.$$

Therefore, in particular we have

$$[H, T(\zeta)] = 0.$$

All these are in the first few chapters of textbook. After that follows the method of the Bethe Ansatz. In fact the said computation of the one-point function by Baxter uses the Bethe Ansatz eigenvectors on a finite-size lattice. The method is hardly extendable to the general case. So we quit the Bethe Ansatz, and switch to another method due to Baxter [4]: Corner transfer matrix.

6. Corner Transfer Matrix

Before going to CTM let us finish the point of the extension of the problem, i.e. to consider the correlation functions for the inhomogeneous $6V$ model. The idea is to associate a variable ζ_j to each line l_j of the $2D$-lattice, and define the Boltzmann weights at the site where two lines cross, by

$R(\zeta_1/\zeta_2) \longleftrightarrow$

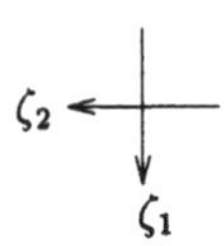

The correlation functions on such inhomogeneous lattice, in general, naturally depend on the parameters ζ_j. Our goal is to show that they satisfy Smirnov-type difference

equation with respect to these variables. The idea of the usual transfer matrix is to cut the $2D$-lattice into $1D$-slices:

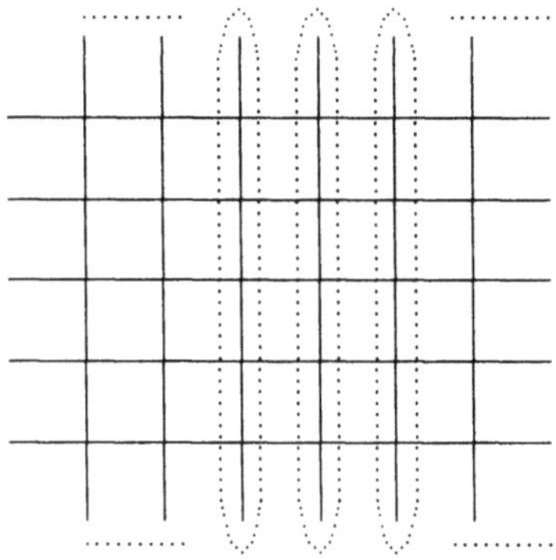

The corner transfer matrix cuts the $2D$-lattice into four quadrants.

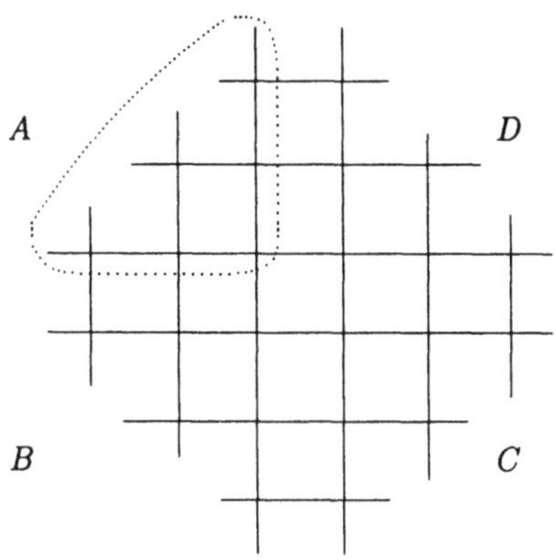

Suppose that all the vertical lines bear the parameter ζ_1 and the horizontal ones ζ_2. Set $\zeta = \zeta_1/\zeta_2$. Consider the CTM at the NW quadrant.

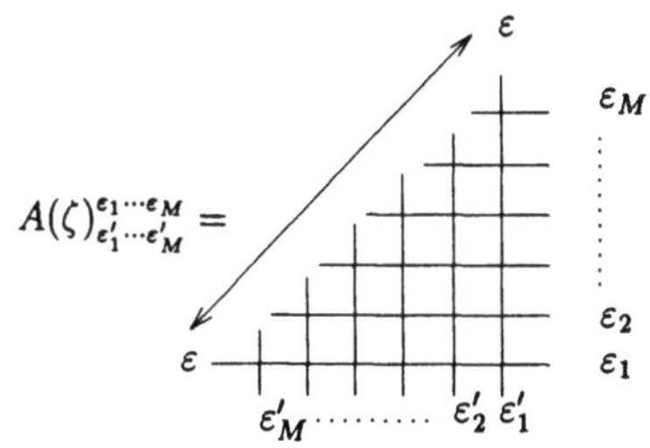

We choose the boundary variables ε are all equal to $(-1)^{M+i}$. With this choice of the boundary condition the CTM $A(\zeta)$ is formally considered to act on the half-infinite space $\mathcal{H}_i$ (Cf. $\mathcal{H}_{i,j}$):

$$A(\zeta) : \mathcal{H}_i \to \mathcal{H}_i.$$

Or rather we define the space $\mathcal{H}_i$ to be the space of the eigenvectors of $A(\zeta)$. This definition is meaningful because up to infinite scalar the CTM has the simple dependence on ζ;

$$A(\zeta) \sim \zeta^{-D}$$

where the operator D is independent of ζ. Therefore our definition of the space $\mathcal{H}_i$ is independent of ζ. The other CTM's are of similar form

$$\begin{aligned} B(\zeta) &\sim XA(-q^{-1}\zeta^{-1}) \\ C(\zeta) &\sim XA(\zeta)X \\ D(\zeta) &\sim A(-q^{-1}\zeta^{-1})X. \end{aligned}$$

Here $X = \bigotimes_{k=1}^{\infty} \sigma_k^x$. Therefore we have

$$Z = \mathrm{tr}_{\mathcal{H}_i} A(\zeta)B(\zeta)C(\zeta)D(\zeta) \sim \mathrm{tr}_{\mathcal{H}_i} q^{2D}.$$

It also follows that

$$\mathrm{Spec} D \subset \{0, 1, 2, \cdots\}.$$

Therefore this expression is quite similar to the character of $\mathcal{H}_i$. In fact, it is equal to the character of the *level one* module of $U_q(\widehat{sl}(2))$ in the case of the XXZ model. This is elaborated later.

7. Vertex Operator

In order to compute the *correlation functions*, we need another machinery. Suppose we are to compute one-point function. We now divide the $2D$-lattice into 6 pieces.

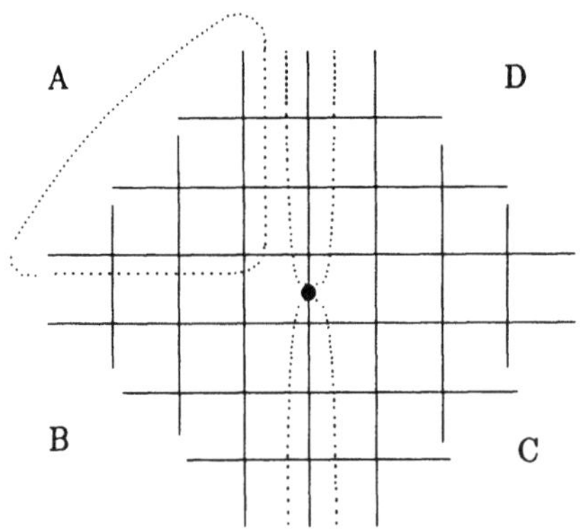

Here • indicate the point 0 on which we are to compute $\langle \sigma_0^z \rangle$. We consider an operator corresponding to the figure

$$\Phi_{\varepsilon_0}(\zeta)^{\varepsilon_1\varepsilon_2\cdots}_{\varepsilon'_1\varepsilon'_2\cdots} =$$

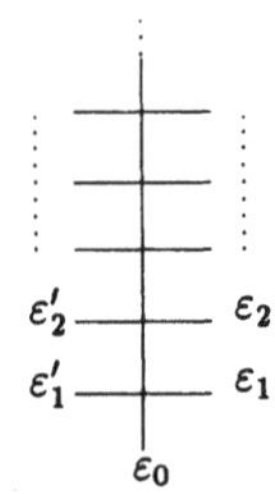

This is called the vertex operator. It acts as

$$\Phi_\varepsilon(\zeta) : \mathcal{H}_i \to \mathcal{H}_{1-i}.$$

Note that the vertex operator is one half of the usual transfer matrix. The other half is equal to $X\Phi_{-\varepsilon}(\zeta)X$. By a graphical argument one can derive the following relations for the vertex operators

$$\xi^{-D}\Phi(\zeta)\xi^{D} = \Phi(\zeta/\xi),$$
$$\sum_{\varepsilon_1'\varepsilon_2'} R(\zeta_1/\zeta_2)_{\varepsilon_1\varepsilon_2}^{\varepsilon_1'\varepsilon_2'}\Phi_{\varepsilon_1'}(\zeta_1)\Phi_{\varepsilon_2'}(\zeta_2) = \Phi_{\varepsilon_2}(\zeta_2)\Phi_{\varepsilon_1}(\zeta_1),$$
$$\sum_{\varepsilon}\Phi_{-\varepsilon}(-q^{-1}\zeta)\Phi_\varepsilon(\zeta) = id.$$

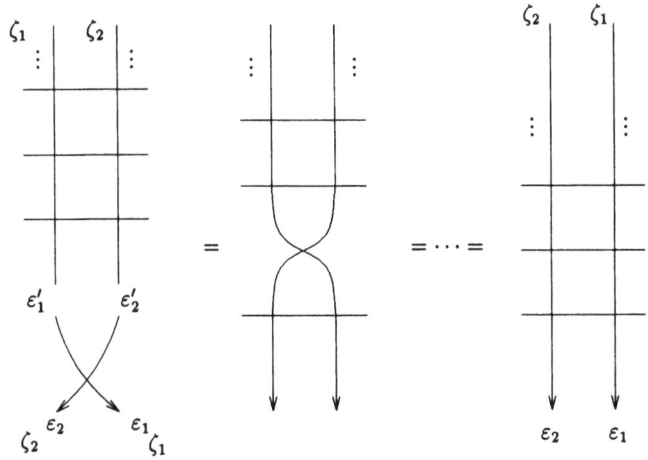

Using these we obtain

$$_0\langle vac|\sigma_0^z|vac\rangle_0 \mathrm{tr}_{\mathcal{H}_0} q^{2D} = \sum_{\varepsilon=\pm} \mathrm{tr}_{\mathcal{H}_0} \varepsilon q^{2D}\Phi_{-\varepsilon}(-q^{-1}\zeta)\Phi_\varepsilon(\zeta).$$

Therefore, in general we consider

$$F^{(i)}(\zeta_1,\cdots,\zeta_n)_{\varepsilon_1\cdots\varepsilon_n} = \mathrm{tr}_{\,\mathcal{H}_i} q^{2D}\Phi_{\varepsilon_1}(\zeta_1)\cdots\Phi_{\varepsilon_n}(\zeta_n).$$

Then the commutation relations immediately imply difference equation for this quantity.

8. Summary of the First Half

Let us summarize what we discussed [5].

(1) We started from the XXZ Hamiltonian

$$H \sim \sum_{k=-\infty}^{\infty} (\sigma_{k+1}^x\sigma_k^x + \sigma_{k+1}^y\sigma_k^y + \Delta\sigma_{k+1}^z\sigma_k^z),$$

which formally acts on the infinite tensor product

$$\cdots V \otimes V \otimes V \otimes V \cdots.$$

(2) We introduced the corner transfer matrix ζ^{-D} where D is given by

$$D \sim \sum_{k=1}^{\infty} k(\sigma_{k+1}^x \sigma_k^x + \sigma_{k+1}^y \sigma_k^y + \Delta \sigma_{k+1}^z \sigma_k^z),$$

which formally acts on the half-infinite tensor product

$$\cdots V \otimes V.$$

(3) We introduced the Hilbert space $\mathcal{H}_i$ spanned by the eigenvectors of the operator D with the boundary condition such that

$$v_{p(k)} = v_{(-1)^{k+i}} \quad k \gg 0.$$

(4) We introduced the vertex operator

$$\Phi(\zeta) : \mathcal{H}_i \to \mathcal{H}_{i-1} \otimes V,$$

being one-half of the usual transfer matrix.
(5) We obtained the commutation relations

$$\begin{aligned} &\xi^{-D}\Phi(\zeta)\xi^D = \Phi(\zeta/\xi) \\ &R^{12}(\zeta_1/\zeta_2)\Phi^1(\zeta_1)\Phi^2(\zeta_2) = \Phi^2(\zeta_2)\Phi^1(\zeta_1). \end{aligned}$$

(6) We reduced the correlation functions to the trace of the product of the vertex operators

$$F^{(i)}(\zeta_1, \cdots, \zeta_n) \underset{\text{def}}{=} \mathrm{tr}_{\mathcal{H}_i} q^{2D} \Phi^1(\zeta_1) \cdots \Phi^n(\zeta_n) \in V^1 \otimes \cdots \otimes V^n.$$

(7) We derived the difference equation for the trace

$$\begin{aligned} &R^{jj+1}(\zeta_j, \zeta_{j+1}) F^{(i)}(\cdots \zeta_j, \zeta_{j+1} \cdots) = F^{(i)}(\cdots \zeta_{j+1}, \zeta_j \cdots), \\ &F^{(i)}(\zeta_2 \cdots \zeta_n q^2 \zeta_1) = F^{(1-i)}(\zeta_1 \zeta_2 \cdots \zeta_n). \end{aligned}$$

In the second half we discuss the following points
(a) The argument given above does not determine the right solution of the difference equation that gives the correlation function. We want to give precise meanings to $\mathcal{H}_i, D, \Phi(\zeta)$.
(b) The whole infinite tensor product are

$$\bigoplus_{i,j=0,1} \mathcal{H}_j \otimes \mathcal{H}_j^* \simeq \bigoplus_{i,j=0,1} \mathrm{Hom}(\mathcal{H}_j, \mathcal{H}_i).$$

Then, how the vacuum and the excited states are realized in this space ?

9. $U_q(\widehat{sl}(2))$

We start from the definition of the quantum affine algebra $U_q(\widehat{sl}(2))$. The algebra is generated by 6 generators

$$e_i, \ f_i, \ t_i \ \ (i = 0, 1).$$

They satisfy the relations:

$$t_i t_j t_i^{-1} = t_j,$$
$$t_i e_j t_i^{-1} = \begin{cases} q^2 e_j & (i = j) \\ q^{-2} e_j & (i \neq j) \end{cases},$$
$$t_i f_j t_i^{-1} = \begin{cases} q^{-2} f_j & (i = j) \\ q^2 f_j & (i \neq j) \end{cases},$$
$$[e_i, f_j] = \delta_{ij} \frac{t_i - t_i^{-1}}{q - q^{-1}},$$
$$e_i^3 e_j - [3] e_i^2 e_j e_i + [3] e_i e_j e_i^2 - e_j e_i^3 = 0,$$
$$f_i^3 f_j - [3] f_i^2 f_j f_i + [3] f_i f_j f_i^2 - f_j f_i^3 = 0,$$

where

$$[n] = \frac{q^n - q^{-n}}{q - q^{-1}}.$$

The Hopf algebra structure provides us with good representation theory.

Counit

$$\varepsilon(e_i) = 0, \quad \varepsilon(f_i) = 0, \quad \varepsilon(t_i) = 1$$

gives us the singlet representation:

$$\varepsilon(x) : \mathbf{C} \to \mathbf{C} \quad x \in U.$$

Antipode

$$a(e_i) = -t_i^{-1} e_i, \quad a(f_i) = -f_i t_i, \quad a(t_i) = t_i^{-1}$$

gives us the dual representation

$$\langle x v^*, v \rangle = \langle v^*, a(x) v \rangle.$$

The property $(xy)v^* = x(yv^*)$ is assured because $a(xy) = a(y)a(x)$.

Coproduct

$$\Delta(e_i) = e_i \otimes 1 + t_i \otimes e_i, \quad \Delta(f_i) = f_i \otimes t_i^{-1} + 1 \otimes f_i, \quad \Delta(t_i) = t_i \otimes t_i$$

gives us the tensor product representation:

$$x(v_1 \otimes v_2) = \Delta(x)(v_1 \otimes v_2).$$

A map

$$\Phi : V_1 \to V_2$$

is called an intertwiner if it commute with the U-action:

$$\begin{array}{ccc} V_1 & \xrightarrow{\Phi} & V_2 \\ \Big\downarrow x & \circlearrowleft & \Big\downarrow x \\ V_1 & \xrightarrow{\Phi} & V_2 \end{array} \quad x \in U.$$

Lemma *Suppose that two maps*

$$\Phi_{12}^{3} : V_1 \otimes V_2 \to V_3$$
$$\Phi_{1}^{32^*} : V_1 \to V_3 \otimes V_2^*$$

are connected by $\Phi_{12}^3(v_1 \otimes v_2) = \langle \Phi_1^{32^*} v_1, v_2 \rangle$. *Then* Φ_{12}^3 *is an intertwiner if and only if* $\Phi_1^{32^*}$ *is an intertwiner.*

Corollary *(The uniqueness of the vacuum.) Suppose that* $\mathcal{H}_i$ *is an irreducible representation of* U. *Then, there exists one and only one singlet in* $\mathcal{H}_i \otimes \mathcal{H}_i^*$.

10. The R-Matrix

We consider doublet, i.e. two-dimensional representation,

$$V = \mathbf{C}v_+ \oplus \mathbf{C}v_-.$$

With respect to this basis the action is represented by the following 2×2 matrices.

$$\begin{aligned} e_0 = \quad f_1 &= \begin{pmatrix} 0 & 0 \\ 1 & 0 \end{pmatrix} \\ e_1 = \quad f_0 &= \begin{pmatrix} 0 & 1 \\ 0 & 0 \end{pmatrix} \\ t_0^{-1} = \quad t_1 &= \begin{pmatrix} q & \\ & q^{-1} \end{pmatrix}. \end{aligned}$$

The coproduct gives a formal action on the infinite tensor product.

$$\begin{aligned} e_0 &= \sum_k \cdots \begin{pmatrix} q^{-1} & \\ & q \end{pmatrix} \otimes \begin{pmatrix} 0 & 0 \\ 1 & 0 \end{pmatrix} \otimes \begin{pmatrix} 1 & \\ & 1 \end{pmatrix} \cdots \\ f_1 &= \sum_k \cdots \begin{pmatrix} 1 & \\ & 1 \end{pmatrix} \otimes \begin{pmatrix} 0 & 0 \\ 1 & 0 \end{pmatrix} \otimes \begin{pmatrix} q^{-1} & \\ & q \end{pmatrix} \cdots \\ e_1 &= \sum_k \cdots \begin{pmatrix} q & \\ & q^{-1} \end{pmatrix} \otimes \begin{pmatrix} 0 & 1 \\ 0 & 0 \end{pmatrix} \otimes \begin{pmatrix} 1 & \\ & 1 \end{pmatrix} \cdots \\ f_0 &= \sum_k \cdots \begin{pmatrix} 1 & \\ & 1 \end{pmatrix} \otimes \begin{pmatrix} 0 & 1 \\ 0 & 0 \end{pmatrix} \otimes \begin{pmatrix} q & \\ & q^{-1} \end{pmatrix} \cdots \\ t_0 &= \cdots \begin{pmatrix} q^{-1} & \\ & q \end{pmatrix} \otimes \begin{pmatrix} q^{-1} & \\ & q \end{pmatrix} \otimes \begin{pmatrix} q^{-1} & \\ & q \end{pmatrix} \cdots \\ t_1 &= \cdots \begin{pmatrix} q & \\ & q^{-1} \end{pmatrix} \otimes \begin{pmatrix} q & \\ & q^{-1} \end{pmatrix} \otimes \begin{pmatrix} q & \\ & q^{-1} \end{pmatrix} \cdots \end{aligned}$$

The fact is the XXZ Hamiltonian commutes with the U-action [6]

$$[H, U_q(\widehat{sl}(2))] = 0.$$

One can deform the representation V_z by putting a complex parameter $z = \zeta^2$:

$$e_0 = z\begin{pmatrix} 0 & 0 \\ 1 & 0 \end{pmatrix}, \quad f_1 = \begin{pmatrix} 0 & 0 \\ 1 & 0 \end{pmatrix},$$
$$e_1 = \begin{pmatrix} 0 & 1 \\ 0 & 0 \end{pmatrix}, \quad f_0 = z^{-1}\begin{pmatrix} 0 & 1 \\ 0 & 0 \end{pmatrix},$$
$$t_0^{-1} = t_1 = \begin{pmatrix} q & \\ & q^{-1} \end{pmatrix}.$$

The R-matrix which gives the Boltzmann weights for the $6V$ model intertwines the two tensor products:

$$PR(z_1/z_2) : V_{z_1} \otimes V_{z_2} \to V_{z_2} \otimes V_{z_1}$$

where $P(v \otimes v') = v' \otimes v$. (To be precise, we need some gauge transformation.)

11. Bosonization

Now we construct $\mathcal{H}_i$ explicitly by using free bosons. Drinfeld [7] gave another set of generators of $U_q(\widehat{sl}(2))$:

$$a_k \quad (k \in \mathbf{Z}\backslash\{0\}), \quad x_k^+, x_k^- \quad (k \in \mathbf{Z}), \quad \gamma, K.$$

The old generators are given by

$$t_1 = K,\ x_0^+ = e_1,\ x_0^- = f_1,$$
$$t_0 = \gamma K^{-1},\ x_1^- = e_0 t_1,\ x_{-1}^+ = t_1^{-1} f_0.$$

The new generators obey the following relations

$$K a_k K^{-1} = a_k,\ K x_k^\pm K^{-1} = q^{\pm 2} x_k^\pm$$
$$[a_k, a_l] = \delta_{k+l,0} \frac{[2k]}{k} \frac{\gamma^k - \gamma^{-k}}{q - q^{-1}}$$
$$[a_k, x_l^\pm] = \pm \frac{[2k]}{k} \gamma^{\pm |k|/2} x_{k+l}^\pm$$
$$x_{k+1}^\pm x_l^\pm - q^{\pm 2} x_l^\pm x_{k+1}^\pm = q^{\pm 2} x_k^\pm x_{l+1}^\pm - x_{l+1}^\pm x_k^\pm$$
$$[x_k^+, x_l^-] = \frac{\gamma^{\frac{k-l}{2}} \psi_{k+l} - \gamma^{\frac{l-k}{2}} \varphi_{k+l}}{q - q^{-1}}$$
$$\sum_{k=0}^{\infty} \psi_k z^{-k} = K \exp\{(q - q^{-1}) \sum_{k=1}^{\infty} a_k z^{-k}\}$$
$$\sum_{k=0}^{\infty} \varphi_k z^{k} = K^{-1} \exp\{-(q - q^{-1}) \sum_{k=1}^{\infty} a_{-k} z^{k}\}.$$

The element γ belongs to the center of U. It implies that on an irreducible representation γ is just a number, say $\gamma = q^l$. The number l is called the level of the representation.

We will construct $\mathcal{H}_i$ as a level one representation. Note that the commutation relation of a_k's tells us that they are free bosons. Prepare the 0 mode: $[\partial, \alpha] = 2$. We set

$$\mathcal{H}_i = \mathbf{C}[a_{-1}, a_{-2}, \cdots] \otimes \Big(\bigoplus_{n \in \mathbf{Z} + \frac{i}{2}} \mathbf{C} e^{n\alpha} \Big).$$

These are the Fock space of the bosons. We set

$$\gamma = q, \ K = 1 \otimes q^{\partial}.$$

The rest of Drinfeld's generators are given as follows [8]: Set

$$X^{\pm}(z) = \sum_{k \in \mathbf{Z}} x_k^{\pm} z^{-k-1}.$$

Then

$$X^{\pm}(z) = \exp\Big(\pm \sum_{n=1}^{\infty} \frac{a_{-n}}{[n]} q^{\mp n/2} z^n \Big) \exp\Big(\mp \sum_{n=1}^{\infty} \frac{a_n}{[n]} q^{\mp n/2} z^{-n} \Big) e^{\pm\alpha} q^{\pm\partial}.$$

This is Frenkel-Jing's result [3].

Now we discuss

$$D = \sum_{k=1}^{\infty} k (\sigma_{k+1}^x \sigma_k^x + \sigma_{k+1}^y \sigma_k^y + \Delta \sigma_{k+1}^z \sigma_k^z).$$

One can check the following commutation relations [9]

$$[D, e_i] = e_i, \quad [D, f_i] = -f_i, \quad [D, t_i] = 0.$$

This relation determines the action of D on $\mathcal{H}_i$.

12. Vertex Operator

The vertex operators are determined by the intertwing relations [10] for

$$\Phi(z) : \mathcal{H}_i \to \mathcal{H}_{1-i} \otimes V_z.$$

The formula reads [11]

$$\Phi_-(z) = \exp\Big(\sum_{n=1}^{\infty} \frac{a_{-n}}{[2n]} q^{7n/2} z^n \Big) \exp\Big(- \sum_{n=1}^{\infty} \frac{a_n}{[2n]} q^{-5n/2} z^{-n} \Big) e^{\alpha/2} (-q^3 z)^{\partial/2}$$

$$\Phi_+(z) = \oint \frac{dw}{2\pi i} \Big(\Phi_-(z) X^-(w) - q X^-(w) \Phi_1(z) \Big).$$

Now it is tedius but straightforward to compute the trace

$$\mathrm{tr}_{\mathcal{H}_i} q^{2D} \Phi_{\epsilon_1}(z_1) \cdots \Phi_{\epsilon_{2n}}(z_{2n}).$$

The obtained formula contains n-fold integrals [11].

13. Space of States

Our understanding of the physical space of states is the $U_q(\widehat{sl}(2))$-module [6]

$$\mathcal{H} = \bigoplus_{i,j=0,1} \mathcal{H}_i \otimes \mathcal{H}_j^* \simeq \bigoplus_{i,j=0,1} \mathrm{Hom}(\mathcal{H}_j, \mathcal{H}_i).$$

If we want to see the *local* tensor product structure, we can open the door by the vertex operator

$$\mathcal{H}_0 \otimes \mathcal{H}_0^* \xrightarrow{\Phi(1)} \mathcal{H}_1 \otimes V \otimes \mathcal{H}_0^* \xrightarrow{\Phi(1)} \mathcal{H}_0 \otimes V \otimes V \otimes \mathcal{H}_0^* \to \cdots.$$

The vacuum vector is realized in $\mathrm{Hom}(\mathcal{H}_i, \mathcal{H}_i)$ as the identity operator. Finally, let us discuss excited states. We want to find finite-dimensional $U_q(\widehat{sl}(2))$-module in $\mathcal{H}$. For example

$$V_z \hookrightarrow \mathcal{H}_i \otimes \mathcal{H}_{1-i}^*.$$

This is equivalent to the vertex operator of the form

$$\Psi^*(z) : V_z \otimes \mathcal{H}_{1-i} \to \mathcal{H}_i.$$

References

[1] R. Kedem and B. M. McCoy, ITP-SB-92-56.

[2] F. A. Smirnov, Adv. Series in Math. Phys. **14**, (1992), World Scientific.

[3] R. J. Baxter, J. Stat. Phys. **9** (1073), 145.

[4] R. J. Baxter, Academic Press, 1982.

[5] M. Jimbo, T. Miwa and A. Nakayashiki, J. Phys. **26A** (1993),2199.

[6] R. Davies, O. Foda, M. Jimbo, T Miwa and A Nakayashiki, Comm. Math. Phys. **151** (1993), 89.

[7] V. G. Drinfeld, Sov. Math. Dokl. **36** (1988),212.

[8] I. B. Frenkel and N. H. Jing, Proc. Nat'l. Acad. Sci. USA **85** (1988) 9373.

[9] O. Foda and T. Miwa, Int. J. Mod. Phys. 7A Suppl. **1A** (1992), 279.

[10] I. Frenkel and N. Yu Reshetikhin, Comm. Math. Phys. **146** (1992), 1.

[11] M. Jimbo, K. Miki, T. Miwa and A. Nakayashiki, Phys. Lett. A **168** (1992), 256.

Conformal-Field-Theory Approach to Quantum-Impurity Problems

I. Affleck

Canadian Institute for Advanced Research and Physics Department, University of British Columbia, Vancouver, BC, V6T1Z1 Canada

Abstract. A brief review is given of a new method for studying the critical behavior of quantum impurity problems, based on conformal field theory techniques, which I developed with Andreas Ludwig. Some results on the overscreened Kondo problem are reviewed. It is shown that the simple open and periodic fixed points, which occur in quantum spin chain impurity models, are related to each other by fusion.

1. Introduction

Quantum impurity problems occur in various areas of physics. Some examples are: an impurity spin in a metal (the Kondo problem), the two-impurity Kondo problem, a defect or impurity spin in a spin chain, tunnelling through a barrier in a quantum wire proton-monopole scattering (the Callan-Rubakov effect). Some common features of these problems are:
1. gapless "bulk" excitations (ie. far from the impurity or in the absence of an impurity the excitations are gapless),
2. essentially harmonic behavior away from the impurity (Fermi or Luttinger liquids),
3. the impurity is localized and may carry quantum-mechanical degrees of freedom (eg. the two spin states of the impurity spin),
4. perturbation theory in the impurity-bulk coupling may be infrared divergent,
5. the problems are either initially defined in one dimension or can be mapped into one dimension (ie. some sort of partial wave expansion can be made and only a finite number of partial waves are important).

The first and last points imply that the systems are equivalent to $(1+1)$-dimensional conformal field theories with the impurity or defect at the origin of the one-dimensional position space. In the imaginary time path-integral formulation, we have a boundary or defect line at $(\tau, 0)$. See Figure 1. John Cardy recently developed [1] a boundary conformal field theory to treat two-dimensional classical critical systems with boundaries. We have extended his technique to deal with the $(1+1)$ dimensional quantum case, allowing for the possibility of a dynamical defect [2-14].

In the next section we review a simple and venerable example to provide some motivation: the local Fermi liquid theory of the Kondo effect, developed by Nozières [15], following earlier ideas of Wilson and Anderson. In Section 3 we discuss our general approach to quantum impurity problems. In Section 4 we present some of our results on the overscreened Kondo effect [2-6, 8-10], in order to demonstrate the power of the new method. In Section 5 we discuss impurities in the spin $s = 1/2$ Heisenberg antiferromagnetic chain [11, 12]. In particular we show that the simple open and periodic fixed points which occur in the problem are related by the "fusion" technique, which we used in our analysis of the Kondo problem.

Springer Series in Solid-State Sciences, Vol. 118
Correlation Effects in Low-Dimensional Electron Systems
Editors: A. Okiji and N. Kawakami

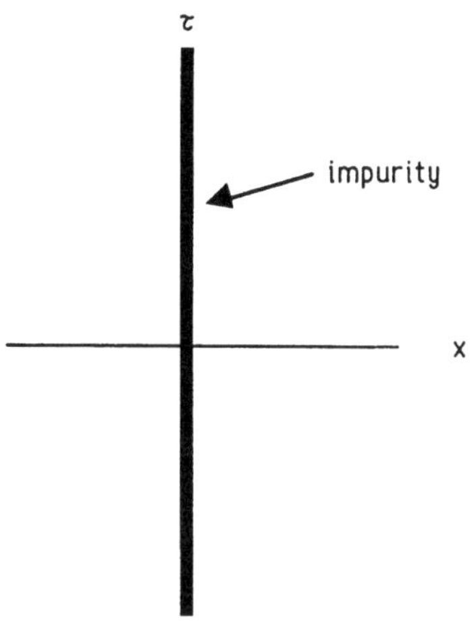

Fig.1. Imaginary time formulation of quantum impurity problems.

2. A Simple Example: Local Fermi Liquid Theory of the Kondo Effect

The continuum form of the Hamiltonian density is:

$$\mathcal{H} = \psi^\dagger \left(-\frac{\nabla^2}{2m}\right)\psi + J\delta^3(\vec{x})\psi^\dagger\frac{\vec{\sigma}}{2}\psi\cdot\vec{S}. \tag{1}$$

Here $\vec{S}$ is an $s = 1/2$ spin operator and the fermion annihilation operator, ψ_α carries a spin index which is not written explicitly. Expanding in spherical harmonics, only the s-wave interacts with the impurity due to the δ-function. Thus we obtain an effective one-dimensional model defined on the half line, $r \geq 0$ with the impurity at the origin. Alternatively, we may reflect the outgoing wave to the negative r-axis, obtaining a theory with a left-mover only on the entire real axis. (See Figure 2.) The corresponding Hamiltonian density, written in terms of a left-moving fermion operator, ψ is:

$$\mathcal{H} = \psi^\dagger i\frac{d}{dx}\psi + \lambda\delta(x)\psi^\dagger\frac{\vec{\sigma}}{2}\psi\cdot\vec{S}. \tag{2}$$

As discovered by Kondo [16], perturbation theory in the Kondo coupling, λ, is infrared divergent. In modern language, this corresponds to the β-function:

$$\frac{d\lambda}{d\ln L} = \lambda^2 + \ldots \quad . \tag{3}$$

It appears that λ renormalizes to ∞ if it is initially positive. (See Figure 3.) What does an infinite effective coupling constant really mean? Nozières made this notion precise, [15], by considering a lattice version of the Kondo problem. Since the dimensionality of the lattice is unimportant we will consider the one-dimensional case:

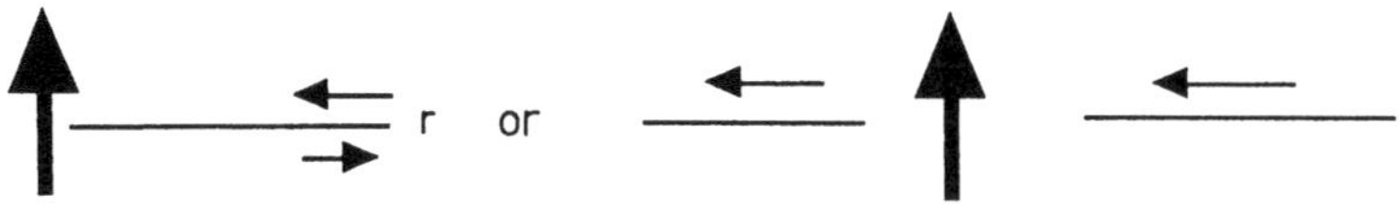

Fig.2. Two possible formulation of quantum impurity problems with left and right-movers on the positive real axis or with left movers only on the entire axis.

Fig.3. Renormalization group flow of the Kondo coupling.

$$H = t\sum_i \left(\psi_i^\dagger \psi_{i+1} + \text{h.c.}\right) + J\psi_0^\dagger \frac{\vec{\sigma}}{2}\psi_0 \cdot \vec{S} \ . \tag{4}$$

This model is easy to study for $J >> t$. For $t = 0$ we must have precisely one electron at the origin, which forms a singlet with the impurity spin, $\vec{S}$. The electronic configuration on the other sites is arbitrary. For a relatively small non-zero t, this degeneracy is broken. The other electrons simply go into a Bloch wave (Slater determinant) state, but the single-particle wave-functions must all vanish at $x = 0$ in order to preserve the spin-singlet condition there. Thus is the even-parity sector, the zero-coupling wave function:

$$\phi(x) = \cos kx \quad (\lambda = 0) \tag{5}$$

gets modified to:

$$\phi(x) = |\sin kx| \quad (\lambda \to \infty). \tag{6}$$

This is a solution of the free-particle Schroedinger equation everywhere except at the origin where the vanishing boundary condition is imposed. On the other hand, the odd-parity wave-function:

$$\phi(x) = \sin kx \tag{7}$$

is the same for zero *or* infinite coupling. As Nozières observed [15], *the strong coupling fixed point is the same as the weak coupling fixed point except that the impurity disappears (screened) and is replaced by a Boundary Condition,*

$$\psi(0) = 0. \tag{8}$$

Equivalently, we have a $\pi/2$ phase shift in the even parity channel.

For finite (or even small) Kondo coupling this picture still holds but only at low energies and long distances. The *boundary condition* is a *fixed point.* ie., we have an *effective boundary condition* which holds in the asymptotic regime. Our main new result is that *All Quantum Impurity Problems work this way!*

3. General Approach

Far from the impurity we expect the low energy physics to be described by a scale-invariant, time-independent boundary condition. This should be true outside a boundary layer whose width, a, is set by the longest microscopic or crossover length scale in the problem. Further from the impurity than this it is reasonable to expect scale invariance. However the critical behavior may still be affected, in a universal way, by the presence of the boundary. Consider, for example, a two-point Green's function. As illustrated in Figure 4 there are two special limits. In the limit where both points are far from the impurity relative to their distance from each other we expect to recover the bulk critical behavior, unaffected by the boundary. However, in the opposite limit where the two points are far apart compared to their distance from the boundary (which is still large compared to the width, a, of the boundary layer), the critical behavior can be quite

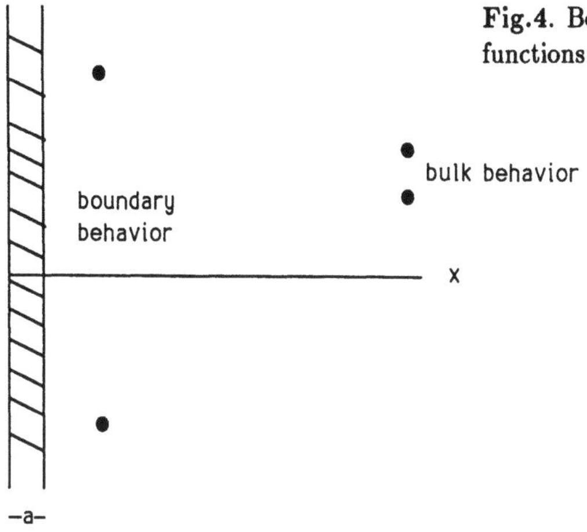

Fig.4. Boundary and bulk limit of correlation functions.

different. ie. boundary critical phenomena can occur. In general, to each bulk universality class corresponds several boundary universality classes. We expect essentially arbitrary boundary interactions should renormalize to one of these. A simple example is provided by the classical two dimensional Ising model, at temperature, $T = T_c$, defined on the half-plane, $x > 0$. There are only three universality classes of boundary conditions, spin up, spin down and free [1]. The correlation exponent, $\eta = 1/4$ far from the boundary, but $\eta = 1$ in the boundary limit, for free boundary conditions. Furthermore, the universal cross-over function, depending only on the ratios of distances from the boundary to distance between the points has been calculated exactly, by Cardy [1]. If we apply a weak magnetic field near the boundary, the system should cross over to the spin up universality class at large distances.

For analogous reasons to the bulk case, we expect not only translational and scale invariance but the infinite-dimensional conformal symmetry to hold at the critical point. We introduce a complex co-ordinate:

$$z \equiv \tau + ix, \tag{9}$$

A general conformal transformation is of the form: $z \to w(z)$ where $w(z)$ is an analytic function. This has the Taylor expansion:

$$w(z) = \sum_{n=1}^{\infty} a_n z^n \quad , \tag{10}$$

where the a_n's are arbitrary complex numbers. With a boundary at $x = 0$ we can, at most, expect the subgroup of the conformal group which leave the boundary invariant to be a symmetry of the problem. ie. we require $w(\tau)^* = w(\tau)$. This implies that the expansion coeffients, a_n are all real. Since the a_n's correspond to generators of the conformal group, we see that one half the conformal group remains in the presence of the conformally invariant boundary condition. From the real-time, Hamiltonian viewpoint, the boundary relates left and right movers. Consequently, the left and right energy density, T and $\bar{T}$ are no longer independently conserved. There is now only one Virasoro algebra, not two as in bulk conformal field theory.

More explicitly, let us assume that the problem is defined on the positive half-plane, $x > 0$ only. (We can always represent the problem this way by reflecting the other half-plane, if neccessary.) We will then assume the condition:

$$T(0,t) = \bar{T}(0,t) \quad . \tag{11}$$

This results from assuming that the momentum density vanishes at the boundary; it is essentially a unitarity condition. Since T is a function of $t+x$ only and $\bar{T}$ is a function of $t-x$ only, Eq. (11) implies that we may regard $\bar{T}$ as the analytic continuation of T to the negative x-axis; ie.

$$T(t,x) \equiv \bar{T}(t,-x) \quad \text{for } x < 0 \quad . \tag{12}$$

Thus we obtain a problem defined on the entire real axis with left-movers only. This identification of left with right leads to a modification of Green's functions near the boundary. An arbitrary operator, O consists of left and right factors, depending on $t+x$ and $t-x$ only:

$$O(x,t) = O_L(t+x)\bar{O}_R(t-x) \quad . \tag{13}$$

Using the left-right identification we obtain (at $t=0$):

$$O(x) \rightarrow O_L(x)\bar{O}_L(-x) \quad , \tag{14}$$

ie. the local operator O becomes effectively bilocal in the presence of a boundary. This is similar to the method of image charges in electrostatics. An immediate consequence of this is that operators pick up non-vanishing one-point functions near the boundary (even if they vanish in the bulk). ie.

$$< O(x) > \rightarrow < O_L(x)\bar{O}_L(-x) > = \frac{C}{(2ix)^{\eta}} \quad , \tag{15}$$

ie. a one-point function becomes a two-point function. Similarly two-point functions becomes four-point functions. In general we can characterize the effects of the boundary by the operator product expansion:

$$O(x) \rightarrow O_L(x)\bar{O}_L(-x) \rightarrow \sum_j \frac{C_j}{(2ix)^{\eta_j}} O_j(0) \quad . \tag{16}$$

The set of operators, O_j and exponents, η_j simply correspond to the left-moving Hilbert Space of bulk operators; they do not depend on the particular boundary conditions. The operator product expansion coefficients, C_j on the other hand, do depend on the boundary condition.

How do we find all possible conformally invariant boundary conditions and calculate Green's functions? Consider the system in a box of length l with arbitrary conformally invariant boundary conditions A and B at $x=0$ and $x=l$, at inverse temperature β. The path integral is defined on a cylinder of circumference β and length l. (See Figure 5.) Letting H^l_{AB} denote the Hamiltonian for the finite system with boundary conditions A and B; the path integral corresponds to the partition function:

$$Z_{AB} = tr e^{-\beta H^l_{AB}} \quad . \tag{17}$$

Alternatively, we may regard l as the time interval and β as the space interval, ie. make a modular transformation. The system is now periodic in space; we write the corresponding Hamiltonian as H^β_P, P denoting periodic. Now the system propogates for a time interval l so the imaginary time evolution operator, $e^{-lH^\beta_P}$ occurs. However, we do not write a trace in this case since the system is not periodic in time. Instead, the system evolves between some initial and final states, $|A>$ and $|B>$; ie.:

$$Z_{AB} = < A|e^{-lH^\beta_P}|B> \quad . \tag{18}$$

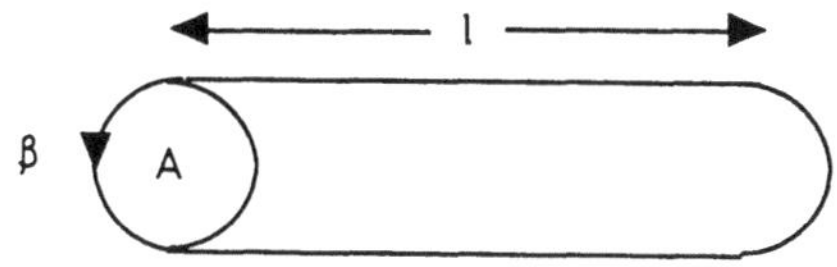

Fig.5. The cylinder geometry: A and B represent conformally invariant boundary conditions.

$|A>$ and $|B>$ are called *boundary states*. To each boundary condition corresponds a boundary state. It turns out to be easier to find all possible boundary states than boundary conditions. Equating the two expressions for Z_{AB} gives an equation (true for all l/β) which determines all conformally invariant boundary states (and hence boundary conditions). The boundary states determine the Green's functions. (Matrix elements of operators between the boundary state and the vacuum give the needed operator product expansion coefficients.) To solve for the critical behavior of an arbitrary quantum impurity problem we "just" have to find (guess?) the corresponding boundary state. Frequently the number of possibilities consistent with the symmetries of a given bulk critical system is very small. We have found boundary states for the various problems listed in the introduction.

4. Multi-Channel Kondo Effect

We generalize the Kondo Hamiltonian of Eq.(2) to include k channels of electrons interacting with a spin-s impurity:

$$\mathcal{H} = \sum_{j=1}^{\infty} \left[\psi^{\dagger j} \frac{id}{dx} \psi_j + \lambda \delta(x) \psi^{\dagger j} \frac{\vec{\sigma}}{2} \psi_j \cdot \vec{S} \right] \quad . \tag{19}$$

Note that we have made the (in general unrealistic) assumption that each channel has identical Fermi velocity (set to 1), identical Kondo coupling and identical potential scattering term (assumed to be 0). Thus the model has an $SU(k)$ symmetry corresponding to interchanging the channels. When the number of channels exceeds twice the impurity spin, $k > 2s$, the impurity is overscreened and the Kondo coupling does not renormalize to ∞. Consider, for example the simplest overscreened case, $k = 2$, $s = 1/2$. See Figure 6. If we assume that the on-site Kondo coupling goes to infinity, then one electron from each channel would sit on the impurity site with spins anti-parallel to that of the impurity. This overscreened complex has spin 1/2. Now consider the electrons on the next site. More correctly we must consider an even parity combination of the two neighboring sites. In the three-dimensional setting we are considering spherical shells around the impurity, as drawn in Figure 6. These electrons feel a weak *antiferromagnetic* Kondo coupling with the spin complex at the origin. This coupling also renormalizes to large values. If we assume it also goes to infinity then we obtain yet another, larger, spin complex with $s = 1/2$. This process would keep on going forever. This basically tells us that the assumption that the Kondo coupling flows to ∞ is not correct.

We have found the boundary states for all values of k and s. From these we can calculate the single-particle Green's function [6, 9]:

$$< \psi_L^{\dagger}(z_1) \psi_R(\bar{z}_2) > \rightarrow \frac{S_{(1)}}{z_1 - \bar{z}_2} \quad . \tag{20}$$

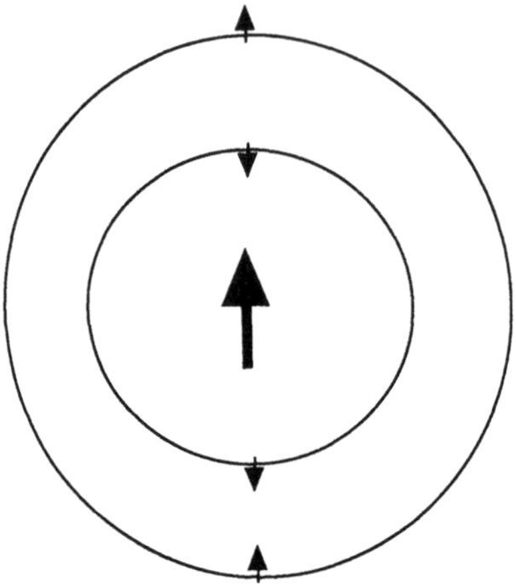

Fig.6. The two-channel, $s = 1/2$ Kondo effect. Successive electron orbitals each over-screen the impurity always leaving behind an effective $s = 1/2$.

$S_{(1)}$ measures the correlation of incoming and outgoing electron. From it we can determine the self-energy for a dilute random array of impurities, and hence the zero-temperature resistivity. This is given by:

$$\rho = \rho_{un} \frac{1 - S_{(1)}}{2} \quad . \tag{21}$$

Here ρ_{un} is the resitivity in the unitary limit. ie., it is the maximum possible resistivity that could occur for potential scattering, corresponding to a $\pi/2$ phase shift. $S_{(1)}$ has the interpretation of the 1 particle $\to$ 1 particle S-matrix at zero energy. For potential scattering, we would have, $S_{(1)} = e^{2i\delta}$. This is also true for the one-channel Kondo effect, with $\delta = \pi/2$. In fact, with a particle-hole symmetric Hamiltonian as in Eq. (19), $S_{(1)}$ must be real. However, in the overscreened case, we find $|S_{(1)}| < 1$!! Unitarity then implies that there must be inelastic scattering even at zero energy, violating the basic assumption of Landau's Fermi liquid theory. Thus we call this a non Fermi liquid fixed point. Our explicit calculation from the boundary state gives:

$$S_{(1)} = \frac{\cos[\pi(2s + 1)/(2 + k)]}{\cos[\pi/(2 + k)]} \quad . \tag{22}$$

We can also calculate the electron pair-operator Green's function [6, 10], for example. For the case, $k = 2$, $s = 1/2$, we define the spin-singlet, flavor-singlet pair operator:

$$O(z) \equiv \psi_{L,\alpha i}(z)\psi_{R,\beta j}(\bar{z})\epsilon^{\alpha\beta}\epsilon^{ij} \quad . \tag{23}$$

We obtain the Green's function:

$$< O(z_1)O^\dagger(z_2) >= \frac{4\eta^{-1/2}}{|z_1 - \bar{z}_2|^2}\left[\frac{1}{1-\eta} + 3\right] \quad . \tag{24}$$

Here,

$$\eta \equiv \frac{4r_1 r_2}{|\tau_1 - \tau_2|^2 + (r_1 + r_2)^2}, \quad z_j \equiv \tau_j + ir_j, \quad (r_j > 0) \ . \tag{25}$$

This universal cross-over function gives the non-interacting result in the bulk limit:

$$< O(z_1)O^\dagger(z_2) > \rightarrow \frac{4}{|z_1 - z_2|^2} \quad (r_1, r_2 >> |z_1 - z_2|) \,, \tag{26}$$

and gives non-trivial singular behavior in the boundary limit:

$$< O(z_1)O^\dagger z_2) > \rightarrow \frac{8}{|\tau_1 - \tau_2|\sqrt{r_1 r_2}} \quad (a << r_1, r_2 << |\tau_1 - \tau_2|) \,. \tag{27}$$

(Here a is the width of the boundary layer, or short-distance cut off.) Note that the $1/\tau$ fall off is more singular than the $1/\tau^2$ behavior in the non-interacting case. Fourier-transforming with respect to time for fixed r_i we obtain a logarithmically divergent local pair susceptibility, $\chi(\omega) \propto \ln \omega$. This has been used as the basis of models of superconductivity [17].

So far, we have calculated the leading temperature dependence of Green's functions, using only properties of the fixed point. The flow to the fixed point is also important and determines various temperature dependent corrections [2, 4, 6, 9]. These are obtained by considering the leading irrelevant operator, O, at the stable fixed point. ie., we write the effective Hamiltonian density:

$$\mathcal{H} = \mathcal{H}_{FP} + g\delta(x)O \,. \tag{28}$$

Here the fixed point Hamiltonian, H_{FP} is defined with the corresponding boundary condition. g is the (non-universal) leading irrelevant coupling constant. For $k = 2$, $s = 1/2$, O has scaling dimension 3/2. It then follows from dimensional analysis that g has dimensions of $(\text{energy})^{-1/2}$. (Note that the δ-function contributes 1 to the dimension.) The corresponding energy scale is, by definition, the Kondo temperature:

$$g \equiv \frac{1}{\sqrt{T_K}} \,. \tag{29}$$

We perform low order perturbation theory in g. This is infrared finite, since g is irrelevant. It gives a series in $(T/T_K)^{1/2}$, by dimensional analysis. We can express the leading temperature dependence of several quantities in terms of one unknown parameter, ratio, exactly for small T. For $k = 2$, $s = 1/2$, we find the resistivity correction [6, 9]:

$$\rho(T) = \frac{1}{2}\rho_{un}[1 + 4g\sqrt{\pi T}] \,, \tag{30}$$

the impurity specific heat [4]:

$$C_{imp}(T) = 9\pi^3 T g^2 ln(T/T_K) \,, \tag{31}$$

and the impurity susceptibility [4]:

$$\chi_{imp}(T) = 18\pi\mu_B^2 g^2 \ln(T_K/T) \,. \tag{32}$$

Notice that $g = 1/\sqrt{T_K}$, drops out of the Wilson ratio, χ_{imp}/C_{imp} and also the ratio of the square of the finite temperature restivity correction to the specific heat. The analogous ratio for the Fermi liquid Kondo fixed point was first calculated by Nozières [15]. In that case the approach to the fixed point is quite different. In particular, the resistivity then takes the form:

$$\rho_{FL}(T) = \rho_{un}[1 - g^2 T^2] \,. \tag{33}$$

Differences occur both in the dimension of the leading irrelevant operator (2 instead of 3/2) and the lowest order in perturbation theory at which various quantities become non-zero. Note the much more singular T-dependence in this non-Fermi liquid case.

5. Impurities in Spin-1/2 Heisenberg Chains: A Fusion Rule Approach

Recently, Sebastian Eggert and I applied the boundary conformal field theory technique to a localized impurity in an s=1/2 antiferromagnetic chain [11]. The impurities we considered were modified exchange couplings on a single link or a pair of links or the coupling of an extra spin (possibly with $s > 1/2$) to the chain. The boundary fixed points that occured were very simple, corresponding to a periodic (ie. unperturbed) chain or an open chain. Unlike in our analysis of the multi-channel or two-impurity Kondo problems [3, 4], we did not use the fusion rules to find the fixed point to which the system renormalized upon adding the impurity. In this section, I wish to show that, in fact, it is possible to pass between these fixed points using fusion, at least in the case of an $SU(2)$ invariant system, ie. the Heisenberg model. We do not know how to do this in the case of the xxz model; it appears likely that fusion is not a general procedure that works in all cases [13, 14]. We hope that this approach to the Heisenberg model may shed light on this general question. It also brings out analogies between the spin chain impurity problem and both two-channel [3, 4] and two-impurity [7] Kondo problems. We begin by briefly reviewing the fusion rules approach to finding boundary states and then turn to the spin chain example.

As mentioned in Sec. 3, the quantum impurity problems that we consider can always be formulated so as to obey the boundary condition $T(t,0) = \bar{T}(t,0)$. In many cases they also obey an analogous condition on some conserved chiral currents:

$$J^a(t,0) = \bar{J}^a(t,0) \, . \tag{34}$$

In the spin chain problem a will run over the three generators of $SU(2)$. As before, we may regard $\bar{J}^a$ as the analytic continuation of J^a to $x < 0$. If we consider the cylinder geometry of Sec. 3, with such boundary conditions at each end, then it follows that

$$J^a(t,-l) = J^a(t,l) \, , \tag{35}$$

and the same for T. Thus all eigenstates of H^l_{AB} will be eigenstates of T defined periodically on an interval of length $2l$. It follows that Z_{AB} can be expanded in Kac-Moody characters of T, corresponding to the algebra of the currents, J^a. These can be written as:

$$\chi_i(q) \equiv \sum_{x;i} q^x \, , \tag{36}$$

where the x label the scaling dimensions in the ith conformal tower, and, $q \equiv e^{-\pi\beta/l}$. ie., in general:

$$Z_{AB} = \sum_i n^i_{AB}\chi_i(q). \tag{37}$$

Only the integer multiplicities, n^i_{AB}, depend on the particular conformally invariant boundary conditions A, B.

Eqs. (11) and (35) correspond to conditions on the boundary states, $|A>$, after the modular transformation:

$$[T(x) - \bar{T}(x)]|A> = [J^a(x) + \bar{J}^a(x)]|A> = 0 \quad . \tag{38}$$

There is one solution of these equations, called an Ishibashi state [18], for each Kac-Moody conformal tower:

$$|i> = \sum_N |i,N>_L \otimes \Omega|i,N>_R \, . \tag{39}$$

Here N is summed over all members of the conformal tower; the anti-unitary operator Ω

takes $\bar{J}^a \to -\bar{J}^a$. All boundary states, $|A>$ can be expanded in Ishibashi states:

$$|A>=\sum_i |i><i,0|A> \quad , \tag{40}$$

where $|i,0>\equiv |i,0>_L \otimes \Omega |i,0>_R$. Thus, the apparently formidable problem of classifying all boundary states and spectra is reduced to finding a finite number of multiplicites, n^i_{AB} or a finite number of matrix elements $<i,0|A>$, one for each conformal tower. By equating the two expressions for Z_{AB}, Eq. (17) and (18) we obtain an important set of equations which can be used to determined these parameters. Using the modular transformation:

$$\chi_i(q) = S_i^j \chi_j(\tilde{q}), \tag{41}$$

where

$$\tilde{q} \equiv e^{-4\pi l/\beta}, \tag{42}$$

and assuming linear independence of the characters, we obtain:

$$\sum_j S^i_j n^j_{AB} =< A|i,0><i,0|B>, \tag{43}$$

Cardy's equation [1].

Cardy discovered an important property of this equation which allows new boundary states to be constructed from old ones. Given any consistent boundary state, $|A>$, and any conformal tower, i, we construct a new boundary state, $|A,i>$ as follows. With some fixed boundary state $|B>$ at the other end of the cylinder, the new spectrum is given by:

$$n^j_{A,i;B} = N^j_{ik} n^k_{AB}. \tag{44}$$

Here the non-negative integers, N^j_{ik}, are the fusion rule coefficients; they give the number of occurrances of the primary field ϕ_j in the operator product expansion of ϕ_i with ϕ_k. The new boundary state is given by:

$$< A,i|j,0>= \frac{S^j_i}{S^j_0} < A|j,0> . \tag{45}$$

The consistency of Eqs. (44) and (45) follows from the Verlinde formula relating the fusion rule coefficients to the modular S-matrix:

$$S^j_k N^k_{il} = \frac{S^j_i S^j_l}{S^j_0}. \tag{46}$$

This strategy for constructing new boundary states was used in our solution for the critical behavior of the multi-channel and two-impurity Kondo problems. We began with a trivial boundary state corresponding to a simple vanishing boundary condition on the free fermions. We then constructed non-trivial boundary states by fusion with particular operators, ϕ_i, motivated by physical considerations.

We now consider the example of the $s=1/2$ Heisenberg spin chain. Our technique can be applied to this problem because, in the absence of a boundary, the system corresponds to a harmonic Luttinger liquid, ie. a free spin boson, or correspondingly a $k=1$, $SU(2)$ Wess-Zumino-Witten (WZW) non-linear σ-model. (See, for example, [19].) The spin operators, in the bulk system are represented in terms of the $SU(2)$ matrix field, g, of the WZW model and the current operators as:

$$\vec{S}_i = (-1)^i \text{constant} \times tr g\vec{\sigma} + (\vec{J}_L + \vec{J}_R). \tag{47}$$

The field g may be regarded as a product of left and right-moving factors:

$$g^{\alpha}_{\beta} = g^{\alpha}_{L} g_{R\beta} \,. \tag{48}$$

Impurity interactions generally involve the field g at the origin. A crucial complication of this Luttinger liquid system, compared to free fermions, is that the impurity interaction involves both left and right-moving fields and cannot be reduced to a single type of field. This is unlike the Kondo problem. If we consider, for example, the Kondo problem in a one-dimensional system of left and right-moving electrons, the Kondo interaction only involves the sum of left and right-moving fermion fields, $[\psi_L(0) + \psi_R(0)]$. We may define even and odd parity channels,

$$\psi_{e,o}(x) \equiv \frac{1}{\sqrt{2}}[\psi_L(x) \pm \psi_R(-x)], \tag{49}$$

and only $\psi_e(0)$ appears in the Kondo interaction. No analgous reduction of the number of degrees of freedom can be made in the Heisenberg model case so we are faced with what is fundamentally a two-channel problem. It is actually convenient to regard the right-movers as being a second channel of left movers. ie., we define:

$$g_{L\alpha 1}(x) \equiv g_{L\alpha}(x), \quad g_{L\alpha 2}(x) \equiv g_{R\alpha}(-x) \,. \tag{50}$$

We now have two left-moving $k = 1$ WZW models, a theory with central charge $c = 2$. This model is known to be equivalent to a single $k = 2$ WZW model ($c = 3/2$) together with an Ising model ($c = 1/2$). This representation is useful because the total spin current, $\vec{J}_1 + \vec{J}_2 \equiv \vec{J}$, is conserved whereas the original left and right-moving currents are not separately conserved in general, with a boundary present. Thus only a finite number of characters and Ishibashi states can occur, corresponding to direct products of $SU(2)_2$ and Ising conformal towers. Each theory has three conformal towers. Those of the WZW model are labeled by the spin of the highest weight state, $j = 0, 1/2, 1$. Those of the Ising model correspond to the identity operator, the order parameter, and the energy operator. It is convenient to also label these by a second index $j_I = 0, 1/2, 1$ respectively because the modular S-matrices are the same with this identification:

$$S = \begin{pmatrix} 1/2 & 1/\sqrt{2} & 1/2 \\ 1/\sqrt{2} & 0 & -1/\sqrt{2} \\ 1/2 & -1/\sqrt{2} & 1/2 \end{pmatrix} .$$

Let us now consider a boundary. It turns out that there are apparently only two relevant boundary conditions corresponding to a periodic system (ie. no boundary at all) or else an open system, ie. a break in the spin chain at one point. As discussed previously, we can always formulate the problem in terms of left-movers only on the entire real axis or left and right-movers on the positive axis. In this problem, in the first formulation, we have two channels of left-movers and in the second formulation we have two channels each of left and right-movers. The latter formulation is convenient for studying the cylinder geometry with two boundary conditions A and B discussed in Sec. 4. We now wish to determine the partition function Z_{AB} for A and B either the periodic or open boundary condition, labelled P and O respectively. Note that Z_{PP} is the partition function for a periodic spin chain of length $2l$; Z_{PO} corresponds to a chain of length $2l$ with open boundary conditions and Z_{OO} corresponds to two decoupled chains with open boundary conditions, each of length l. Although we ignore corrections of $O(1/l)$ to the energies, we must carefully distinguish the case of even or odd l. These spectra were worked out in [11] by mapping the boundary conditions on the spins into boundary conditions on the abelian boson which is equivalent to the $k = 1$ WZW model. Periodic boundary conditions on the spins map into periodic boundary conditions on the boson and open map into vanishing boundary conditions. From this we determined the partition functions, written in terms of $k=1$ WZW characters, which are labelled by highest weight states, $j=0$ and $j=1/2$. We label the corresponding characters, $\chi_j^{(1)}$; the superscript (1) labels the Kac-Moody central charge $k=1$. The partition functions depend on whether the chains have even or odd length. For a periodic chain of even length $2l$ the partition function is given by $[\chi_0^{(1)}(q)]^2+[\chi_{1/2}^{(1)}(q)]^2$. For odd length $2l+1$ it is given by $2\chi_0^{(1)}(q)\chi_{1/2}^{(1)}(q)$. For an open chain of even length, l it is given by $\chi_0^{(1)}(q)$ and for odd length, $l+1$ it is given by $\chi_{1/2}^{(1)}(q)$.

We can reformulate these spectra in terms of the $SU(2)_2\times$ Ising representation. We now give the spectra for the three possible pairs of boundary conditions (P or O at each end) and for even or odd length of the line segments (denoted by a superscript e or o.). These are given by:

$$\begin{aligned}
Z^e_{PP}(q) &= [\chi_0^{(2)}(q)+\chi_1^{(2)}(q)]\cdot[\chi_0^I(q)+\chi_1^I(q)]\ ,\\
Z^o_{PP}(q) &= 2\chi_{1/2}^{(2)}(q)\cdot\chi_{1/2}^I(q)\,,\\
Z^e_{PO}(q) &= [\chi_0^{(2)}(q)+\chi_1^{(2)}(q)]\cdot\chi_{1/2}^I(q)\,,\\
Z^o_{PO}(q) &= \chi_{1/2}^{(2)}(q)\cdot[\chi_0^I(q)+\chi_1^I(q)]\ ,\\
Z^{ee}_{OO}(q) &= \chi_0^{(2)}(q)\cdot\chi_0^I(q)+\chi_1^{(2)}(q)\cdot\chi_1^I(q)\ ,\\
Z^{eo}_{OO}(q) &= \chi_{1/2}^{(2)}(q)\cdot\chi_{1/2}^I(q)\ ,\\
Z^{oo}_{OO}(q) &= \chi_0^{(2)}(q)\cdot\chi_1^I(q)+\chi_1^{(2)}(q)\cdot\chi_0^I(q)\ .
\end{aligned}\tag{51}$$

It is straightforward to check all these identities explicitly since the $SU(2)_2$ and Ising characters all have simple expressions in terms of θ-functions and the $SU(2)_1$ characters are given by simple free boson formulas.

We now wish to demonstrate that we can pass between the different spectra by fusion. The fusion rules are the same for $SU(2)_2$ and for the Ising model using the j_I representation. They are:

$$\begin{aligned}
0\times j &\to j\ ,\\
1/2\times 1/2 &\to 0+1\ ,\\
1\times 1/2 &\to 1/2\ .
\end{aligned}\tag{52}$$

Let us consider starting with two open sections, both of even length, corresponding to Z^{ee}_{OO}. Suppose we now couple one extra $s=1/2$ variable infinitesimally with equal strength to the two open chains at $x=0$. We expect this weak link to "heal"; ie. to renormalize to a periodic boundary condition. We should then obtain the partition function, Z^o_{PO}. (The total length is now odd since we have added one extra spin.) It is natural to suspect that this process should correspond to fusion with the $j=1/2$ primary field in the $SU(2)_2$ sector. It can be verified explicitly from Eqs. (51) and (52) that this is the case. If we now couple the remaining open ends to another $s=1/2$ variable then we obtain the Z^e_{PP} spectrum. This again can be obtained by fusion with the $j=1/2$ primary field. Alternatively, we may start with two open sections both of odd length or one odd and one even. Adding extra $s=1/2$'s, coupled with equal strength at both ends, producing cross-over to Z_{PO} and then Z_{PP} corresponds to fusion with $j=1/2$ in all cases. This process is formally identical to a spin-only version of the two-channel, $s=1/2$ Kondo effect [4]. In that problem the non-trivial fixed point is also obtained by fusion with $j=1/2$ in the $SU(2)_2$ sector. The only difference lies in the presence of charge excitations in the Kondo problem; but these play a passive role anyway.

Alternatively, we may again start with two open even length sections and now couple them together with *two* intervening spins. This is analogous to the two-impurity Kondo problem; the coupling of the two extra spins to the ends of the open chains is analogous to the Kondo coupling and the self-coupling of the two extra spins is analogous to the RKKY coupling. (The two Kondo couplings are assumed equal.) The stable fixed points in this problem correspond to open chains, with the two extra spins either coupling together to form a singlet or else with one fastening onto the end of each chain. In between these two stable fixed points we expect an unstable one, which can be obtained by adjusting the ratio of the inter-impurity coupling to the Kondo coupling. This fixed point corresponds to the periodic boundary condition, with spectrum Z^e_{PO}. It can be checked from Eq. (51) and (52) that this is obtained from fusion with the Ising primary, $j_I = 1/2$. Critically coupling the two remaining free ends to two more impurity spins produces Z^e_{PP}, again from fusion with $j_I = 1/2$. Again the same process occurs if we start with two open chains of odd length or one even and one odd. This fusion process, with the $j_I = 1/2$ field is again identical to the one that determines the non-trivial unstable critical point in the two-impurity Kondo problem [7].

Rather remarkably, both two channel one impurity and one channel two impurity processes occur in this simple model, in a simplified form with no charge degrees of freedom present. In both cases the "non-trivial" fixed point is simply the periodic one, the "healing" process discovered in [11] corresponds to the flow to the non-trivial fixed point in both types of Kondo problem, a stable flow for one impurity but unstable for two impurities.

We may also write down boundary states corresponding to periodic and open boundary conditions, related to each other by fusion. Subtleties arise because of the distinction between even and odd length chains and we find it neccessary to introduce (at least) two types of periodic boundary states and three types of open ones, in order to obtain all seven partition functions. Ishibashi states can be labelled either in terms of the two $SU(2)_1$ algebras or in terms of $SU(2)_2\times$ Ising. In the latter formulation ($|jj_I>$ basis), the boundary states are:

$$\begin{aligned}
|per\ 1> &= |0,0> -|0,1> +|1,0> -|1,1>\ , \\
|per\ 2> &= |0,0> +|0,1> -|1,0> -|1,1>\ , \qquad (53)\\
|open\ 1> &= \frac{1}{\sqrt{2}}[|0,0> +|0,1> +|1,0> +|1,1> +|1/2,1/2>_1 +|1/2,1/2>_2]\ , \\
|open\ 2> &= \frac{1}{\sqrt{2}}[|0,0> +|0,1> +|1,0> +|1,1> -|1/2,1/2>_1 -|1/2,1/2>_2]\ , \\
|open\ 3> &= \frac{1}{\sqrt{2}}[|0,0> -|0,1> -|1,0> +|1,1> +|1/2,1/2>_1 -|1/2,1/2>_2]\ .
\end{aligned}$$

Two different Ishibashi states $|1/2,1/2>_i$, $i = 1,2$ occur in this formulation, corresponding to the $SU(2)_1 \otimes SU(2)_1$ Ishibashi states $|1/2,0>$ and $|0,1/2>$. Using the ratio of S-matrix elements:

$$\frac{S^j_{1/2}}{S^j_0} = (\sqrt{2},0,-\sqrt{2}), \qquad (54)$$

we see from Eq. (45) that each of the $|open\ i>$ states goes into one of the $|per\ j>$ states under fusion with either the $j = 1/2$ or $j_I = 1/2$ field. The 15 matrix elements of the form $< A|e^{-lH^\beta_P}|B>$ obtained from these five states are all equal to one of the seven different partition functions given in Eq. (51). For instance, $Z_{per\ i,\ per\ j} = Z^e_{PP}$ for $i = j$, and $= Z^o_{PP}$ for $i \neq j$.

To conclude this section, the simple problem of an impurity in a Heisenberg spin chain can be understood using the fusion rules. It represents a spin-only version of both the two channel and two impurity Kondo problems.

I would like to thank Andreas Ludwig for his collaboration in much of this work and Eugene Wong for checking the identities in Eq. (51). I would also like to thank my other collaborators in this field: Sebastian Eggert, Jacob Sagi and Erik Sørensen. This research was supported in part by NSERC of Canada.

References

[1] J.L. Cardy, *Infinite Lie Algebras and Conformal Invariance in Condensed Matter and Particle Physics* (ed. K. Dietz and V. Rittenberg, World Scientific, Singapore, 1987), 81; Nucl. Phys. **B324**, 581 (1989); D. Lewellen and J.L. Cardy, Phys. Lett. **B259**, 274 (1991).

[2] I. Affleck, Nucl. Phys. **B336**, 517 (1990).

[3] I. Affleck and A.W.W. Ludwig, Nucl. Phys. **B352**, 849 (1991).

[4] I. Affleck and A.W.W. Ludwig, Nucl. Phys. **B360**, 641 (1991).

[5] I. Affleck and A.W.W. Ludwig, Phys. Rev. Lett. **67**, 161 (1991).

[6] A.W.W. Ludwig and I. Affleck, Phys. Rev. Lett., **67**, 3160 (1991).

[7] I. Affleck and A.W.W. Ludwig, Phys. Rev. Lett. **68**, 1046 (1992).

[8] I. Affleck, A.W.W. Ludwig, H.-B. Pang and D.L. Cox, Phys. Rev. **B45**, 7918 (1992).

[9] I. Affleck and A.W.W. Ludwig, Phys. Rev. **B48**, 7297 (1993).

[10] A.W.W. Ludwig and I. Affleck, in preparation.

[11] S. Eggert and I. Affleck, Phys. Rev. **B46**, 10866 (1992).

[12] E. Sørensen, S. Eggert and I. Affleck, preprint, UBCTP-93-07 (cond-mat@babbage.sissa.it,no. 9306003) to appear in J. Phys. **A**.

[13] I. Affleck and J. Sagi, preprint, UBCTP-93-18 (hep-th@babbage.sissa.it, no. 9311056).

[14] E. Wong and I. Affleck, preprint, UBCTP-93-20 (cond-mat@babbage.sissa.it, no. 931140).

[15] P. Nozières, J. Low Temp. Phys. **17**, 31 (1974); P. Nozières, Proceedings of the 14th Int.'l Conf. on Low Temp. Phys. (ed. M. Krusius and M. Vuorio, North Holland, Amsterdam, 1974) V.5, p. 339.

[16] J. Kondo, Prog. Theor. Phys. **32**, 37 (1964).

[17] D.L. Cox, Phys. Rev. Lett. **67** 2883 (1987); V.J. Emery and S. Kivelson, Phys. Rev. **B46**, 10812 (1992).

[18] N. Ishibashi, Mod. Phys. Lett. **A4**, 251 (1989); T. Onogi and N. Ishibashi, Nucl. Phys. **B318**, 239 (1989).

[19] I. Affleck, *Fields, Strings and Critical Phenomena* (ed. E. Brézin and J. Zinn-Justin, North-Holland, Amsterdam), 563 (1990).

Elementary Excitations for Impurity Models

A. Okiji[1], *S. Suga*[1], *M. Yamashita*[1,*], *and N. Kawakami*[2]

[1]Department of Applied Physics, Osaka University, Suita, Osaka 565, Japan

[2]Yukawa Institute for Theoretical Physics, Kyoto University, Kyoto 606, Japan

*Present Address: NTT Basic Research Laboratories, Atsugi, Kanagawa 243-01, Japan

Abstract. Elementary excitations for impurity models are investigated with the use of the Bethe-Ansatz solution. We study three kinds of Hamiltonians concerning the Kondo problem: the ordinary Anderson model, the orbitally degenerate Anderson model and the multi-channel Kondo model. The calculations for the multi-channel Kondo model are further extended to the case at finite temperatures by means of Yang and Yang's method developed in one-dimensional interacting boson systems.

1. Introduction

Thermodynamic quantities in the Kondo problem have been investigated precisely at arbitrary temperatures by the renormalization-group method [1,2] and also by the Bethe-Ansatz solution [3]. Although the Bethe-Ansatz method is quite powerful, physical quantities calculated were restricted to only static ones. This problem has been resolved to some extent by combining the exact solution with the local Fermi-liquid theory [1,4,5], which enables us to exactly calculate some transport quantities [3,6] and low-frequency susceptibilities [7]. More recently, dynamical quantities have been fully calculated by the numerical renormalization-group technique [8]. Furthermore, conformal-field-theory approach has made it possible to discuss the critical behavior of correlation functions [9,10].

In this article, we discuss elementary excitation spectra in the Kondo problem based on the Bethe-Ansatz solution. We introduce *dressed particles* (or *dressed holes*) in the thermodynamic limit [11,12], which are the simplest elementary excitations for the spin and charge degrees of freedom. Excitation spectra of these *dressed particles* determine the low-energy behavior of bulk quantities completely, such as the spin and the charge susceptibilities. In the following, we study three kinds of Hamiltonians concerning the Kondo problem: the ordinary Anderson model, the orbitally degenerate Anderson model and the multi-channel Kondo model. We will observe the formation of the sharp resonance at the Fermi level and the deep-lying charge excitation spectrum in the symmetric Anderson model. We then discuss the orbitally degenerate Anderson model in connection with the Kondo effect for rare-earth impurity systems. The multi-channel Kondo model is investigated for the orbital singlet case as well as for the over- and under-screening cases [13]. Finally, we extend the calculation of the multi-channel Kondo model to the case at finite temperatures, based on Yang and Yang's method for one-dimensional boson systems [12,14].

Springer Series in Solid-State Sciences, Vol. 118
Correlation Effects in Low-Dimensional Electron Systems
Editors: A. Okiji and N. Kawakami

2. Finite U Anderson Model

Let us consider the Anderson impurity model,

$$H = \sum_{k,\sigma} \epsilon_k c_{k\sigma}^\dagger c_{k\sigma} + V \sum_{k,\sigma} [c_{k\sigma}^\dagger d_\sigma + h.c.] + \epsilon_d \sum_{\sigma} d_\sigma^\dagger d_\sigma + U d_\uparrow^\dagger d_\uparrow d_\downarrow^\dagger d_\downarrow, \tag{2.1}$$

which describes the electron correlation problem for dilute alloy systems. The Bethe-Ansatz solution of this model results in the basic algebraic equation for two kinds of the quasi-momenta k and Λ [15],

$$\exp(ik_j L) = \prod_{\beta=1}^{M} e[\Lambda_\beta - g(k_j)] e[-U(k_j + U/2)], \tag{2.2}$$

$$-\prod_{j=1}^{N} e[-(\Lambda_\alpha - g(k_j))] = \prod_{\beta=1}^{M} e[(\Lambda_\beta - \Lambda_\alpha)/2], \tag{2.3}$$

where we have introduced $e[x] = (x - iU\Delta)/(x + iU\Delta)$ and $g(k) = (k - U/2 - \epsilon)^2$. Here $N(M)$ is the number of total (up-spin) electrons and L is the diameter of the sphere system around the impurity site. The density of states for conduction electrons is assumed to be constant so that the width of the virtual bound state Δ is independent of the energy. The total energy and the z-component of the total spin are given by $E = \sum k_j$ and $S_z = (N - 2M)/2$. In contrast to ordinary repulsive models solved exactly, such as the Hubbard model, the ground state of the Anderson model is expressed by the bound-state solutions of two electrons (complex k), and the magnetic excitation is expressed by unpaired electrons (real k) [16]. Hence, the ground state is given by the sea of two-electron pairs with anti-parallel spins. Hereafter, we restrict our consideration to the symmetric case where the localized electron number is set to unity.

Let us begin with the simplest excitation concerning the charge degrees of freedom. As mentioned above, the ground state of the Anderson model is constructed by the sea of charged bound states. We remove a bound state from the ground state. Although the bare change of the electron charge is 2e, the back-flow effect induced by the many-body effect partially compensates the change of the electron charge. As a result, the spinless *dressed hole* with charge e is formed. In Fig.1(a), the charge excitation spectrum is shown for several values of the Coulomb interaction. For $U = 0$, the shape of the spectrum is given by the Lorentzian. In the *s-d* limit ($U \gg \Delta$), the hump structure around the impurity level is described by the universal function,

$$D_c(\epsilon) = \frac{1}{2\pi\Delta} \int_{-\infty}^{\infty} dx \frac{1}{\cosh(\pi x/2\Delta)} \frac{\Delta}{(\epsilon - \epsilon_d - x)^2 + \Delta^2}. \tag{2.4}$$

It is noteworthy that $D_c(0)$ gives the exact value of the charge susceptibility, χ_c.

We next consider the spin excitation [11]. First, we add an electron to the local-singlet of the ground state. In contrast to the charge excitation, the bare increase of the electron charge is exactly canceled by the back-flow effect. Consequently, the *dressed particle* of the spin excitation with the effective spin $s^* = 1/2$ and no charge is formed. In Fig.1(b), the spin excitation spectrum is shown for several values of the Coulomb interaction. Note that the spin susceptibility (in the unit of $g\mu_B$), χ_s, is given by $D_s(0)$. It is observed that the shape becomes narrower with the increase of the Coulomb interaction. In the *s-d* limit, this spectrum describes the sharp many-body resonance. In this limit, the spin excitation spectrum takes the universal form,

$$D_s(\epsilon) = \frac{T_K/\pi}{\epsilon^2 + (T_K)^2}, \tag{2.5}$$

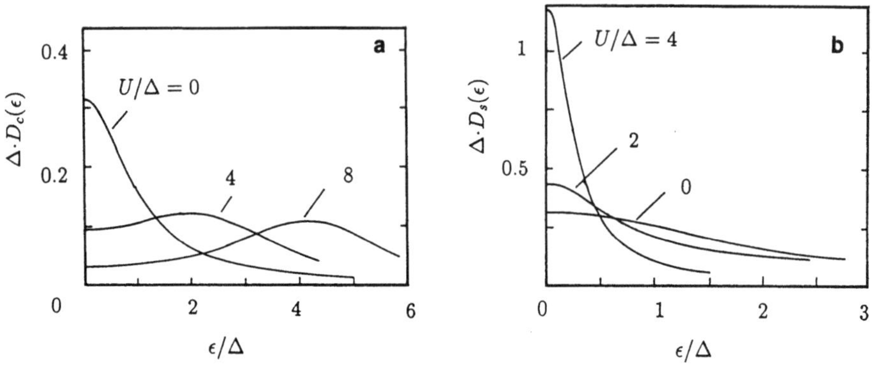

Fig.1(a) Charge excitation spectrum for the symmetric Anderson model;(b) spin excitation spectrum.

with the Kondo temperature, $T_K = (1/\pi)(2U\Delta)^{1/2}\exp[-\pi(U/(8\Delta) - \Delta/(2U))]$. It is noted that the density of states for the dressed particle has a constant weight irrespective of the Coulomb interaction, hence it is distinguished from the density of states determined by the one-particle Green function which possesses the renormalization factor of the order T_K/Δ.

3. Highly Correlated Degenerate Anderson Model

The orbitally degenerate Anderson model in the limit of infinite Coulomb repulsion has been often used to investigate the rare-earth impurity problem (*e.g.* Ce, Yb). The Hamiltonian is given by

$$H = \sum_{k,m} \epsilon_k c_{km}^\dagger c_{km} + V \sum_{k,m} [c_{km}^\dagger f_m + h.c.] + \epsilon_f \sum_m f_m^\dagger f_m + U \sum_{m,n} f_m^\dagger f_m f_n^\dagger f_n, \tag{3.1}$$

with $U \to \infty$, where the f-electron states are ν-fold degenerate and the corresponding partial-wave representation is used for the conduction electron state. The ν-fold f-states are assumed to be classified by the spin-orbit coupling scheme. We thus set $\nu = 2J + 1$ with the total angular momentum J. This model has been diagonalized by the nested Bethe-Ansatz method [17]. In this case, we have to introduce ν kinds of quasi-momenta in order to express the internal spin degrees of freedom. It is noted that the ground state of the ν-fold degenerate Anderson model is given by the bound state out of ν-electrons [17], and there exist ν kinds of elementary excitations. These excitations are classified into the charge excitation and the $\nu - 1$ kinds of spin excitations. Here, we show the results for the Coqblin-Schrieffer model where $\epsilon_f \to -\infty$. The excitation spectrum for this model is obtained in the universal form [11],

$$D_c(\epsilon) = \frac{1}{2\pi\Delta} \int_{-\infty}^{\infty} dx \frac{\sin(\pi/\nu)}{\cosh(\pi x/\nu\Delta) + \cos(\pi/\nu)} \frac{\Delta}{(\epsilon - \tilde{\epsilon}_f - x)^2 + \Delta^2}, \tag{3.2}$$

where $\tilde{\epsilon}_f$ is the effective f-level position introduced by Haldane [18]. For $\nu = 2$ case, this expression coincides with that for the Anderson model. Similarly, the spin excitation spectrum for the Coqblin-Schrieffer model is obtained in the universal function [11],

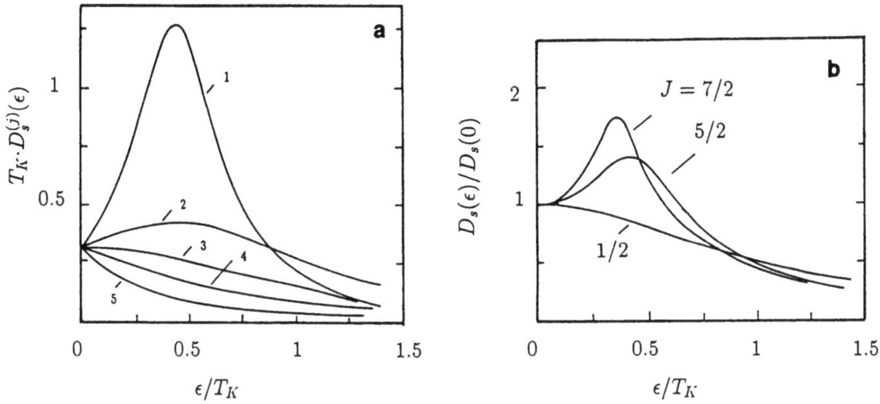

Fig.2 (a) Spin excitation spectrum for the highly correlated degenerate Anderson model in the *s*-*d* regime; (b)excitation spectrum $\Sigma D_s^{(j)}(\epsilon)$.

$$D_s^{(j)}(\epsilon) = \frac{1}{\pi} \frac{\Delta_j^*}{(\epsilon - \epsilon_j^*)^2 + (\Delta_j^*)^2}, \qquad 1 \leq j \leq \nu - 1, \tag{3.3}$$

where the renormalized f-level ϵ_j^* and the renormalized resonance width Δ_j^* are respectively given by $\epsilon_j^* = T_K \sin(2\pi j/\nu)$ and $\Delta_j^* = 2T_K \sin^2(\pi j/\nu)$. We note that there are $\nu - 1$ kinds of spin excitations in contrast to the charge excitation. We recall that the ground state of the present system is the local singlet in which the impurity moment is completely screened by the ν channels of conduction electrons. The $j = 1$ spin excitation corresponds to the excitation in which the $(\nu - 1)$ f-channels of the impurity state are screened by conduction electrons. Generally, the j-th spin excitation describes the excited state in which $(\nu - j)$ channels are screened. In Fig.2(a), the spin excitation spectrum is shown for the case of Ce impurities with $J = 5/2$. One can see the characteristic hump for the excitation spectrum, for $j = 1, 2$, at finite excitation energies. Physical quantities in rare-earth compounds often exhibit the maximum structure in their temperature and magnetic-field dependences, which may reflect this hump structure in the spin excitation. In the present classification, this hump is caused by the $j = 1, 2$ spin excitations for the $J = 5/2$ case, in which the impurity spin is partially compensated by the corresponding conduction-electron channels. As a reference, we show the total density of states (spin-excitation spectrum) in Fig.2(b), which is consistent with the results obtained by Rajan [19] who deduced the hump structure from the static quantities.

4. Multi-Channel Kondo Model

The multi-channel Kondo model is given by

$$H = \sum_{k,m,\sigma} \epsilon_k c_{km\sigma}^\dagger c_{km\sigma} + J \sum_{k,k',m,\sigma,\sigma'} c_{km\sigma}^\dagger (\sigma_{\sigma\sigma'} \cdot \mathrm{S}) c_{k'm\sigma'}, \tag{4.1}$$

where the impurity has the spin moment S, and n kinds of the screening channels are classified by the index m. This model has been used to study the Kondo effect for transition-metal impurities, such as Mn impurities in the Cu matrix, in which Mn atoms are expected to be in so-called high spin state S with no angular momentum. That is, $n = 2S$. Another interesting feature of the model has been demonstrated by Nozières and Blandin with the scaling arguments: a peculiar screening effect of the impurity spin occurs for the over-screening case ($n > 2S$) [13]. More recently, it has been found experimentally that the specific heat, the resistivity and the residual entropy of uranium heavy-fermion materials exhibit characteristic behaviors expected for the two-channel Kondo problem [20]. This behavior may be attributed to quadrupolar degrees of freedom on the uranium sites (quadrupolar Kondo effect), which was firstly proposed by Cox [21].

We now consider the spin excitation spectrum for the multi-channel Kondo model. The basic algebraic Bethe-Ansatz equation has been given for the Hamiltonian (4.1) [22,23],

$$\exp(ik_jL) = \prod_{\alpha=1}^{M} e_n(\Lambda_\alpha), \tag{4.2}$$

$$[e_n(\Lambda_\alpha)]^N e_{2S}(\Lambda_\alpha + 1/J) = \prod_{\beta=1}^{M} e_2(\Lambda_\alpha - \Lambda_\beta), \tag{4.3}$$

where $e_n(x) = (x - in/2)/(x + in/2)$. We note that Tsvelick-Wiegmann and Andrei-Destri investigated static quantities with the aid of the Bethe-Ansatz solution [22,23].

4.1 Dressed Particles at Zero Temperature

The ground state of the n-channel model is constructed by the bound state of n particles [22,23]. Therefore, the simplest spin excitation is described by introducing the hole state of the n-particle bound-state solution [11]. Note that there is only one species of dressed particles in the multi-channel Kondo model in contrast to the degenerate Anderson model. This result reflects that the present system is in the orbital singlet state. The spin excitation spectrum behaves quite differently according to the value of $2S/n$, because the ground state is different from each other. In case of $n = 2S$, the ground state is the local Fermi-liquid. The excitation spectrum in this case is given by [11],

$$D_s(\epsilon) = \frac{1}{\pi}\frac{T_K}{\epsilon^2 + T_K^2}, \tag{4.4}$$

which contains neither the quantities n nor S and takes the same form as in the $S = 1/2$ case. This is consistent with the calculated magnetization which describes quite similar curves irrespective of the values of n and $2S$ [22,23]. So, the essential properties of the multi-channel Kondo model in case $n = 2S$ may be described by the usual s-d model ($S = 1/2$). In the under-screening case, $n < 2S$, the excitation spectrum in the low energy region shows the singular behavior as [11],

$$D_s(\epsilon) \sim \frac{1}{4}(2S - n)\frac{1}{\epsilon[\ln(\epsilon/T_K)]^2}, \tag{4.5}$$

which is consistent with low-field magnetizations $M \sim (S - n/2)(1 - \beta/\ln H)$. In contrast to above cases, the screening effect is known to be rather peculiar in the case of $n > 2S$, as claimed by Nozières and Blandin. In terms of the renormalization-group language, the former is characterized by the strong coupling fixed point, while the latter by the non-trivial fixed point. We obtain the excitation spectrum for $n > 2S$ as [11],

$$D_s(\epsilon) \sim \frac{\sin(2S\pi/n)}{n\pi\cos(\pi/n)T_K}\left(\frac{\epsilon}{T_K}\right)^{2/n-1}. \tag{4.6}$$

This power-law behavior is consistent with low-field magnetizations: $M \sim \alpha H^{2/n}$ [22,23].

4.2 Elementary Excitation Spectra at Finite Temperatures

At finite temperatures, string solutions are introduced in order to describe excited states. In the thermodynamic limit, the distribution of the solutions is expressed by the density function $\sigma_l(\Lambda)$ for l-string solutions and $\sigma_l^h(\Lambda)$ for holes, which are determined so as to minimize the free-energy functional at thermal equilibrium. As a result, the infinite set of non-linear integral equations is obtained in terms of the pseudo-energy defined by $\varepsilon_l(\Lambda) = T\ln[\sigma_l^h(\Lambda)/\sigma_l(\Lambda)]$ [22]

$$\begin{aligned}\varepsilon_l(\Lambda)/T = 2J\mathrm{sech}(2\pi J\Lambda)*\{\ln[1+\exp(\varepsilon_{l-1}/T)] + \ln[1+\exp(\varepsilon_{l+1}/T)]\}\\ -2(E_F/T)\exp(2\pi J\Lambda)\delta_{ln},\end{aligned} \tag{4.7}$$

where the asterisk denotes convolution. We now consider the elementary excitation in this system at finite temperatures, applying Yang and Yang's method developed in one-dimensional interacting boson systems [12,14]. A simple spin excitation is given by removing one of l-string states at thermal equilibrium (the real part is denoted as Λ_0). The back-flow effect due to the electron correlation rearranges the distribution of the spin rapidities, hence the energy deviates from the value at thermal equilibrium. We note that the excitation energy renormalized by the back-flow effect is given by the corresponding pseudo-energy in thermal equilibrium, $\Delta\varepsilon_l(\Lambda_0) = -\varepsilon_l(\Lambda_0) + \varepsilon_l(-\infty)$. One readily finds that, at zero temperature, $\varepsilon_n(\Lambda_0)$ coincides with the energy of the spin excitation from the ground state, whereas $\varepsilon_l(\Lambda_0)$ vanishes for $l \neq n$. Accordingly, it is considered that $\varepsilon_n(\Lambda_0)$ (i.e. n-string solution) mainly controls the low-energy behavior of physical quantities at low temperatures. We introduce the density for spin excitations by counting the number of states in the range of the excitation energy, $[\varepsilon, \varepsilon + \mathrm{d}\varepsilon]$. The excitation spectrum of the *dressed hole* takes the form as,

$$D_s^{(l)}(\varepsilon) = \frac{\sigma_{lI}(\Lambda_0)}{|\mathrm{d}\varepsilon_l(\Lambda_0)/\mathrm{d}\Lambda_0|}, \tag{4.8}$$

where $\sigma_{lI}(\Lambda_0)$ denotes the impurity part of the density function in thermal equilibrium, and each excitation spectrum is given as a function of the excitation energy, $\varepsilon = -\varepsilon_l(\Lambda_0) + \varepsilon_l(-\infty)$. Similarly, we obtain the excitation spectrum for the *dressed particle* as $\sigma_{lI}^h(\Lambda_0)/|d\varepsilon_l(\Lambda_0)/d\Lambda_0|$. In order to discuss the temperature dependence of the excitation spectrum, it is helpful to consider the following 'whole' excitation spectrum, $\rho_s^{(l)}(\varepsilon) = [\sigma_{lI}(\Lambda_0) + \sigma_{lI}^h(\Lambda_0)]/|d\varepsilon_l(\Lambda_0)/d\Lambda_0|$, which counts the whole number of states including the dressed particle and dressed hole excitations. These excitation spectra satisfy the relation,

$$D_s^{(l)}(\varepsilon) = \frac{1}{1+\exp[(-\varepsilon+\varepsilon_l(-\infty))/T]}\rho_s^{(l)}(\varepsilon). \tag{4.9}$$

Let us begin discussions with the case of $n = 2S$. In Fig.3(a), we show the spin excitation spectrum in case of $n = 1$ and $S = 1/2$, for several temperatures. It is seen that the resonant peak structure developes in the low energy region as the temperature is decreased, which gives rise to the characteristic temperature dependence of the physical quantities such as the maximum structure of the specific heat around T_K [3,24]. It is noted that the weight at $\varepsilon = 0$ is independent of temperatures and is given by $D_s^{(n)}(0) = (1/2\pi)/(n+1)^2$ in the case $n = 2S$. One can see the above temperature dependence more clearly by observing the 'whole' excitation spectrum, $\rho_s^{(1)}(\varepsilon)$. It is seen in Fig.3(b) that $\rho_s^{(1)}(\varepsilon)$ exhibits the Lorentzian-like shape whose temperature dependence is not so strong: the width of the spectrum becomes slightly larger as the temperature is increased. These results indicate that the excitation spectrum, $D_s^{(n)}(\varepsilon)$, is well described by the

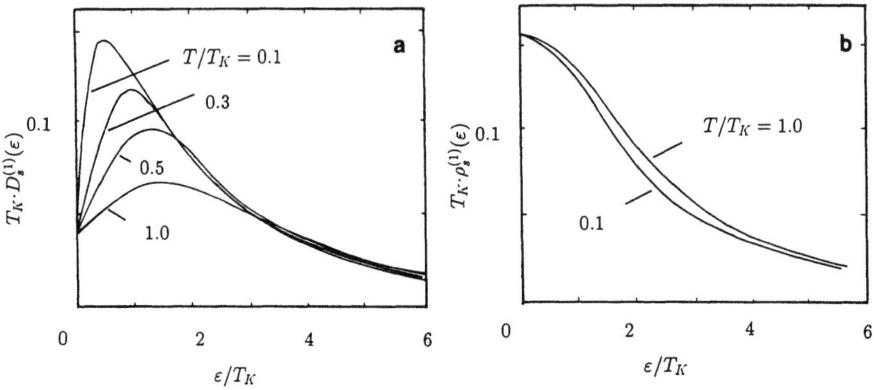

Fig.3(a) Spin excitation spectrum $D^{(1)}(\varepsilon)$ in case of $n = 1$ and $S = 1/2$; (b) whole excitation spectrum $\rho^{(1)}(\varepsilon)$.

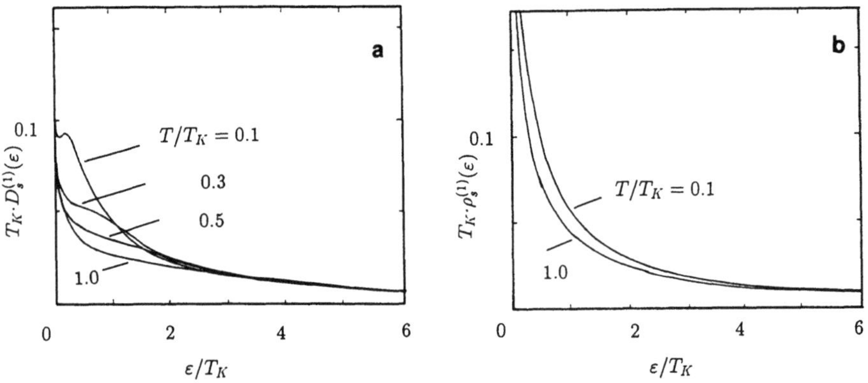

Fig.4(a) Spin excitation spectrum $D^{(1)}(\varepsilon)$ in case of $n = 1$ and $= 1$; (b) whole excitation spectrum $\rho^{(1)}(\varepsilon)$.

renormalized resonance-level model at low temperatures [25]. Let us next see the spin excitation spectra for the over- and under-screening cases at finite temperatures. In the case of $n < 2S$, screening channels of the conduction electrons are not enough to quench the impurity spin completely, leading to the residual spin $S - n/2$. The excitation spectra are shown in Figs.4(a) and 4(b), in case of $n = 1$ and $S = 1$. It is seen that the 'whole' excitation spectrum may diverge at zero energy, irrespective of the temperature. This divergence property reflects the unquenched spin. In the over-screening case, the screening is rather peculiar in contrast to the above two cases: the residual entropy in this case is given by $\ln\{\sin[\pi(2S+1)/(n+1)]/\sin[\pi/(n+2)]\}$, for zero-magnetic field [26]. The excitation spectra $D_s^{(2)}(\varepsilon)$ and $\rho_s^{(2)}(\varepsilon)$ are shown in Figs.5(a) and 5(b), respectively, for $n = 2$ and $S = 1/2$. In this case, it is seen that the peak structure in $D_s^{(2)}(\varepsilon)$ developes in the low energy region as the temperature is decreased. In contrast to the case of $n = 2S$, the 'whole' excitation spectrum does not behave like Lorentzian and its zero-energy value increases with decreasing the temperature. It is noted that the $\rho_s^{(2)}(\varepsilon)$ approaches the shape obtained analytically in eq.(4.6) as the temperature is decreased.

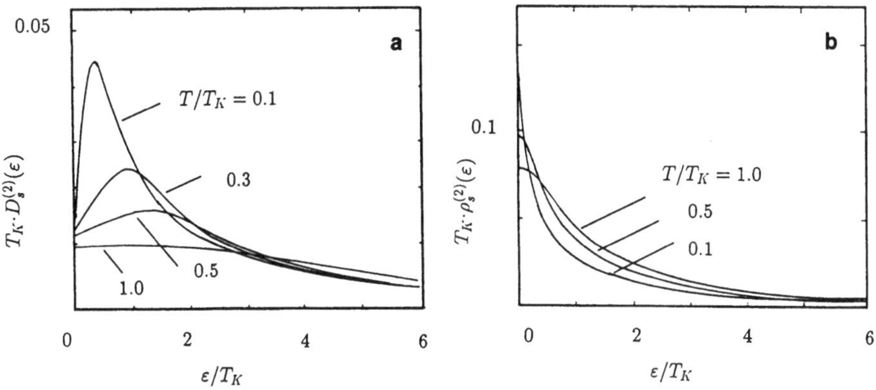

Fig.5(a) Spin excitation spectrum $D^{(2)}(\varepsilon)$ in case of $n = 2$ and $S = 1/2$; (b) whole excitation spectra $\rho^{(2)}(\varepsilon)$ in case of $n = 2$ and $S = 1/2$.

5. Summary

We have investigated the elementary excitation spectra for the Kondo problem with the use of dressed particles, which are the local analogues of the holon and spinon excitations for one-dimensional correlated electron systems. Hence, these dressed particles are considered to describe fundamental excitations for the impurity models.

Acknowledgements

This work was partly supported by Grant-in-Aid from the Ministry of Education, Science and Culture.

References

[1] K. G. Wilson, Rev. Mod. Phys. **47**, 773(1975). 483(1981).

[2] H.R. Krishna-murthy, J. W Wilkins and K. G. Wilson, Phys. Rev. B **21**, 1003(1980); H.R. Krishna-murthy, J. W Wilkins and K. G. Wilson, *ibid*, B **21**, 1044(1980).

[3] For reviews see,*e.g.*, N. Andrei, K. Furuya and J. H. Lowenstein, Rev. Mod. Phys. **55**. 331(1983); A. M. Tsvelick and P. B. Wiegmann, Adv. Phys. **32**, 453(1983); A. Okiji and N. Kawakami, *Springer Series in Solid State Science* (Springer, Berlin, 1984), Vol. 62, p. 46; A. Okiji, *Springer Series in Solid State Science* (Springer, Berlin, 1989), Vol. 77, p. 63; P. Schlottmann, Phys. Rep. **181**, 1(1989).

[4] P. Nozières, J. Low Tem. Phys. **17**, 31(1974).

[5] K. Yosida and K. Yamada, Prog. Theor. Phys. Suppl. No.46, 244(1970); K. Yamada, Prog. Theor. Phys. **53**, 970(1975); K. Yosida and K. Yamada, *ibid.* **53**, 1286(1975); K. Yamada, *ibid.* **54**, 316(1975).

[6] N. Kawakami and A. Okiji, Phys. Lett. A **86**, 301(1986); N. Kawakami, T. Usuki and A. Okiji, J. Phys. Soc. Jpn. **56**, 1539(1987); N. Kawakami, A. Nakamura and A. Okiji, J. Phys. Soc. Jpn. **57**, 4359(1988).

[7] A. Nakamura, N. Kawakami and A. Okiji, J. Phys. Soc. Jpn. **56**, 3667(1987).

[8] O. Sakai, Y. Shimizu and T. Kasuya, J. Phys. Soc. Jpn. **58**, 3666(1989); O. Sakai, Y. Shimizu and T. Kasuya, Prog. Theor. Phys. Suppl. No.108, 73(1992).

[9] A. M. Tsvelick, J. Phys. C **2**, 2833(1990).

[10] A. W. W. Ludwig and I Affleck, Phys. Rev. Lett. **67**, 3160(1991); I. Affleck and A. W. W. Ludwig, Nucl. Phys. **B360**, 641(1991).

[11] N. Kawakami and A. Okiji, Phys. Rev. B **42**, 2383(1990).

[12] M. Yamashita, A. Okiji and N. Kawakami, J. Phys. Soc. Jpn. **61**, 180(1992).

[13] P. Nozières and A. Blandin, J. Phys. (Paris) **41**, 193(1980); D. M. Cragg, P. Lloyd and P. Nozières, J. Phys. C **13**, 803(1980).

[14] C. N. Yang and C. P. Yang: J. Math. Phys. **10**, 1115(1969).

[15] P. B. Wiegmann, Phys. Lett. A **80**, 163(1980).

[16] N. Kawakami and A. Okiji, Phys. Lett. A **86**, 483(1981).

[17] P. Schlottmann, Phys. Rev. Lett. **50**, 1697(1983).

[18] F. D. M. Haldane, J. Phys. C **11**, 5015(1978).

[19] V. T. Rajan, Phys. Rev. Lett. **51**, 308(1983).

[20] C. L. Seaman, M. B. Maple, B. W. Lee, S. Ghamaty, M. S. Torikachvilli, J.-S. Kang, L. Z. Liu, J. W. Allen and D. L. Cox, Phys. Rev. Lett. **67**, 2882(1991); B. Andraka and A. M. Tsvelick, Phys. Rev. Lett. **67**, 2886(1991).

[21] D. L. Cox, Phys. Rev. Lett. **59**, 1240(1987).

[22] N. Andrei and C. Destri, Phys. Rev. Lett. **52**, 364(1984).

[23] A. M. Tsvelick and P. B. Wiegmann, Z. Phys. B **54**, 201(1984); A. M. Tsvelick and P. B. Wiegmann, J. Stat. Phys. **38**, 125(1985).

[24] H-U. Desgranges, J. Phys. C **18**, 5481(1985); P. D. Sacramento and P. Schlottmann, Phys. Rev. B **40**, 431(1989); P. D. Sacramento and P. Schlottmann, *ibid*, B **43**, 13294(1991).

[25] D. M. Newns and A. C. Hewson, J. Phys. F **10**, 2429(1980).

[26] A. M. Tsvelick, J. Phys. C **18**, 159(1985).

Part III

Phase Diagram for 1D Correlated Electron Systems

Phase Diagram of the One-Dimensional Kondo-Lattice Model

K. Ueda

Institute of Materials Science, University of Tsukuba,
Tsukuba 305, Japan

Abstract. The goundstate phase diagram of the one dimensional Kondo lattice model is discussed based on exact analyses in several limiting cases and numerical results interpolating those limits. In one dimension there are three phases: a ferromagnetic metallic phase, a paramagnetic metallic phase and an insulating spin-liquid phase. Emphasis is placed upon the strong coupling limits which are important to understand the properties of these phases.

1. Introduction

Heavy fermion systems are a class of rare-earth and actinide compounds which show very large electronic specific heat at low temperatures, several hundreds to a thousand times bigger than ordinary metals. One of the characteristic features of the heavy fermion systems is variety of different types of groundstates. Some of the well known examples are normal heavy fermion states, the heavy fermion superconductivity and antiferromagnetically ordered states. Recent interests in this class of materials extend also to semiconducting states with a small gap and low-density-carrier systems [1].

In the heavy fermion systems two types of electrons are involved: one is a set of itinerant conduction electrons which have extended orbitals, while the other are f-electrons which have orbitals more localized in each of rare earth or actinide ions. One of the canonical models for the heavy fermion systems is the periodic Anderson model (PAM). The PAM in its simplest version which neglects the orbital degeneracy for the localized electrons is given by

$$\mathcal{H}_{PAM} = -t\sum_{\langle ij\rangle}\sum_{s} c_{is}^{\dagger}c_{js} + \sum_{is}\varepsilon_f n_{is} + V\sum_{is}(c_{is}^{\dagger}f_{is} + f_{is}^{\dagger}c_{is}) + U\sum_{i} n_{i\uparrow}n_{i\downarrow}, \quad (1)$$

where $n_{is} = f_{is}^{\dagger}f_{is}$ is the number operator for the f-orbital at site i.

The heavy mass of quasiparticles originates from strong electron-electron interaction, U. When the f-level is deep enough, charge fluctuations in the f-shells are suppressed as the Coulomb interaction is increased. In this limit, an effective model to the PAM is give by the Kondo lattice model (KLM), which is given by

$$\mathcal{H}_{KLM} = -t\sum_{\langle ij\rangle}\sum_{s} c_{is}^{\dagger}c_{js} + J\sum_{i}\sum_{s,s'} \vec{S}_i \cdot \vec{\sigma}_{ss'}c_{is}^{\dagger}c_{is'} \quad (2)$$

Springer Series in Solid-State Sciences, Vol. 118
Correlation Effects in Low-Dimensional Electron Systems
Editors: A. Okiji and N. Kawakami

where $\vec{S}_i = \sum_{s,s'} \vec{\sigma}_{ss'} f^\dagger_{is} f_{is'}$ and the exchange coupling is antiferromagnetic ($J > 0$).

In this article we will discuss the groundstate phase diagram of the KLM in one dimension. The reason to discuss mainly the KLM is just its simplicity compared with the PAM. If the hopping processes are restricted to nearest neighbor pairs, the KLM has only two parameters: one is the exchange coupling constant in units of the hopping matrix element, J/t, and the other is the density of conduction electrons, $\rho = N/L$. Since there is an electron-hole symmetry for the conduction band, it is sufficient to consider from the low density limit to the half-filled band, $0 < \rho < 1$.

In connection with heavy fermions, we are mostly concerned with the weak coupling region, $J/t < 1$, since the strong Coulomb interaction for the PAM means a small $J/t < 1$ in the KLM. However in the next section we will start our discussions from strong coupling cases. As will be seen in the following, the strong coupling cases are not only easier to treat than the weak coupling cases but also they define fixed points which are relevant to weak coupling regions.

2. Approaches from Strong Coupling Limits

2.1 Half-Filled Case

At half-filling there is one conduction electron per site on the average. In the strong coupling limit each localized spin may capture one conduction electron to form a local singlet. As a collection of singlets is a singlet the groundstate is trivially a global singlet. In this limit elementary excitations are also simple to obtain. One can create a spin excitation by breaking a local singlet into a spin triplet. Energy cost for this is just the exchange energy, J. To create a charge excitation, one electron has to be removed from a local singlet and placed on a neighboring site. Since two doublets are left by this process, the energy cost is $3J/2$. From these simple arguments, it is easy to see that at half-filling the groundstate in the strong coupling limit is a singlet with excitation gaps both for the spin- and charge-excitations, a spin-liquid state. In this limit the charge gap is 50% larger than the spin gap.

2.2 Away from Half-Filling

Away from half-filling, a local basis set contains only three states in the strong coupling limit rather than eight states of the original KLM: a singlet state composed of an f-spin and a conduction electron and up- and down-states of an unpaired f-spin. By identifying the singlet as a vacant state and the unpaired f-spin as a mobile particle, the KLM can be mapped to the $U = \infty$ Hubbard model. From the mapping the effective Hamiltonian for the KLM is given by

$$\mathcal{H}_t = +\frac{t}{2} \sum_{j\sigma} \left(\tilde{f}^\dagger_{j\sigma} \tilde{f}_{j+1\sigma} + \text{H.c.} \right)$$

where the double occupancy of the $\tilde{f}$-particles is forbidden (t-model). The hopping matrix element is reduced by half since the initial and the final states are the local singlets. The change in the sign of the hopping term comes from the fermionic sign of the original f-electron which is a component of the local singlet.

In one dimension there is a macroscopic 2^{L-N}-fold degeneracy in the spin space, because a sequential order of spins is not changed by the nearest neighbor hopping. The wave functions in the groundstate multiplet are given by those for spinless fermions [2],

$$|\Psi\{\sigma_j\}\rangle = \sum_{j_1<\cdots<j_{\tilde{N}}} \det|\phi_\alpha(j)|\, \tilde{f}^\dagger_{j_1\sigma_1}\cdots\tilde{f}^\dagger_{j_{\tilde{N}}\sigma_{\tilde{N}}}|0\rangle, \tag{3}$$

where the one-particle eigenfunctions $\{\phi_\alpha\}$ are chosen to be the lowest $\tilde{N}$ levels of the nearest-neighbor hopping term.

The spin degeneracy is lifted up when the second order processes are considered [3]. In the second order there are two types of contributions. In one type of processes the charge configuration given by the wave function of the spinless fermion is modified, while the other type of processes lead to spin exchange processes in the groundstate multiplet. Since the degeneracy is macroscopic, the latter effect is expected to be dominant in the limit of small t/J. The effective spin interaction in the groundstate multiplet is given by $\langle\sigma'_1\cdots\sigma'_{\tilde{N}}|J_{\text{eff}}\sum_j \mathbf{S}_j\cdot\mathbf{S}_{j+1}|\sigma_1\cdots\sigma_{\tilde{N}}\rangle$ with

$$J_{\text{eff}} = -\frac{t^2}{J}\frac{1}{2\pi}\left(\frac{2}{\pi\tilde{\rho}}\sin^2\pi\tilde{\rho} - \sin 2\pi\tilde{\rho}\right), \tag{4}$$

where $\tilde{\rho} = 1-\rho$ is the density of the $\tilde{f}$-particles. This effective exchange interaction is ferromagnetic for any density. From this we can conclude that in the strong coupling limit the groundstate is ferromagnetic for any density. In this ferromagnetic phase, magnetic moment per site decreases towards half-filling as $g\mu_B(1-\rho)/2$.

3. Phases of the 1D Kondo Lattice Model

3.1 Groundstate Phase Diagram

Recently we have determined the groundstate phase diagram of the one dimensional KLM by exact diagonalization (Figure 1) [4]. There are three phases in one dimension. First there is the ferromagnetic metallic phase which extends from the low density limit [5] to strong coupling region [3]. Second one is a paramagnetic metallic phase which extends from the weak coupling region towards half-filling. The third one is a paramagnetic insulating phase at half-filling [6,7,8].

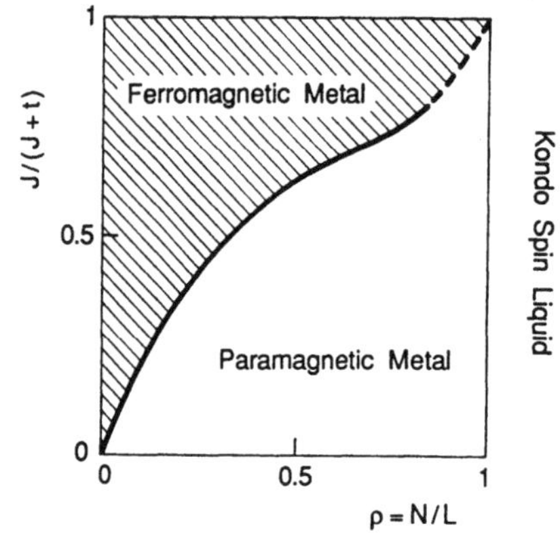

Figure 1. The phase diagram of the one dimensional Kondo lattice model.

3.2 Kondo Spin-Liquid Phase

In the strong coupling limit at half-filling the groundstate is a spin-liquid state as we have shown in the section **2.1**. In this case spin correlation functions decay exponentially. Now we consider the weak coupling limit. If the conventional RKKY picture is valid, the effective model is a Heisenberg model of spin 1/2 with long range antiferromagnetic interaction. There is, however, some pathological aspect in one dimension for this long range model: the exchange constants decay only as $1/r$ and there is no lower energy bound even for the classical Néel state. To save the pathology it may be essential to take account of the screening effect for the localized moments by the conduction electron spins. However to clarify the nature of the problem let us assume that the Heisenberg model is valid as an effective model. For the Heisenberg model with nearest neighbor interaction, but not the long range one, it is well known that there is a gapless magnon-like modes found by des Cloizeau and Pearson [9]. When there are gapless modes spin correlation functions decay only algebraically. In short this problem of gapful or gapless excitations is a one dimensional analogue to the problem of existence or absence of magnetic long range order in higher dimensions.

To look at the problem, we studied the half-filled case for the entire range of exchange coupling constant by exact numerical diagonalization [7]. By using a finite size scaling we have concluded that there is a spin gap always. The spin gap vanishes with essentially singular form at $J = 0$.

The Kondo spin-liquid state is a theoretical prototype of Kondo insulators. Concerning the Kondo insulators the most fundamental question would be what is the difference between ordinary semiconductors and the Kondo insulators except for the smallness of the gap. In this respect too the strong coupling limit is instructive. In the ordinary semiconductors the spin and charge gaps have the same value, i.e. the both are given by the band gap, while the charge gap in the KLM is 50% larger than the spin gap in the strong coupling limit. Numerical study by the exact diagonalization shows that the charge gap decreases much more slowly than the spin gap as J is decreased and that the ratio of the charge gap to the spin gap diverges in the weak coupling limit. Therefore the ratio bigger than unity is a characteristic feature of the Kondo insulators compared with ordinary semiconductors [8].

3.3 Ferromagnetic Metallic Phase

Figure 1 shows that the boundary between the ferromagnetic phase and the paramagnetic phase comes down from the strong coupling region to the weak coupling limit as the density of conduction electrons decreases. In the low density limit one can consider a special case of single electron in the KLM. In this case the following theorem is proven [5].

Theorem. The groundstate of the KLM with one conduction electron has the total spin quantum number $S_{\text{tot}} = (L-1)/2$ and is unique apart from the $(2S_{\text{tot}}+1)$-fold spin degeneracy.

The proof is based on the Perron-Frobenius theorem and holds in any dimension. Translational symmetry is not necessary for this proof and the hopping matrix elements and the exchange interactions may depend on sites and also may go beyond nearest neighbor pairs, as long as the hopping matrix elements are negative and the exchange couplings are antiferromagnetic.

Once the spin quantum number is known it is easy to obtain the wave function of the groundstate. Single electron in a magnetic medium is a magnetic polaron. The present case is a first example that a magnetic polaron problem is solved exactly.

One electron in a lattice does not define a proper low density limit in the thermodynamic sense. However the spin quantum number, $S_{\text{tot}} = (L-1)/2$, is consistent with the ferromagnetism in the strong coupling limit. Therefore the present theorem is relevant for the thermodynamic limit. It is also confirmed numerically that an effective interaction between the two magnetic polarons is ferromagnetic and therefore the ferromagnetic state is stabilized [10].

Elementary spin excitations in the ferromagnetic phase are the spin waves. Since the magnons are Goldstone modes, they are gapless. The charge excitations are given by particle-hole pair excitations and they are also gapless. Both the magnons and the particle-hole pair excitations are gapless at $q = 0$ and $q = 2\pi\rho$. The latter momentum corresponds to the $2k_{\text{F}}$ of the completely spin-polarized conduction electrons [3].

3.4 Paramagnetic Metallic Phase

Properties in the paramagnetic metallic phase are not well understood compared with those in either the spin-liquid phase or the ferromagnetic phase. One of the reason for this is that there is no simple limit for the paramagnetic metallic phase, see Fig. 1. A controlled perturbation with respect to J in a paramagnetic phase is very difficult due to the huge spin degeneracy at $J = 0$. On the other hand, towards the strong coupling region the paramagnetic phase is bounded by the ferromagnetic phase typically at intermediate coupling. Finally, towards the half filled line, once the spin-liquid phase is doped the groundstate is not trivial, although the spin-liquid phase is relatively well understood.

Concerning the paramagnetic phase, there are several fundamental questions to be answered. First concern is whether it is a Luttinger liquid [11] like many other interacting electron systems in one dimension. If it is the case, the next question is where is the Fermi momentum which is defined by the singularity in momentum distribution function. These questions are not solved yet. One possible approach to address these questions is discussed in an article in this volume [12].

Acknowledgments I would like to thank my colleagues, Manfred Sigrist, Hirokazu Tsunetsugu and Tomotoshi Nishino who are the key persons in the collaboration of the present subject. I am also grateful to Maurice Rice and Yasuhiro Hatsugai for many helpful discussions.

References

[1] For recent developments see *Proceedings of the international Conference on Strongly Correlated Electron Systems*, San Diego (1993).
[2] M. Ogata and H. Shiba, Phys. Rev. B **41**, 2326 (1990).
[3] M. Sigrist, H. Tsunetsugu, K. Ueda, and T.M. Rice, Phys. Rev. B **46**, 13838 (1992).
[4] H. Tsunetsugu, M. Sigrist, and K. Ueda, Phys. Rev. B **47**, 8345 (1993).
[5] M. Sigrist, H. Tsunetsugu, and K. Ueda, Phys. Rev. Lett. **67**, 2211 (1991).
[6] K. Ueda, H. Tsunetsugu, and M. Sigrist, Phys. Rev. Lett. **68**, 1030 (1992).
[7] H. Tsunetsugu, Y. Hatsugai, K. Ueda, and M. Sigrist, Phys. Rev. B **46**, 3175 (1992).
[8] T. Nishino and K. Ueda: Phys. Rev. B **47**, 12451 (1993).
[9] J. des Cloizeaux and J.J. Pearson, Phys. Rev. **128**, 2131 (1962).
[10] M. Sigrist, K. Ueda, and H. Tsunetsugu, Phys. Rev. **B46**, 175(1992).
[11] F. D. M. Haldane, J. Phys. C **14**, 2585 (1981).
[12] T. Nishino, K. Ueda, and H. Tsunetsugu, in this volume.

Fermi Volume of the Kondo-Lattice Model

T. Nishino[1], *K. Ueda*[2], *and H. Tsunetsugu*[3,4]

[1]Department of Physics, Tohoku University, Kawauchi, Sendai 980, Japan
[2]Institute of Materials Science, University of Tsukuba, Tsukuba 305, Japan
[3]Theoretische Physik, ETH-Zürich, CH-8093 Zürich, Switzerland
[4]Interdisziplinäres Projektzentrum für Supercomputing, ETH-Zürich, CH-8092 Zürich, Switzerland

Abstract. It is shown that the one-dimensional Kondo lattice model has a new Luttinger-liquid fixed point in the strong coupling limit when frustrating next-nearest-neighbor hoppings are introduced. This new phase is characterized by a large Fermi surface whose area is determined by the total number of electrons including the localized f-electrons.

1. Introduction

The Luttinger's Fermi surface (FS) sum rule[1] is one of the most fundamental properties of Fermi liquid state. It tells us that the Fermi volume is given by the total number of electrons independent of type of interactions and their strength. Even for Luttinger liquids in one-dimension, which are typical non-Fermi liquids, the "Fermi volume" defined by the singularities in the momentum distribution function is independent of the strength of the coupling constant.

The heavy electron materials are a typical highly correlated system that contains both itinerant conduction electrons and localized f-electrons. In the Kondo lattice model (KLM), which is a canonical model for the heavy electron systems, the localized f-electrons are treated as spins neglecting their charge degrees of freedom completely. In this limit it is difficult to tell whether the localized electrons contribute to the Fermi volume or not, because there is no adiabatic path from extended states to localized states.

To address the question of the FS sum rule for the KLM, in this paper we will study a one-dimensional KLM with a next-nearest-neighbor hopping term

$$H = -t\sum_{i\sigma}(c^{\dagger}_{i\sigma}c_{i+1\sigma} + h.c.) - t'\sum_{i\sigma}(c^{\dagger}_{i\sigma}c_{i+2\sigma} + h.c.) + J\sum_{i} \boldsymbol{S}^{c}_{i} \cdot \boldsymbol{S}^{f}_{i}, \quad (1)$$

where $\boldsymbol{S}^{c}_{i}$ and $\boldsymbol{S}^{f}_{i}$ are the spin of the conduction electrons $\sum_{\tau\tau'} c^{\dagger}_{i\tau}(\frac{\boldsymbol{\sigma}}{2})_{\tau\tau'}c_{i\tau'}$ and that of the f-electrons $\sum_{\tau\tau'} f^{\dagger}_{i\tau}(\frac{\boldsymbol{\sigma}}{2})_{\tau\tau'}f_{i\tau'}$, respectively. We shall call this model the t-t'-KLM hereafter. The sign of t does not matter for bi-

Springer Series in Solid-State Sciences, Vol. 118
Correlation Effects in Low-Dimensional Electron Systems
Editors: A. Okiji and N. Kawakami

partite lattices and we will set t=1 as units of energy. One has to keep in mind the local constraint that no charge fluctuation is allowed in any f-orbital, $\sum_\sigma f_{i\sigma}^\dagger f_{i\sigma} = 1$. Recently the phase diagram of the one-dimensional KLM, the t-t'-KLM at t'=0, was determined[2,3]. It consists of three different phases: (a) an insulating singlet spin-liquid phase at half-filling[4,5], (b) a ferromagnetic metallic phase in the low-density[6] or strong coupling region[7], and (c) a paramagnetic metallic phase in the weak coupling region.

Among the three phases, the least understood is the last one. Even we do not know the size of Fermi volume of the singlet phase. Two different view points are possible. One is based on the RKKY picture. In this picture spin and charge correlations of the whole system are determined by those of conduction electrons. Therefore this picture is consistent with the small FS, the volume being $(2\pi)^D N/(2L)$ (D: dimensions, N: the number of conduction electrons, L: the number of sites). On the other hand, we can regard the KLM as an effective model for the periodic Anderson model in the strong correlation limit U=∞. From this point of view, it may be natural to expect that the paramagnetic metallic phase has a large FS, where both conduction and localized f-electrons contribute to the Fermi volume, $(2\pi)^D(N+L)/(2L)$.

Since only very few are known about the weak coupling region, a possibly nice alternative is to start from the opposite limit, the strong coupling region, which is more tractable as we will explain later, and to use the continuity argument. However, the next-nearest-neighbor hoppings are necessary instead to suppress the undesirable Ferromagnetic long range order which otherwise exists. In the next section, we will prove that a new paramagnetic Luttinger liquid phase appears in the strong coupling limit under the presence of negative t'. The momentum distribution of conduction electron in the t-t'-KLM will be calculated in §3.

2. Luttinger-Liquid State in the Strong Coupling Limit

A great advantage of starting from the strong coupling limit $J \to \infty$ is that the Hamiltonian becomes much simpler, because the spin and charge degrees of freedom are substantially reduced there. This is because each conduction electron forms a local singlet pair with the f-electron at the same site and the double occupancy of conduction electrons is excluded, leaving only three local configurations allowed out of the original eight. One of the three is a spin singlet, $(c_{i\uparrow}^\dagger f_{i\downarrow}^\dagger - c_{i\downarrow}^\dagger f_{i\uparrow}^\dagger)/\sqrt{2}|0\rangle$, and the others are an unpaired f-spin with the z-projection $\pm 1/2$, $f_{i\uparrow}^\dagger|0\rangle$, $f_{i\downarrow}^\dagger|0\rangle$. By mapping the local singlets to vacancies and the unpaired f-spins to new Fermions (say $\tilde{f}$-particles), the KLM is shown to be now isomorphic to the U=∞ Hubbard model, whose hopping terms describe the motion of the original c-electrons in the KLM:

$$c_{i\sigma}^{\dagger}c_{j\sigma} \rightarrow \begin{cases} -\frac{1}{2}\tilde{f}_{j-\sigma}^{\dagger}\tilde{f}_{i-\sigma}, & i \neq j \\ \frac{1}{2}(1 - \tilde{f}_{i\uparrow}^{\dagger}\tilde{f}_{i\uparrow} - \tilde{f}_{i\downarrow}^{\dagger}\tilde{f}_{i\downarrow}), & i = j. \end{cases} \tag{2}$$

Thus, in the limit $J{=}\infty$ the t-t'-KLM (1) is rewritten as

$$\tilde{H} = +\frac{t}{2}\sum_{i\sigma}(\tilde{f}_{i\sigma}^{\dagger}\tilde{f}_{i+1\sigma} + h.c.) + \frac{t'}{2}\sum_{i\sigma}(\tilde{f}_{i\sigma}^{\dagger}\tilde{f}_{i+2\sigma} + h.c.), \tag{3}$$

to which we shall refer as the t-t'-model. It is important to remember that the number of $\tilde{f}$ particles is given by $\tilde{N} = L - N$.

Any correlation functions of the t-t'-KLM at $J{=}\infty$ are accordingly given by those of the t-t'-model. The momentum distribution of the c-electrons is, for example, obtained as

$$n_{\sigma}^{c}(k) \equiv c_{k\sigma}^{\dagger}c_{k\sigma} = \tfrac{1}{2}\big(1 - \tfrac{1}{2}\tilde{\rho} - n_{-\sigma}^{\tilde{f}}(-k)\big), \tag{4}$$

where $\tilde{\rho}$ is the density of the $\tilde{f}$-particles, $\tilde{N}/L \equiv \sum_{i,\sigma}\tilde{f}_{i\sigma}^{\dagger}\tilde{f}_{i\sigma}/L$. This relation gives an upper bound $n_{\sigma}^{c}(k) \leq (1 - \frac{1}{2}\tilde{\rho})/2$, since $n_{\sigma}^{\tilde{f}}(k)$ is non-negative.

All the eigenstate wave functions can be written explicitly at $t'{=}0$, because the spin configuration of $\tilde{f}$-particles, $\{\sigma_i\}{=}(\sigma_1,\ldots,\sigma_{\tilde{N}})$, remains unchanged during the hopping processes. This is a special character of the one dimensional system, although the mapping to the $U{=}\infty$ Hubbard model is always exact independent of the dimensionality. Therefore the spin degrees of freedom are trivially decoupled from the charge part, leaving the complete $2^{\tilde{N}}$-fold spin degeneracy for each eigenstate. A wave function in the ground-state multiplet is given by[8,9]

$$|\Psi\{\sigma_i\}\rangle = \sum_{1\leq j_1<\cdots<j_{\tilde{N}}\leq L} \begin{vmatrix} \phi_1(j_1) & \cdots & \phi_1(j_{\tilde{N}}) \\ \vdots & \ddots & \vdots \\ \phi_{\tilde{N}}(j_1) & \cdots & \phi_{\tilde{N}}(j_{\tilde{N}}) \end{vmatrix} \tilde{f}_{j_1\sigma_1}^{\dagger}\cdots\tilde{f}_{j_{\tilde{N}}\sigma_{\tilde{N}}}^{\dagger}|0\rangle, \tag{5}$$

where the one-particle states $\phi_1,\cdots,\phi_{\tilde{N}}$ are the lowest $\tilde{N}$ levels of the nearest neighbor hopping term.

The spin degeneracy will be lifted as the t'-term is switched on, and we can show the ground state is singlet by using the first-order perturbation as long as the t' is a small negative value. In the ground-state multiplet, the degenerate perturbation in t' gives spin exchange process

$$\langle\Psi\{\sigma_i'\}|\tilde{H}|\Psi\{\sigma_i\}\rangle = \langle\sigma_1'\ldots\sigma_{\tilde{N}}'|J_{\text{eff}}\sum\nolimits_j \mathbf{S}_j\cdot\mathbf{S}_{j+1}|\sigma_1'\ldots\sigma_{\tilde{N}}'\rangle + \text{const.}, \tag{6}$$

where the effective coupling is $J_{\rm eff} = -\frac{t'}{2\pi}(\frac{2}{\pi\tilde{\rho}}\sin^2\pi\tilde{\rho} - \sin 2\pi\tilde{\rho})$. If $t' > 0$ the exchange coupling is ferromagnetic, leading to a ground state with a maximal spin, $S=\frac{1}{2}\tilde{N}$, for any concentrations[10]. On the other hand, when $t' < 0$ the ground state is spin singlet. For sufficiently small $-t'$, the ground-state wave function is given by a linear combination

$$|\Phi\rangle = \sum_{\{\sigma_i\}} \chi(\sigma_1 \dots \sigma_{\tilde{N}}) |\Psi(\sigma_1 \dots \sigma_{\tilde{N}})\rangle, \tag{7}$$

where the amplitude $\chi(\sigma_1 \dots \sigma_{\tilde{N}})$ is the ground-state wave function of the squeezed spin chain of $\tilde{N}$ sites with the nearest-neighbor antiferromagnetic couplings $J_{\rm eff}$. The state $|\Phi\rangle$ is essentially the same as the Luttinger-liquid ground state of the $U=\infty$ Hubbard model studied by Ogata and Shiba[8,9]. Note that since in Eq.(3) the nearest-neighbor hoppings have a positive matrix element, the $\tilde{f}$-particles are distributed around $k=\pm\pi$ in the momentum space, and the Fermi momentum is $\pm\pi(1-\tilde{\rho}/2)$. Correspondingly, the $J=\infty$ t-t'-KLM has also a singularity at the same momentum $\pm\pi(1-\tilde{\rho}/2)=\pm\pi(1+\rho)/2$, as is seen in Eq.(4). *Thus, the t-t'-KLM has a large FS at least in the double limit $J=\infty$ and $t'\to 0$.* (Be careful to the order of the limits.)

The Luttinger liquids have power-law singularities in their correlation functions[11-14]. For example, the momentum distribution has the form

$$n^c(k) = n^c(k_{\rm F}) - \text{const.} \times \text{sign}\big(|k| - k_{\rm F}\big)\big||k| - k_{\rm F}\big|^{\alpha}. \tag{8}$$

The $U=\infty$ Hubbard model is shown to have the exponent of $\alpha=1/8$[13,14], so does for the t-t'-model in the limit $t'\to 0$ with $n^c(k_{\rm F})=\rho/4$. Figures 1a and 1b show the momentum distribution of the $J=\infty$ t-t'-KLM at $t'=-0.25$ and -1, respectively. Two cases of filling, $\rho=1/2$ and $2/3$, are shown. It is clear that the FS is large. The $n^c(k)$ for $t'=-1$ has a maximum value at a nonzero momentum, since the one-particle energy of the conduction electron has two additional minima besides $k=0$ when $|t'|/t > 1/4$. The critical exponent α decreases with increasing $|t'|$. Fitting the data at $\rho=1/2$ by Eq.(8), we get $\alpha=0.095$ with $n^c(k_{\rm F})=0.11$ at $t'=-0.25$, and $\alpha=0.024$ with $n^c(k_{\rm F})=0.075$ at $t'=-1$ with about 20% of numerical uncertainty.

In the Luttinger liquids, both the spin and charge correlation functions show characteristic singularities. In the limit of $t'\to -0$, the $J=\infty$ t-t'-KLM has a cusp at $2k_{\rm F}$ in the spin structure factor

$$S(k) \equiv \Big\langle \frac{1}{L}\sum_{lm} e^{ik(l-m)} (S_l^c + S_l^f)^z (S_m^c + S_m^f)^z \Big\rangle, \tag{9}$$

which is the same as the spin structure factor of the t-t'-model. In Fig.2a we

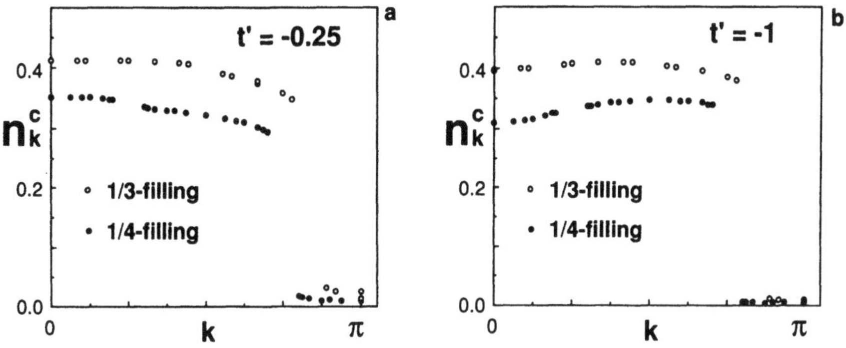

Fig.1. Momentum distribution of the $J=\infty$ t-t'-KLM up to $L=20$ (**a**) at $t'=-0.25$ and (**b**) at $t'=-1$.

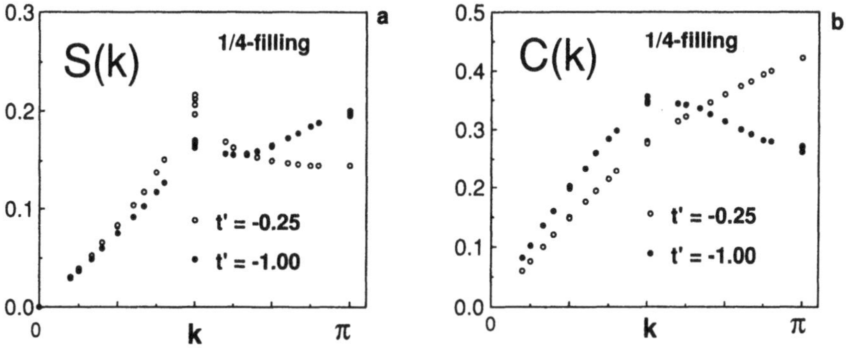

Fig.2. Correlation functions of the $J=\infty$ t-t'-KLM at $\rho=1/2$: (**a**) spin structure factor $S(k)$: (**b**) charge structure factor $C(k)$.

clearly observe the $2k_{\mathrm{F}}$ singularity, and the peak height decreases with $|t'|$. On the other hand, a kink appears at $4k_{\mathrm{F}}=\pi$ for non-zero $|t'|$. Concerning the charge correlations

$$C(k) \equiv \left\langle \frac{1}{L} \sum_{lm} e^{ik(l-m)} n_l^c n_m^c \right\rangle \tag{10}$$

at $t'=0$, the $C(k)$ is linear in k for $|k|<4k_{\mathrm{F}}$ and stays constant for $|k|>4k_{\mathrm{F}}$. Numerical data in Fig.2b shows that the $C(k)$ additionally shows a $2k_{\mathrm{F}}$ structure for $t' \neq 0$. This $2k_{\mathrm{F}}$ singularity is expected, for example, the g-ology suggests a possibility that $C(k)$ has the same degree of singularity as that of $S(k)$.

3. Singlet state for arbitrary Kondo coupling

Now that we have established the new Luttinger-liquid ground state in the $J = \infty$ limit, we will proceed to the study for finite J's. Figure 3 shows the ground state phase diagram determined numerically for the t-t'-KLM at t'=-0.1. Finite clusters with the open boundary conditions from L=4 to 9 have a ground state with a large spin $S = (L - N)/2$ in the shaded region between open circles and squares, and have a ground state with a small spin outside. The upper phase boundary shown by open circles is consistent with the result of the second-order perturbation. If J is finite but sufficiently large, a degenerate perturbation up to the order of t^2/J gives the same effective Hamiltonian as Eq.(6) with a modified exchange coupling[8]

$$J_{\rm eff} = -\frac{1}{2\pi}(t' + \frac{t^2}{J})(\frac{2}{\pi\tilde{\rho}}\sin^2\pi\tilde{\rho} - \sin 2\pi\tilde{\rho}). \tag{11}$$

The sign change in $J_{\rm eff}$ gives the phase boundary at $J \sim -t^2/t'$=10, consistent with the numerical result in Fig.3.

In Fig.3 there is another singlet phase near $J\sim 0$. An important question is if the two singlet phases in the strong and weak coupling regions have the same characters, particularly the same Fermi volume. Energy level structures of the finite clusters give a hint to the question. In the upper paramagnetic region, there are 2^{L-N} low-lying levels nearly degenerated, the lowest of which is continuously connected to the wave function $|\Phi\rangle$ in Eq.(7). At the upper phase boundary, these 2^{L-N} levels reverse their order because of the sign change in $J_{\rm eff}$. These 2^{L-N} states have almost the same energy down to $J\sim 1$. At the lower phase boundary, the singlet state which is the continuation of the ground state in the strong coupling region comes down again to be the ground state. This behavior of the level crossing shows that there is an "adiabatic path" between the ground states in upper and lower singlet phases for the finite systems studied. Since the Fermi volume is

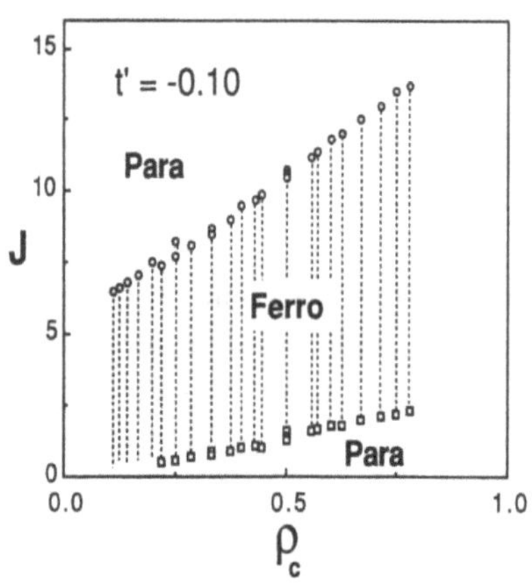

Fig.3. Phase diagram of the t-t'-KLM.

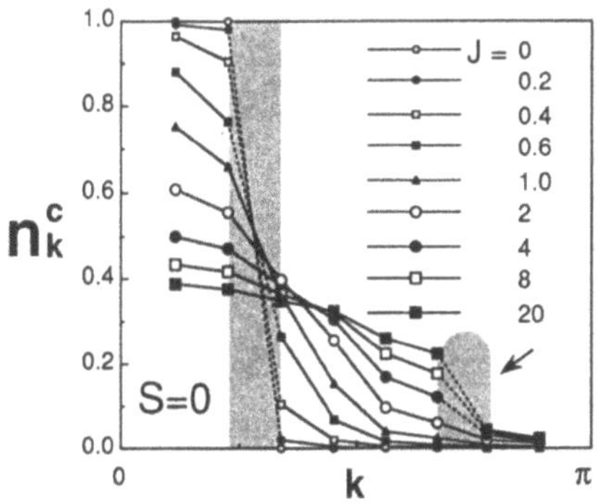

Fig.4. The momentum distribution $n^c(k)$ on the adiabatic path.

expected to be independent of the Kondo coupling J, this continuity means that one important necessary condition is fulfilled for having the large FS in the lower singlet phase in the thermodynamic limit.

Figure 4 shows the change in momentum distribution along the adiabatic path, tracing the lowest singlet state with decreasing J. Here is shown the case of L=8 and t'=-0.1 at ρ=1/2 with the open boundary conditions. In order to reduce the boundary effect, the momentum distribution is here evaluated as $n^c(k) = \sum_{i,j,\sigma} \phi_k^*(i)\phi_k(j)c_{i\sigma}^\dagger c_{j\sigma}$, where $\phi_k(i)$ is the one-particle wave function of the conduction electron. The effective momentum k_l (l=1...L) is defined by the relation, $\epsilon_l = -2t\cos k_l - 2t'\cos 2k_l$, where ϵ_l is the one particle eigenvalue for ϕ_{k_l}.

It is difficult to draw a conclusion directly from Fig.4 whether the Fermi surface is large or small, since $n^c(k)$ is given only at discrete momenta. A finite size scaling of the "jump" of $n^c(k)$ around k_F is necessary to determine the Fermi momentum k_F. For each system size L, let us define a crossover coupling constant J_c^L where the difference in $n^c(k)$ around large-FS (k_F=$\pi/4$) and small-FS (k_F=$3\pi/4$) are the same. If the system has a large-FS in the limit $L \to \infty$, J_c^L must become zero as the system size increases to $L \to \infty$. From Fig.4 we get $J_c^8 \sim 3.8$; for a smaller system $J_c^4 \sim 4.7$. The decreasing J_c^L is consistent with the idea of the large FS in the weak coupling region in the limit of $L \to \infty$. However it is premature to draw a definitive conclusion on this issue, since the system sizes available are just two.

4. Conclusion

We have studied the Fermi surface sum rule of the Kondo lattice model with the frustrating next-nearest-neighbor hoppings. A new metallic paramagnetic phase is found in the strong coupling limit, whose fixed point is a Luttinger liquid. It is rigorously shown that it has a large Fermi surface whose area is determined by the total number of the electrons including the localized

f-electrons. To the best of our knowledge, this is the first case where local spins are shown to contribute to the Fermi surface sum rule. The relevance of this limit to the paramagnetic phase in the weak coupling region is still an open question, although there are some encouraging results of numerical simulation.

Acknowledgment: The authors thank T. Maurice Rice, Manfred Sigrist and Masao Ogata for helpful discussions. This work was supported in part by the Ministry of Education, Science and Culture of Japan and the Swiss National Science Foundation Grant No. NFP-304030-032833. The numerical calculations were mainly done on the NEC SX-3/22R at Centro Svizzero di Calcolo Scientifico and SX-3/14R at the Computation Center, Osaka University.

References

[1] J.M.Luttinger, Phys. Rev. **119**, 1153 (1960).
[2] M.Troyer and D.Würtz, Phys. Rev. B **47**, 2886 (1993).
[3] H.Tsunetsugu, M.Sigrist, and K.Ueda, Phys. Rev. B **47**, 8345 (1993).
[4] H.Tsunetsugu, Y.Hatsugai, K.Ueda, and M.Sigrist, Phys. Rev. B **46**, 3175 (1992).
[5] T.Nishino and K.Ueda: Phys. Rev. B **47** 12451 (1993).
[6] M.Sigrist, H.Tsunetsugu, and K.Ueda, Phys. Rev. Lett. **67**, 2211 (1991).
[7] M.Sigrist, H.Tsunetsugu, K.Ueda, and T.M.Rice, Phys. Rev. B **47**, 13838 (1993).
[8] M.Ogata and H.Shiba, Phys. Rev. B **42**, 10553 (1990).
[9] H.Shiba and M.Ogata, Int. J. Mod. Phys. B **5**, 31 (1991).
[10] M.Ogata, private communication.
[11] F.D.M.Haldane, J. Phys. C **14**, 2585 (1981).
[12] H.J.Schulz, Phys. Rev. Lett. **64**, 2831 (1990).
[13] H.Frahm and V.E.Korepin, Phys. Rev. B **42**, 10553 (1990).
[14] N.Kawakami and S.-K.Yang, Phys. Lett. A **148**, 359 (1990).

Variational Wave Functions in the One-Dimensional t-J Model for Strongly Correlated Regimes

M. Ogata

Institute of Physics, University of Tokyo, Komaba,
Meguro-ku, Tokyo 153, Japan

Abstract. The one-dimensional t-J model is investigated by using variational wave functions for strongly correlated regimes. The Gutzwiller-Jastrow-type trial state has been studied extensively but it does not have spin-charge separation explicitly. On the other hand, another type of variational wave function is considered, which has the spin-charge separation and is easy to extend to higher dimensions. This trial state has spinless fermions as the charge degrees of freedom and a resonating-valence-bond state as the spin degrees of freedom.

1. Introduction

The t-J model is an important model to study highly correlated electron systems for its simplicity and close relationship to high temperature superconductivity. In the strongly correlated regime, the spin-charge separation will be playing an essential role. Another interesting phenomenon in high-T_c superconductors is an appearance of (pseudo) spin gap. Stimulated by this, strongly correlated one-dimensional (1D) electron systems have been studied extensively using various methods, and many properties have been clarified. Actually in 1D systems, the spin-charge separation is realized [1] and in some models, spin gaps open with charge-type excitations being kept gapless. The important question is whether we can extend the idea of 1D correlated electron systems to higher dimensions [2]. For this purpose it is important to understand the ground-state wave functions which are realized in strong correlation regimes. However, the wave function itself is not known so well except for a few cases. Although the Bethe ansatz solution gives us the explicit form of the wave function for 1D Hubbard model [3] and for the supersymmetric t-J model [4,5], they are still too complicated for the purpose of extension to higher dimensions.

In this paper we study the 1D t-J model using variational wave functions including the Gutzwiller-Jastrow-type wave functions and resonating-valence-bond type wave functions. In the t-J model, the low-lying energy spectrum of finite systems was studied [6,7] and the exponents, which characterize the power-law decays of correlation functions, were obtained for general values of J/t and the electron density n. In the phase diagram of J/t and n, the Tomonaga-Luttinger (TL) liquid theory holds in small J/t below $J_c/t = 2.5 \sim 3.5$ depending on n. It is also found that a phase separation takes place for larger J/t. However the ground-state wave function is understood only for three special cases:

(1) For $J = 0$, equivalent to $U \to \infty$ Hubbard model, the Bethe ansatz wave function has a very simple form owing to a decoupling of charge and spin degrees of freedom [8].
(2) Near $J/t = 2$ (supersymmetric case [4,5]), the Gutzwiller wave function (GWF) is an extremely good trial state [9,10]. This GWF can be modified by an introduction of Jastrow correlation factors for different values of J.
(3) In the region of phase separation, the ground-state consists of a vacuum and an island of a Heisenberg spin chain.

Springer Series in Solid-State Sciences, Vol. 118
Correlation Effects in Low-Dimensional Electron Systems
Editors: A. Okiji and N. Kawakami

The spin-charge separation is apparent in the wave function (1), but this function is difficult to extend to higher dimensions. On the other hand, an extension to higher dimensions is easy for the wave function (2), but the spin-charge separation is not clear. As a result, it is difficult to reproduce the density-density correlation functions which are same as those for free spinless fermions for $J/t \to 0$. In §2 and 3, we summarize the results for the wave functions (1) and (2). In §4 we study another type of variational state which contains a resonating-valence-bond (RVB) [11] spin wave-function. This wave function has an explicit spin-charge separation and is relatively easy to extend to higher dimensions. As is expected, there is also a possibility of a spin gap in this state.

2. $J \to 0$ Limit

In the subspace with no double occupancy, the 1D t-J model is defined as

$$H = -t \sum_{i\sigma}(c_{i\sigma}^{\dagger}c_{i+1\sigma}+\text{h.c.})+J\sum_{i}(\mathbf{S}_i\cdot\mathbf{S}_{i+1}-\frac{1}{4}n_i n_{i+1}), \tag{1}$$

where $t, J \geq 0$, and spin operators vanish when they are applied to empty sites.

First we summarize the ground-state wave function for $J \to 0$ limit [8] which can be obtained from the Lieb-Wu's Bethe ansatz solution for the Hubbard model. Let $f(x_1,\cdots,x_N)$ be the amplitude in the wave function when M down spins are located at the sites $x_1,\cdots,x_M$, and $N-M$ up spins at $x_{M+1},\cdots,x_N$. For $J/t \to 0$, f can be expressed as

$$f(x_1,\cdots,x_N) = \det[\exp(ik_ix_j)]\Phi(y_1,\cdots,y_M), \tag{2}$$

where $y_1,\cdots,y_M$ represent the "coordinates" of M down spins in the squeezed spin configuration with vacant sites omitted. In Eq. (2), the determinant depends only on the sites of particles and not on their spins. Thus it is the same as the Slater determinant of spinless fermions with momenta k_j's. This part represents the charge degrees of freedom. Furthermore it can be shown that the remaining spin degrees of freedom, $\Phi(y_1,\cdots,y_M)$, is just the same as the Bethe's exact solution of the 1D $S=\frac{1}{2}$ Heisenberg spin system. The equations to determine k_j's and the spin rapidities Λ_α's are completely decoupled, which means that the spin-charge separation is perfectly takes place in the ground-state wave function. This notable feature holds for any electron density.

The form of the wave function (2) is, however, still complicated to extend in higher dimensions, because its spin degrees of freedom are represented by a Bethe ansatz solution. There is not an exact solution for the two-dimensional Heisenberg model and furthermore it is easily seen that the ordering in one-dimensional systems is important in (2).

3. Gutzwiller-Jastrow Type Variational Wave Functions

Next we study the variational wave functions which have been considered by modifying the Gutzwiller wave function (GWF). The GWF is defined as

$$P_d\Phi_F = \prod_j(1-n_{j\uparrow}n_{j\downarrow})\Phi_F, \tag{3}$$

where Φ_F is a simple Fermi sea and P_d is a Gutzwiller projection operator. This wave function was studied numerically [12,13] and analytically [14,15] without using secondary approximations. It was found that the GWF is excellent for the Heisenberg model, but it is not satisfactory in describing the properties of the strong coupling Hubbard model or the small J/t region of the t-J model. For example, the GWF does not reproduce the $2k_F$ peak in the spin correlation function. However it was shown later that GWF is an extremely good trial wave function for larger values of J, i.e. for the supersymmetric case ($J/t = 2$) [9]. The energy difference is only 0.06% at the quarter filling, for example, and various correlation functions are reproduced fairly well.

In order to modify GWF for small J/t, we have to take account of the density correlations which can be introduced by a Jastrow-type two-body correlation factors. We used a spin-independent density correlation like [9]

$$\Psi = \prod_{j\ell}\prod_{\sigma\sigma'}\left[1-(1-\eta(r_{j\ell}))n_{j\sigma}n_{\ell\sigma'}\right]\Phi_F, \qquad (4)$$

where $r_{j\ell} = |r_j - r_\ell|$. For $\eta(r)$, simple forms are desirable, satisfying the condition of the t-J model: $\eta(0) = 0$ and $\eta(\infty) = 1$. The former condition is to project out the double occupancies. Here we consider two cases in addition to the GWF:

$$\eta(r) = \begin{cases} \frac{2}{\pi}\arctan\frac{r}{\zeta} & \text{(a)} \quad \text{(RJWF)} \\ 1 & \text{(b)} \quad \text{(GWF)} \\ 1+\frac{\alpha}{r^\beta} & \text{(c)} \quad \text{(AJWF)} \end{cases}, \qquad (5)$$

where $r \neq 0$ and ζ,α and β are positive variational parameters.

The form (a), which we call the repulsive Jastrow wave function (RJWF), includes repulsive correlation. It prefers configurations with electrons mutually apart. In the limit $\zeta \to 0$, it is reduced to the GWF (b). This wave function was previously introduced to study the repulsive Hubbard model in strong coupling regimes [16]. An attractive Jastrow wave function (AJWF) with the correlation factor (c) favors local configurations with electrons close to each other. Comparing with the RJWF and the AJWF, we may regard the GWF as a "free-electron" state in that there is no amplitude modification from the non-interacting state except for the exclusion of doubly occupied sites.

We evaluate the variational expectation values using the variational Monte Carlo (VMC) method [12,13]. In the VMC calculations we use systems of $N = 4I + 2$ (I: integer) with the periodic boundary conditions. We find that, by optimizing the variational parameters, ζ, α, and β, the variational energy is considerably improved upon the GWF. It becomes very close to the estimated ground-state energy which is obtained from the thermodynamic extrapolation of the small-cluster results [9]. For $J/t < 2$, the RJWF is a good variational state, while for $J/t > 2$ the AJWF is a good variational state. The latter also describes a fully phase-separated state for large values of J as well. The global features of various correlation functions are also reproduced by these states. For example, Figure 1 shows $n(k)$, $S(k)$ and $N(k)$ of the optimized RJWF ($\zeta = 0.7$) for $J/t = 1$. They are compared with the exact diagonalization results. Except for the vicinity of the singularities, two results are close to each other. The deviation around the singularities is caused by the same reason with the GWF, i.e. the variational states in Eq. (4) are always so-called Fermi-liquid states in that they have a jump at k_F in $n(k)$.

Hellberg and Mele [10] have introduced another type of Jastrow factor and succeeded in representing the power-law singularity which is characteristic of the TL liquid. Their variational state is written as

$$\Psi = P_d|F(r_i^\uparrow, r_j^\downarrow)|^\nu \Phi_F, \qquad (6)$$

where $|F(r_i^\uparrow, r_j^\downarrow)|$ is a determinant of all electron coordinates. In this wave function, the correlation functions have power-law decays and the correlation exponent, K_ρ, is analytically related to the variational parameter, ν, in a relation, $K_\rho = \frac{1}{2\nu+1}$ [17].

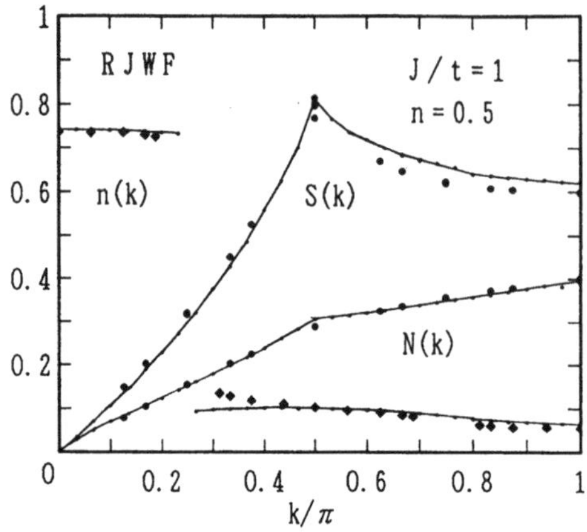

Fig.1 Momentum distribution function $n(k)$, spin and charge density structure factors, $S(k)$ and $N(k)$ for the RJWF ($\zeta = 0.7$, solid line) and the exact diagonalization (solid diamond and circle) for $J/t = 1$. In the VMC calculations, 5×10^4 samples are collected for a system with 60 sites.

It is easily shown that state (6) is rewritten in the form, Eq. (4), with

$$\eta(r) = \left[\frac{L}{\pi}\sin(\frac{\pi}{L}r)\right]^{\nu}. \tag{7}$$

Therefore the state (6) is another type of RJWF. Comparing (5) and (7), we can see that the long-distance behavior of the Jastrow factor is different from each other. In (5) $\eta(r)$ goes to 1 as $r \to \infty$. On the other hand, in (7), $\eta(r)$ is size-dependent and the long-distance correlation, $\eta(\frac{L}{2})$, scales as $(\frac{L}{\pi})^{\nu}$. This indicates that the long-ranged Jastrow correlation is necessary in order to obtain the power-law behavior of the correlation functions.

If we compare the optimized value of ν, however, we notice that the actual values of the correlation exponent do not coincide with those obtained in the t-J model. For example at $J = 0$, the exact value is $\nu = 0.5$ (corresponding to $K_\rho = \frac{1}{2}$), while the optimized state of (6) gives $\nu \sim 1$ for $n \sim 0$, $\nu \sim 0.75$ for $n \sim 0.5$ and $\nu \sim 0.5$ for $n \sim 1$, respectively. This difference becomes apparent if we see the density-density correlation functions; the state (6) gives a small peak at $2k_F$ in $N(k)$, which should not show up in the exact ground-state at $J = 0$. This is because the charge degrees of freedom in the exact ground-state are represented by free spinless fermions (1), while it is difficult to resemble this behavior in (6) because of the Fermi sea Φ_F.

To reproduce the exact $N(k)$ is easy, since it is the density-density correlation of the free spinless fermions. This can be done if we use a variational state like Eq. (2) which has a Slater determinant of spinless fermions as charge degrees of freedom. In the next section we try to construct a variational state which has this character of the spin-charge separation.

4. RVB-type Variational Wave Functions

We can consider a RVB-type wave function which may be a good trial state for small J/t. In order to have a spin-charge separation in the wave function, it is most convenient to use the state like Eq. (2). In (2), the spin part of the wave function is same as the

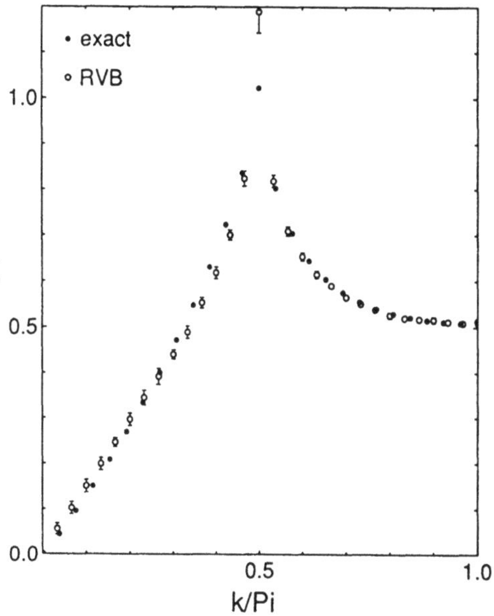

Fig.2 The spin correlation function $S(k)$ for the RVB-spinless fermion type variational wave function (open circles) compared with the exact results for $J/t \to 0$ using the Bethe ansatz [8] (closed circles).

squeezed Heisenberg chain, so that it is difficult to extend to higher dimensions. Instead we introduce a RVB-type spin wave function in place of the Heisenberg chain. A similar wave function is derived in the slave-fermion mean-field theory [18].

Instead of the Heisenberg solution, $\Phi(y_1, \cdots, y_M)$, we introduce

$$P_d P_N \exp(i \sum_k \tilde{A}_k S^\dagger_{k\uparrow} S^\dagger_{-k\downarrow}) = P_d (\sum_{i,j} A_{i,j} S^\dagger_{i\uparrow} S^\dagger_{j\downarrow})^{\frac{N}{2}}, \tag{8}$$

where $S^\dagger_{i\uparrow}$ is a Schwinger boson creation operator, $\tilde{A}_k$ is a function which appears in the Bogoliubov transformation in the mean-field theory, $\frac{v_k}{u_k}$, and $A_{i,j}$ is its Fourier transform. Generally $A_{i,j} = -A_{j,i}$ so that $A_{i,j}$ represents the amplitude of the RVB singlet with a distance $r_i - r_j$.

For small J/t, it is appropriate to choose $|A_{i,j}|$ = constant even if there are holes between the singlet pair at r_i and r_j. (In principle, a candidate for $|A_{i,j}|$ is obtained in the mean-field approximation. Here we regard it as a variational parameter.) In this case, the spin part of the wave function does not depend on the charge configuration, so that the spin-charge separation is explicitly realized. The density-density correlation function is determined solely from the free spinless fermions.

Actually there are additional degrees of freedom to choose the RVB amplitude, when some RVB bonds cross or jump over other bonds. Figure 2 shows the spin-spin correlation function which is given by an optimized RVB state. ($|A_{i,j}| = 0$ when it crosses the other bond and $A_{i,j}$ changes its sign when it jumps over another bond.) Although it seems to give a stronger peak at $2k_F$, the global feature of the correlation function is fairly close to the exact result. Note that the appearance of the peak at $2k_F$ in this variational state is non-trivial.

5. Conclusion

In this paper we have examined various variational wave functions for one-dimensional t-J model. The variational expectation values are estimated exactly by the Monte Carlo procedure and compared with those obtained from the Bethe Ansatz and exact diagonalization of small clusters. Results are summarized as follows:

1) The ground-state energies and various correlation functions are reproduced by using variational states with repulsive Jastrow correlations for $J/t < 2$. The global features of correlation functions can be reproduced by short-range Jastrow correlations. However, in order to reproduce the power-law behavior around the singularities, the long-ranged Jastrow correlation is essential. This type of variational state is easy to extend to higher dimensions [19] but the spin and charge degrees of freedom are not decoupled in these states, so that it is difficult to reproduce the density-density correlation functions which is simply given by the free spinless fermions.

2) In order to construct a variational state with an explicit spin-charge separation, we have examined a state with free spinless fermions for the charge degrees of freedom and a RVB-type spin wave-function for the spin degrees of freedom. The spin-charge separation shows up explicitly in the wave function and it is relatively easy to extend to the higher dimensions.

Recently it has been shown that Schwinger-boson mean-field theory for the two-dimensional Heisenberg spin system gives a RVB state which gives a very good variational energy [20]. Therefore the present state with a RVB-type spin wave-function will be a good trial state in the strongly correlated regimes. Moreover, since the spin part is written in a RVB state, it has a possibility to open a spin gap depending on the parameter J/t and the electron density n.

Part of the present work is carried out in collaboration with H. Yokoyama.

References

[1] For reviews see, J. Sólyom, Adv. Phys. **28**, 201 (1979), F. D. M. Haldane, J. Phys. C **14**, 2585 (1981), and H. Fukuyama and H. Takayama, *Electronic Properties of Inorganic Quasi-One-Di-mensional Materials, I*, ed. by P. Monceau (Reidel, 1985) p. 41.

[2] P. W. Anderson, Phys. Rev. Lett. **64**, 1839 (1990), **65** ,2306 (1990) and Prog. Theor. Phys. Supple. **107**, 41 (1992).

[3] Lieb and Wu: Phys. Rev. Lett. **20**, 1445 (1968).

[4] B. Sutherland: Phys. Rev. B **12**, 3795 (1975), P. Schlottmann: Phys. Rev. B **36**, 5177 (1987).

[5] P. -A. Bares and G. Blatter: Phys. Rev. Lett. **64**, 2567 (1990) P. -A. Bares, G. Blatter and M. Ogata: Phys. Rev. B **44**, 130 (1991).

[6] N. Kawakami and S. -K. Yang: Phys. Rev. Lett. **65**, 2309 (1990).

[7] M. Ogata, M. U. Luchini, S. Sorella and F. F. Assaad: Phys. Rev. Lett. **66**, 2388 (1991).

[8] M. Ogata and H. Shiba: Phys. Rev. B **41**, 2326 (1990); H. Shiba and M. Ogata: Prog. Theor. Phys. Supple. **108**, 265 (1993).

[9] H. Yokoyama and M. Ogata: Phys. Rev. Lett. **67**, 3610 (1991).

[10] C. S. Hellberg and E. J. Mele: Int. J. Mod. Phys. **5**, 1791 (1991) and Phys. Rev. Lett. **67**, 2080 (1991).

[11] P. W. Anderson: Science **235**, 1196 (1987).

[12] C. Gros, R. Joynt and T.M. Rice: Phys. Rev. B **36**, 381 (1987).

[13] H. Yokoyama and H. Shiba: J. Phys. Soc. Jpn. **56**, 1490 (1987).

[14] W. Metzner and D. Vollhardt: Phys. Rev. B **37**, 7382 (1988).

[15] F. Gebhard and D. Vollhardt: Phys. Rev. B **38**, 6911 (1988).

[16] H. Yokoyama and H. Shiba: J. Phys. Soc. Jpn. **59**, 3669 (1990).

[17] N. Kawakami and P. Horsch: Phys. Rev. Lett. **68**, 3110 (1992).

[18] D. Yoshioka: J. Phys. Soc. Jpn. **58**, 1516 (1989).

[19] R. Valenti and C. Gros: Phys. Rev. Lett. **68**, 2402 (1992).

[20] D. Yoshioka: J. Phys. Soc. Jpn. **58**, 32 (1989), and private communication.

Strong Correlation and Strong Electron-Phonon Interaction – Peierls Transition with Mott Transition

H. Fukuyama

Department of Physics, Faculty of Science, University of Tokyo, 7-3-1 Hongo, Bunkyo-ku, Tokyo 113, Japan

It is well-known that the electron-phonon interaction in quasi-one-dimensional electron systems leads to the Peierls transition with resultant suppression of spin susceptibility. In a class of organic conductors, $(DCNQI)_2Cu$, however, the Peierls transition leads to the Mott transition associated with the Cu$-d$ band resulting in the localized magnetic moments; *i.e.* the Mott transition and the Peierls transition go hand in hand.

The theoretical understanding of this unique system is presented.

1. Introduction

The family of one-dimensional organic conductors, $(DCNQI)_2Cu$, is unique [1-4]. It has the phase diagram in the plane of pressure (p) and temperature (T) as schematically shown in Fig.1, where M and I indicate the metallic and the insulating states, respectively. In the insulating state there exists a three-fold lattice distortions along the chain. Hence this metal-insulator transition could naively be associated with the Peierls transition [5]. It is not, however, the case since, in contrast to the ordinary Peierls transition, the spin susceptibility (χ) in the insulating state is enhanced over the temperature-independent Pauli-like paramagnetism seen in the metallic state. Moreover the temperature dependence of χ in the insulating stage is close to that of localized magnetic moments with $S = 1/2$ and total number of spins about one-third of Cu atoms. These facts can be understood simply by the assumption of the charge ordering of Cu^{++}-Cu^{+}-Cu^{+}-Cu^{++} $\cdots$ along the chain in the insulating state as was indicated from the early stage of the studies. However the relationship of the electronic states in the insulating and the metallic states, i.e. the nature of the metal-insulator transtion, has been left unresolved.

Recently the theoretical works [6-10] based on the periodic Anderson model have developed to study basic features of the metal-insulator transition of this family. The theory is based on the experimental findings by photoemission spectroscopy [11] that the average charge transfer is close to $(DCNQI)_2^{-2/3}Cu^{+4/3}$ both in the metallic and

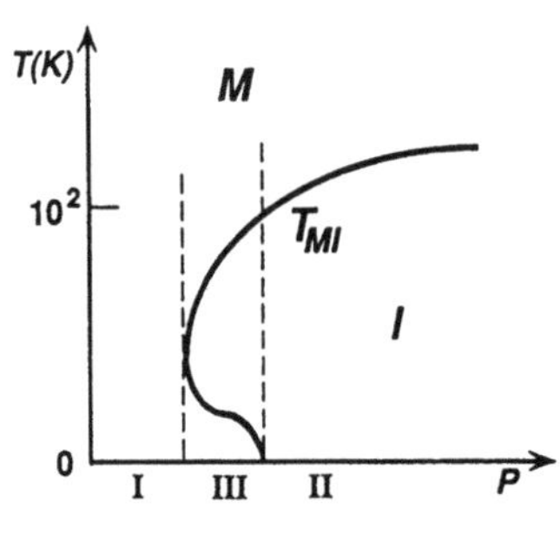

Fig. 1 The phase diagram of $(DCNQI)_2Cu$ family.

Springer Series in Solid-State Sciences, Vol. 118
Correlation Effects in Low-Dimensional Electron Systems
Editors: A. Okiji and N. Kawakami

insulating states. It has been shown [7,8] that in this case the Peierls transition of the DCNQI π-bands is accompanied by the Mott transition of the Cu d-bands and the transition can be of the first order as seen experimentally. (Intrestingly there is a reentrant metallic state for some particular range of the pressure, which has been proposed as due to the quantal-classical crossover of proton-tunneling [9].) Actually the hysterisis associated with this metal-insulator transition is very large; i.e. in deuterated DME–DCNQI the resistivity jumps up about eight orders at around $T = 50K$ and jumps down by nine orders at arouond $T = 10K$ [4].

The purpose of this paper is to briefly discuss about the origins of the stability of the charge transfer of $Cu^{+4/3}$ on one hand and the large hysterisis at the transition on the other hand. We will see that both of these are consequences of the strong correlation.

We take unit $k_B = \hbar = 1$.

2. Why is $Cu^{+4/3}$ stable?

Our model Hamiltonian is given by

$$H = \sum_{\ell=1}^{2} \sum_{k} \sum_{\sigma} \varepsilon_k \, a^\dagger_{\ell k \sigma} \, a_{\ell k \sigma} + \frac{V}{\sqrt{N}} \sum_{\ell,k,\sigma} \sum_{j=1}^{N} (a^\dagger_{\ell k \sigma} \, d_{j\sigma} \, e^{-ikj} + h.c.) + E_d \sum_{j,\sigma} n_{j\sigma} + U \sum_{i} n_{i\uparrow} n_{i\downarrow} \,. \quad (1)$$

Here for simplicity the three-dimensionality due to interchain transfer integrals has been ignored and the two π–bands associated with DCNQI molecules and the one d–state due to Cu atoms are taken into account in the unit cell in the plane perpendicular to the conducting direction (the c–axis). In eq.(1) the first term represents the kinetic energy of the π–electrons, which is doubly degenerate, and the second and the third terms express the mixing integral between the DCNQI π–orbitals and the Cu d–orbitals and the atomic levels of the d–orbitals, respectively. The last term denotes the Coulomb repulsion in the atomic d levels.

The band structure in the metallic state in the hole picture is schematically shown in Fig.2, where the Cu d–level, ε_d, is close to the Fermi level, ε_F, resulting in the charge transfer leading to $Cu^{+4/3}$. If the Coulomb interaction in the d–orbital , U, is negligible the states with $1/6 < |k| < 1/3$ in the figure are occupied, which are essentially d–like. However in the presence of the strong U as in the present case the spectrum weight is distributed over entire "d–band" ($0 < |k| < 1/3$); i.e. the lower Hubbard band is just completely filled [7].

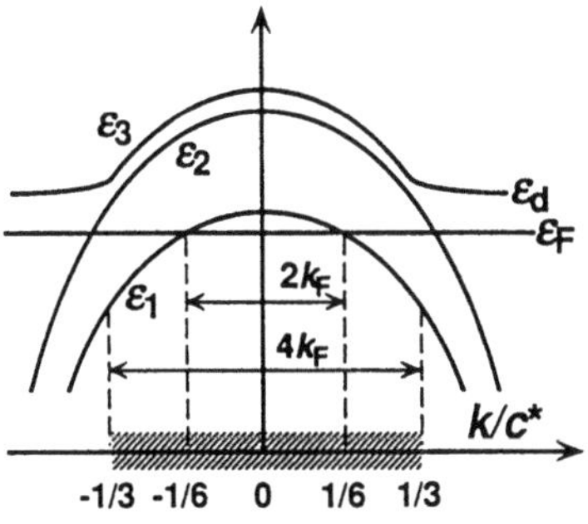

Fig. 2 The expected band structure in the case of $(DCNQI)_2^{-2/3}Cu^{+4/3}$.

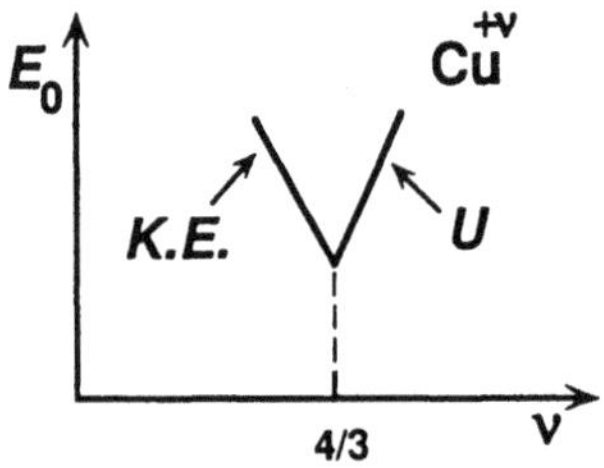

Fig. 3 Schematic representation of the ground state energy, E_0, as a function of the average valence, ν, of Cu.

If the average valence of Cu is less than 4/3, there exist less $d-$holes so that the number of holes in the $\pi-$band is more leading to the appreciable increase of the total kinetic energy. On the other hand, if the average valence of Cu is more than 4/3, the extra $d-$holes have to occupy the upper Hubbard band separated by the lower Hubbard band by U. Consequently the ground state energy, E_0, will have a cusp at $\nu=4/3$ as a function of average valence, ν, of Cu as shown in Fig.3. This will lead to the locking of the charge transfer ν at around $4/3$.

3. Why is the hysterisis so large?

Next we will study why the hysterisis associated with the first order metal-insulator transition is so large in this system. First of all it is to be noted that the existence of the first, instead of the second, order transition itself in the presence of the three-fold lattice distortion can be understood as due to the third order commensurability energy [12] as is experimentally the case in TTF-TCNQ under pressure [13]. The real interest here is in the extraordinary large change through this first order transiton, which we will study by use of the Ginzburg-Landau theory.

The free energy in terms of the Peierls gap, Δ, is given as follows for the ordinary Peierls transition [5]. Here Δ is refered to the gap in the $\pi-$band (ε_2-band).

$$\Delta F_0 = N(0)[-\varepsilon|\Delta|^2 + A|\Delta|^3 + \frac{b}{2}|\Delta|^4] , \tag{2}$$

where $\varepsilon = \ell n(T_p/T), b=7\zeta(3)/(8\pi^2 T_p^2)$, T_p being the Peierls transition temperature and A is due to the third order commensurability energy [12]. In eq.(2) N(0) is the density of state at the Fermi energy and is of the order of N/ε_F where N is the number of the total sites. In the present case with the $d-$band having strong correlations we take account of the following extra contribution, ΔF_d, due to the onset of localized magnetic moments in the insulating state,

$$\Delta F = \Delta F_0 - \frac{N}{3} T\, \ell n(1+2S) , \tag{3.a}$$

$$\equiv \Delta F_0 + \Delta F_d, \tag{3.b}$$

where one third in the prefactor corresponds to the threefold lattice distortions. The magnitude of spin S is determined by the energy gap, E_g, of the $d-$band (ε_1-band) at $|k|/c^* = 1/3$ in Fig.2 introduced by the lattice distortion i.e. $E_g \propto \Delta$. Though it will be studied in more detail elsewhere, we phenomenologically express S to be (under the assumption that the width of the ε_1-band is much smaller than U)

$$2S = f(\frac{E_g}{U}) , \tag{4}$$

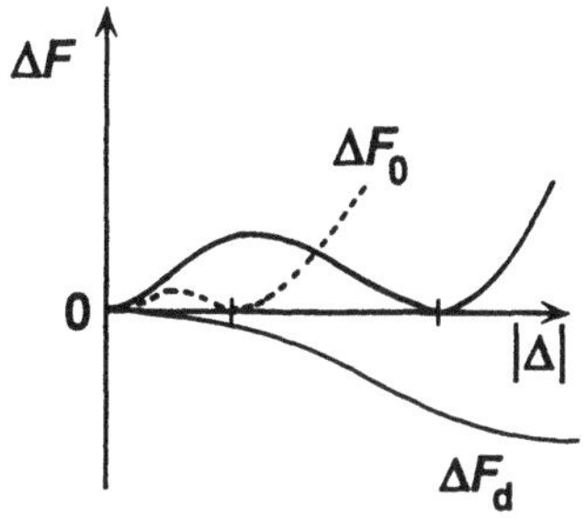

Fig. 4 The free energy as a function of the Peierls gap energy, Δ; with π−electrons only (dotted line) and with d−electrons as well (solid line).

where $f(x) \sim x^2$ as $x << \Phi$ and $f(x) \to 1$ as $x >> \Phi$. By noting the large prefactor $NT/3 \sim NT_p/3$ due to the large number of localized moment compared to the conventional one in eq.(2), $N(0)\Delta^2 \sim NT_p^2/\varepsilon_F$, we expect the dependences on Δ of ΔF_0 and ΔF as schematically shown in Fig.4.

The onset of magnetic moments in the insulating state caused by the strong correlations at Cu sites is seen to lead to large sudden change of the lattice distortions at the transition.

4. Summary and Discussions

Based on the recently proposed theoretical model for the metal-insulator transition in $(DCNQI)_2Cu$, the stability of the charge transfer leading to $Cu^{+4/3}$ as an average and the extraodinaty large hysterisis through the transition are explained as natural consequences of Peierls transition in the presence of the strong correlation in the Cu d−band.

The system is a unique example where the Peierls transition goes hand in hand with the Mott transition.

References

[1] A. Aumüller, P.Erk, G. Klebe, S. Hunig, J.U. von Schütz and H.-P. Werner, Angew. Chem. Int. Ed. Engl., **25**, 740 (1986).

[2] R. Kato, H. Kobayashi, A. kobayashi, T. Mori and H. Inokuchi, Chem. Lett. **1987**, 1579.

[3] S. Tomic, D. Jerome, A. Aumüller, P. Erk, S. Hünig and J.U. von Schütz, J. Phys. C. **21**, L203 (1988)

[4] R. Kato, H. Sawa, A. Aonuma, M. Tamura, M. Kinoshita and H. Kobayashi, Solid State Commun. **85**, 831 (1993).

[5] For example, L.J. Sham, *Highly Conducting One-Dimensional Solids*, ed. by J.T. Devreese, R.P. Evrard and V.E. van Doren (Plenum Press, 1977) p.227.

[6] Y. Suzumura and H. Fukuyama, J. Phys. Soc. Jpn. **61**, 3322 (1992).

[7] H. Fukuyama, J. Phys. Soc. Jpn. **61**, 3452 (1992).

[8] H. Fukuyama, to appear in Prog. Theor. Phys. Suppl. **106**, (1993).

[9] H. Fukuyama, J. Phys. Soc. Jpn. **62**, 1436 (1993).
[10] Y. Suzumura and Y. Ono, J. Phys. Soc. Jpn. **62**, 3244 (1993).
[11] I.H. Inoue, A. Kakizaki, AS. Kobayashi, R. Kajita, H. Kobayashi, H. Namatame and A. Fujimori, Physica C **185-189**, 2692 (1991).
[12] Y. Suzumura and H. Fukuyama, J. Phys. Soc. Jpn. **49**, 2081 (1980).
[13] R.H. Friend, M. Miljak and D. Jerome, Phys. Rev. Lett **40**, 1048 (1978).

Part IV

Edge States of the Quantum Hall Effect

Edge Excitations in Abelian and Non-Abelian FQH States

X.G. Wen

Department of Physics, MIT, 77 Massachusetts Avenue,
Cambridge, MA 02139, USA

Abstract. We summarize edge excitations in some abelian and non-abelian FQH states using partion construction and conformal field theory.

1. Introduction

Fractional quantum Hall effects (FQHE) discovered by Tsui, Stormer and Gossard[1] open a new era in theory of strongly correlated system. For the first time we have to completely abandon the theories based on the single-body picture (such as Fermi liquid theory) and use an intrinsic many-body theory proposed by Laughlin[2] and others[3] to describe FQHE. Due to the repulsive interaction and strong correlation between the electrons, a FQH liquid is an incompressible state despite the first Landau level is only partially filled. All the bulk excitations in the FQH states have finite energy gaps. The FQH states and insulators are very similar in the sense that both states have finite energy gap and short ranged electron propagators. Because of this similarity, people were puzzled by the fact that the FQH systems apparently have very different transport properties than ordinary insulators. Halperin first pointed out that the integral quantum Hall (IQH) states contain gapless edge excitations.[4] Although the electronic states in the bulk are localized, the electronic states at the edge of the sample are extended (*i.e.*, the electron propagator along the edge is long ranged).[5] Therefore the nontrivial transport properties of the IQH states come from the gapless edge excitations.[4,6] The edge transport picture has been supported by many experiments.[7]

We know that different FQH states were generally labeled by their filling fractions. However now it becomes clear that FQH states contain extremely rich internal structures (called topological orders) that the filling fraction alone is not enough to classify all the different universality classes of FQH states.[8,9] One can easily construct different FQH states with the same filling fraction.[10,11,12,13,14] Now the question is that how to experimentaly prob those rich internal structures, or topological orders, in FQH states. We will see later that the edge states [15,16] open up a practical window through which we can look into the internal structures in FQH states. The measurements of the edge states can provide us new quantum numbers, in addition to the filling fractions, to characterize different quantum Hall states.

2. The Edge Excitations of Abelian FQH States - Parton Construction

In this section we will use the parton construction to obtain the edge excitations of some abelian FQH states.[17] Let us first review the parton construction of the bulk FQH states, using $\nu = \frac{n}{mn+1}$ FQH state as an example.[10] The trick is to split the electrons into m charge $\frac{ne}{mn+1}$ partons $\psi_\alpha|_{\alpha=1}^{m}$ and a charge $\frac{e}{mn+1}$ parton ψ_{m+1}. All the partons have fermionic statistics and m is even. Since each kind of partons has the same density as the electrons: $n_\alpha = \psi_\alpha^\dagger \psi_\alpha = n_e$, the parton ψ_{m+1} have a filling fraction $\nu_{m+1} = n$ and the parton $\psi_\alpha|_{\alpha=1}^{m}$ have a filling fraction $\nu_\alpha = 1$. Thus if the partons

Springer Series in Solid-State Sciences, Vol. 118
Correlation Effects in Low-Dimensional Electron Systems
Editors: A. Okiji and N. Kawakami

were independent, the partons ψ_{m+1} will form a $\nu = n$ IQH state described by the wave function $\chi_n(z_i^{(m+1)})$ and the partons $\psi_\alpha|_{\alpha=1}^m$ will form $\nu = 1$ states described by $\chi_1(z_i^{(\alpha)})$. But in reality the partons are not independent. By introducing the fictitious particles, the partons, we introduce some unphysical degrees of freedom, namely the density fluctuations of the form $n_\alpha - n_{\alpha'}$. In a physical electron states, the densities of the partons always satisfy the constraint $n_1 = n_2 =, ..., = n_{m+1}$, or

$$\sum_\alpha C_\alpha n_\alpha = 0, \quad \text{for any } \sum_\alpha C_\alpha = 0.$$

To use the parton picture to describe the real electron states and to obtain the correct physics for the electrons, we need to make a projection to project away all the unphysical degrees of freedom. The ground state (trial) wave function Ψ that satisfies the above constraint can be obtained by doing a projection $z_i^{(1)} = z_i^{(2)} = ... = z_i$, where z_i are the electron coordinates:

$$\Psi(z_i) = \chi_n(z^{(m+1)}) \prod_{\alpha=1}^m \chi_1(z_i^{(\alpha)})|_{z_i^{(1)}=...=z_i^{(m+1)}=z_i}. \tag{1}$$

The parton construction is very convenient for the construction of the edge excitations. This is because the projection can be done at the effective theory level.

Let us first discuss the edge excitations in the $\nu = \frac{n}{mn+1}$ FQH state (m is an even integer). We assume that the FQH state has a disk-like geometry. The edge of the disk is parameterized by x.

Before the projection, all the partons are independent. The charge $\frac{n}{mn+1} = q_2$ partons $\psi_\alpha|_{\alpha=1}^m$ form a $\nu_\alpha = 1$ IQH state and support a single branch of edge excitations for each α. Those excitations are described by chiral fermion theory $\sum_{\alpha=1}^m \lambda_\alpha^\dagger(\partial_t - v\partial_x)\lambda_\alpha$. After bosonization they are described by the following Kac-Moody (K-M) algebra[18]

$$[\rho_{\alpha,k}, \rho_{\alpha',k'}] = \frac{k}{2\pi}\delta_{k+k'}\delta_{\alpha\alpha'}, \tag{2}$$

where $\rho_\alpha = \lambda_\alpha^\dagger\lambda_\alpha$ is the edge density of the αth partons. The parton creation operator on the edge is given by $\psi_\alpha = e^{i\phi_\alpha}$ for $\alpha = 1, ..., m$, where $\partial_x\phi_\alpha = 2\pi\rho_\alpha$. The charge $\frac{1}{mn+1} = q_1$ partons ψ_{m+1} form a $\nu = n$ IQH states that support n branches of edge excitations. Those edge excitations are described by

$$[\rho_{i,k}, \rho_{i',k'}] = \frac{k}{2\pi}\delta_{k+k'}\delta_{i,i'}, \quad i, i' = 1, 2, \ldots, n, \tag{3}$$

where ρ_i is the edge density of the partons ψ_{m+1} in the ith Landau level. The parton creation operators on the edge are given by $\psi_i^\dagger = e^{i\phi_i}$ with $\partial_x\phi_i = 2\pi\rho_i$, $i = 1, \ldots, n$. They carry an electric charge q_1. $\psi_i^\dagger$ create a parton in the ith Landau level. The coupling between the edge densities and the external electric potential is given by

$$e(q_2 \sum_{\alpha=1}^m \rho_\alpha + q_1 \sum_{i=1}^n \rho_i)A_0. \tag{4}$$

Before the projection, the Hilbert space of the edge excitations is generated by ρ_α, ρ_i, ψ_α and ψ_i with $\alpha = 1, ..., m$ and $i = 1, \ldots, n$, which contains $n + m$ branches.

Because the fluctuations associated with $\tilde{\rho}_C = \sum_{\alpha=1}^m C_\alpha\rho_\alpha + C_{m+1}\sum_{i=1}^n \rho_i$ are unphysical for any C_a satisfying $\sum_{a=1}^{m+1} C_a = 0$, we should remove all such fluctuations to obtain the correct edge excitations for electrons. To accomplish this, we will first specify the physical operators. A physical operator must not create any fluctuations associated with $\tilde{\rho}_C$. Hence a physical operator must commute with $\tilde{\rho}_C$:

$$[\hat{O}_{phy}, \tilde{\rho}_C] = 0 \tag{5}$$

for any C_a that satisfy $\sum_{a=1}^{m+1} C_a = 0$. One can easily check that the following edge density operators are physical

$$\begin{aligned} j_0 &= \sqrt{\nu}(\rho_0 + \frac{1}{n}\sum_{i=1}^{n} \rho_i) , \\ j_i &= \sum_{j=1}^{n} a_i^j \rho_j, \quad i = 1, \ldots, n-1 , \end{aligned} \tag{6}$$

where $\rho_0 = \sum_{\alpha=1}^{m} \rho_\alpha$. a_i^j in (6) are orthogonal vectors satisfy $\sum_{j=1}^{n} a_i^j = 0$ and $\sum_{j=1}^{n} a_i^j a_{i'}^j = \delta_{i,i'}$. Similarly the charged physical operators (with the minimum charge) are given by $\Psi_i = e^{i(\phi_i + \sum_{\alpha=1}^{m} \phi_\alpha)}$, $i = 1, \ldots, n$ The operators Ψ_i carry an electric charge e. They are just the electron creation operators on the edge. The Hilbert space of the physical edge excitations is generated by j_i and Ψ_i, and contains n branches. Thus the edge excitations of the $\nu = \frac{n}{mn+1}$ FQH state (obtained from the parton construction) have n branches.

From (6), (2) and (3) we see that the physical edge density operators satisfy the following K-M algebra

$$[j_{i,k}, j_{i',k'}] = \frac{k}{2\pi}\delta_{k+k'}\delta_{i,i'} \quad i = 0, ..., n-1. \tag{7}$$

From (4) we find that only j_0 couples to the electric potential $e\sqrt{\nu} j_0 A_0$.

Using the algebra (2) and (3) we can easily calculate the equal time correlations between Ψ_i and $\Psi_i^\dagger$: $\langle \Psi_i^\dagger(x)\Psi_j(y)\rangle \propto (x-y)^{-m-1}\delta_{i,j}$. The electronic state on the edge is definitely not a Fermi liquid due to the anomalous exponent in the correlation functions. We can also show that $\{\Psi_i(x), \Psi_j(y)\} = 0$. Therefore Ψ_i are indeed fermionic operators.

Notice that when $n = 1$, (1) becomes the Laughlin wave functions. In this case the above results reduce to the results obtained in Ref. [19]

3. The Edge Excitations of Non-Abelian FQH States

It is now known that there are two classes of QH states, abelian QH states whose quasiparticles all have abelian statistics and non-abelian QH states[20,21,22] that contain some quasiparticles with non-abelian statistics. In the following we will briefly study the edge excitations of some non-abelian QH states.

A class of non-abelian QH states is represented by wave functions[21]

$$\chi_1^p(z_i)[\chi_m(z_i)]^n \tag{8}$$

whose filling fraction is $\nu = (p + \frac{m}{n})^{-1}$. In (8), $p + n$ is an odd integer and χ_m is the fermion wave function with m filled Landau levels. The electrons in this wave function stay within the first $n(m-1)+1$ Landau levels. There are local Hamiltonians such that (8) is an exact ground state.

The non-abelian QH state represented by (8) contain a quasiparticle excitation with charge $\frac{n}{pnm+m^2}$. Such a quasiparticle can be shown to carry a non-abelian statistics described by the $SU(n)$ level m, or in short $SU(n)_m$, Chern-Simons theory. For more detailed discussions see Ref. [21] and [22].

To understand the edge states of the non-abelian QH states, let us first concentrate on the non-abelian state described by $[\chi_m]^n$. Note such a state can be obtained through the parton construction[10] by "splitting" electrons into n charge $1/n$ partons. Therefore it is convenient to use the method discussed in section 2 to construct the edge states of non-abelian states. Let us use $\alpha = 1, ..., n$ to label different partons. We first assume that all partons are independent, and thus the ground state wave function is given by

$$\prod_{\alpha=1}^{n} \chi_m(z_i^{(\alpha)}), \tag{9}$$

where $z_i^{(\alpha)}$ is the coordinates of the αth kind of the partons. The original electron wave function $[\chi_m(z_i)]^n$ is obtained by doing the projection $z_i^{(1)} = ... = z_i^{(n)} = z_i$. Before the projection the state (9) is just n independent IQH states each with filling fraction m and contain mn branches of edge excitations that are described by the following low energy effective theory in 1D:

$$\mathcal{L} = \sum_{\alpha,a} i\psi^{\alpha a\dagger}(x,t)(\partial_t - v\partial_x)\psi^{\alpha a}(x,t), \tag{10}$$

where $\psi^{\alpha a}$, $a = 1, ..., m$, are charge $1/n$ fermion fields that describe the ath edge branch of the αth kind of the partons. (Note that each kind of the partons forms the $\nu = m$ IQH state that contains m branches of edge excitations.)

Within the effective theory (10) we can define three densities $J(x) = \frac{1}{n}\sum_{\alpha a}\psi^{\alpha a\dagger}\psi^{\alpha a}$, $j^s(x) = \sum_{\alpha,a,b}\psi^{\alpha a\dagger}T^s_{ab}\psi^{\alpha b}$ and $\tilde{j}^{\tilde{s}}(x) = \sum_{\alpha,\beta,a}\psi^{\alpha a\dagger}\tilde{T}^{\tilde{s}}_{\alpha\beta}\psi^{\beta a}$, where T^s_{ab} ($\tilde{T}^{\tilde{s}}_{\alpha\beta}$) is generators of the $SU(m)$ ($SU(n)$) Lie algebra. J is the electric charge density of the edge excitations and j^s and $\tilde{j}^{\tilde{s}}$ represent the densities of non-abelian charges associated with the Landau level index and the parton index respectively. The above three densities satisfy the $U(1) \times SU(m)_n \times SU(n)_m$ K-M algebra:[23,18]

$$\begin{aligned}
[J_k, J_{k'}] &= \frac{m}{2\pi n} k\delta_{k+k'} , \\
[j^r_k, j^s_{k'}] &= \frac{n}{2\pi} k\delta_{r,s}\delta_{k+k'} + L^{-\frac{1}{2}} f^{rst} j^t_{k+k'} , \\
[\tilde{j}^{\tilde{r}}_k, \tilde{j}^{\tilde{s}}_{k'}] &= \frac{m}{2\pi} k\delta_{\tilde{r},\tilde{s}}\delta_{k+k'} + L^{-\frac{1}{2}} \tilde{f}^{\tilde{r}\tilde{s}\tilde{t}} \tilde{j}^{\tilde{t}}_{k+k'} ,
\end{aligned} \tag{11}$$

where f^{rst} ($\tilde{f}^{\tilde{r}\tilde{s}\tilde{t}}$) is the structure constant of the $SU(m)$ ($SU(n)$) Lie algebra. J, j and $\tilde{j}$ commute with each other. It has been shown that the excitations in (10) are completely described by the above K-M algebra.[23]

Now we are ready to do the projection and to discuss the edge excitations of the electron system. We know the independent parton model contains unphysical degrees of freedom. To remove the unphysical fluctuation, we need to recombine the partons into electrons. The electron operator $\psi_e = \prod \psi^\alpha$ is a $SU(n)$ singlet under the $SU(n)$ transformation $\psi^\alpha \to U^{\alpha\beta}\psi^\beta$. (Here ψ^α is the operator of the αth kind of the partons.) Therefore any excitations of the electron system must be $SU(n)$ singlets and $\tilde{j}^{\tilde{s}}$ are identically zero for physical excitations. Therefore we can recombine partons into electron by removing all excitations with non-trivial $SU(n)$ quantum numbers. A physical operator must not creat any $SU(n)$ fluctuations and therefore must commute with the $SU(n)$ density operator $\tilde{j}$:

$$[\hat{O}_{phy}, \tilde{j}^{\tilde{s}}] = 0. \tag{12}$$

The above equation is the analogue of the equation (5) in the section 2.

We can see that all the operators in the $U(1)$ and the $SU(m)_n$ K-M algebra commute with $\tilde{j}$, therefore edge excitations of the electronic state $(\chi_m)^n$ are described by the $U(1) \times SU(m)_n$ K-M algebra.

The additional factor χ_1^p in the electron wave function (8) will only modify the $U(1)$ K-M algebra. One can show that the edge excitations in the state (8) are described by the following $U(1) \times SU(m)_n$ K-M algebra:

$$\begin{aligned}
[J_k, J_{k'}] =& \frac{\nu}{2\pi} k\delta_{k+k'} , \\
[j_k^r, j_{k'}^s] =& \frac{n}{2\pi} k\delta_{k+k'} + L^{-1/2} f^{rst} j_{k+k'}^t , \\
\nu =& \frac{m}{pm+n} .
\end{aligned} \tag{13}$$

(More precisely, the Hilbert space of the low energy edge excitations forms a representation of the above algebra.) The low energy effective Hamiltonian takes the following general form

$$H = \sum_k V J_k J_{-k} + \sum_{k;r,s} V_{r,s} j_k^r, j_{-k}^s + \sum_{k;s} V_s (j_k^s J_{-k} + J_k j_{-k}^s). \tag{14}$$

(13) and (14) give us a complete description of the low energy

dynamics of the edge excitations of non-abelian QH states. Note that although the Hilbert space is generated by the $U(1) \times SU(m)_n$ K-M algebra, the Hamiltonian in general does not respect the $SU(m)$ symmetry. So when we say the edge excitations of a non-abelian state are described by non-abelian K-M algebra, we only mean that the Hilbert space of the low lying excitations forms a representation of the non-abelian algebra.

Now let us study the electron operators and the quasiparticle operators. First let us assume $p = 0$. From the parton construction and requirement that the electron operators must satisfy (12), we find that the electron operators on the edge are given by

$$\begin{aligned}
\Psi_e^M =& \epsilon_{\alpha_1,\ldots,\alpha_n} \psi^{\alpha_1 a_1} \ldots \psi^{\alpha_n a_n} S^M_{a_1,\ldots,a_n} \\
=& e^{i\frac{n}{m}\phi} V_{\Lambda_n,M} ,
\end{aligned} \tag{15}$$

where $\epsilon_{\alpha_1,\ldots,\alpha_n}$ are rank n antisymmetric tensors (labeled by M), $S^M_{a_1,\ldots,a_n}$ is the rank n symmetric tensor, and $V_{\Lambda_n,M}$ is the primary field of the $SU(m)_n$ K-M algebra in the representation of the rank n symmetric tensor. The field ϕ in (15) is determined through $J(x) = \frac{1}{2\pi}\partial_x\phi(x)$ and $e^{i\frac{n}{m}\phi}$ is a primary field of the $U(1)$ K-M algebra. The electron operators in (15) satisfy (12) and are $SU(n)$ singlet. They belong to the $U(1) \times SU(m)_n$ K-M algebra and can be written as a product of the primary fields in the $U(1)$ and the $SU(m)_n$ K-M algebra.[24] Note that there are many different electron operators just as in the IQH states (with $\nu > 1$) and the hierarchical FQH states. Those different electron operators form a representation of the rank n symmetric tensors of the $SU(m)$ group. (15) also implies Ψ_e^M anti-commute (commute) when n is odd (even). The equal time correlation has a form

$$\langle \Psi_e^{\dagger M}(x) \Psi_e^M(y) \rangle = \langle e^{-i\frac{n}{m}\phi(x)} V^\dagger_{\Lambda_n,M}(x) e^{i\frac{n}{m}\phi(y)} V_{\Lambda_n,M}(y) \rangle \propto (\frac{1}{x-y})^n. \tag{16}$$

When $p \neq 0$, we only need to modify the $U(1)$ vertex operator. We find in this case the electron operators are given by

$$\Psi_e^M = e^{i\frac{1}{\nu}\phi} V_{\Lambda_n,M} , \tag{17}$$

where ν is given in (13). Using (13) one can easily check that Ψ_e^M in (17) create an unit localized charge. We can also show that

$$\frac{e^{i\frac{1}{\nu}\phi(x)}e^{i\frac{1}{\nu}\phi(y)}}{e^{i\frac{n}{m}\phi(x)}e^{i\frac{n}{m}\phi(y)}} = (-)^p \frac{e^{i\frac{1}{\nu}\phi(y)}e^{i\frac{1}{\nu}\phi(x)}}{e^{i\frac{n}{m}\phi(y)}e^{i\frac{n}{m}\phi(x)}}. \tag{18}$$

Therefore Ψ_e^M anti-commute when $n+p$ =odd. In this case Ψ_e^M is a fermionic operator as expected. The equal time correlation is given by

$$\langle \Psi_e^{\dagger M}(x)\Psi_e^M(y)\rangle = \langle e^{-i\frac{1}{\nu}\phi(x)} V^\dagger_{\Lambda_n,M}(x) e^{i\frac{1}{\nu}\phi(y)} V_{\Lambda_n,M}(y)\rangle \propto (\frac{1}{x-y})^{n+p}. \tag{19}$$

The exponent $n+p$ in the electron propagator can be measured experimentally through tunneling between edge states.[25] Such a measurement combining with the knowledge of the filling fraction may allow us to determine experimentally whether a FQH state is an abelian or a non-abelian state.

Now let us consider the quasiparticle operators. A generic quasiparticle operator is a product of a primary field in the $U(1)$ K-M algebra and a primary field V_Λ in the $SU(m)$ K-M algebra:

$$\Psi_q = e^{iq_\Lambda \phi} V_\Lambda \ , \tag{20}$$

where Λ labels the representation of the primary field. The value of the q_Λ is determined by reqiring the electron wave function to be single valued. A calculation of q_Λ can be found in Ref. [22]. The electric charge of the quasiparticle is given by

$$Q = q_\Lambda \nu. \tag{21}$$

Within the edge theory, the single-valueness of the electron wave function is equivalent to the requirement that the electron operators Ψ_e and the quasiparticle operator Ψ_q are mutually local to each other. This means the correlation function

$$\langle T(\Psi_e(x_1,t_1)\Psi_q(x_2,t_2)...)\rangle \tag{22}$$

to be single valued as the electron operator goes around the quasiparticle operator in the space-time. In (22) "..." represents other operators that make the correlation function non-zero.

In the following we will attempt to calculate q_Λ within the edge theory. For simplicity let us consider the quasiparticles in the fundamental representation (labeled by Λ_1) of the $SU(m)$ and assume the Hamiltonian to respect the $SU(m)$ symmetry, *i.e.*, $V_{rs} \propto \delta_{rs}$ and $V_r = 0$ in (14). In this case we can use the conformal field theory to calculate the correlations. The primary fields V_{Λ_1} and V_{Λ_n} satisfy the following operator product expansion:

$$V_{\Lambda_1}(z_1)V_{\Lambda_n}(z_2) \propto (z_1 - z_2)^{h_0 - h_1 - h_n} V_{\Lambda_0} \ , \tag{23}$$

where h_0, h_1 and h_n are the conformal dimensions of the primary fields V_{Λ_0}, V_{Λ_1} and V_{Λ_n} respectively. z in (23) is given by $x + i\tau$ and τ is the imaginary time it. One can show that[26]

$$h_0 - h_1 - h_n = \frac{1}{m}. \tag{24}$$

From the operator product expansion

$$\begin{aligned} e^{i\frac{1}{\nu}\phi(z_1)} e^{iq_{\Lambda_1}\phi(z_2)} &= (z_1 - z_2)^\lambda e^{i(\frac{1}{\nu} + q_{\Lambda_1})\phi(z_2)} \ , \\ \lambda &= \frac{\nu}{2}[(\frac{1}{\nu} + q_{\Lambda_1})^2 - \frac{1}{\nu^2} - q_{\Lambda_1}^2] = q_{\Lambda_1} \ , \end{aligned} \tag{25}$$

we see that the single valueness of the correlation requires that $\lambda + h_0 - h_1 - h_n$ =integer, or

$$q_{\Lambda_1} = -\frac{1}{m} + l \ , \tag{26}$$

where l is an integer. We find the quasiparticle operator to have a form

$$\Psi_q^{(l)} = e^{i(l-\frac{1}{m})\phi} V_{\Lambda_1}. \tag{27}$$

Such a quasiparticle operator carries a charge $(l-\frac{1}{m})\nu$. Thus the quasiparticle mentioned at the beginning of this section correspond to $\Psi_q^{(0)}$. The above quasiparticle operator has a conformal dimension $h = \frac{\nu}{2}(l - \frac{1}{m})^2 + \frac{m^2-1}{2m(m+n)}$.[26] Thus the propagator of the quasiparticle is given by

$$\langle \Psi_q^{(l)\dagger}(x,t)\Psi_q^{(l)}(0)\rangle \propto (x - v_J t)^{-\nu(l-\frac{1}{m})^2}(x - v_j t)^{-\frac{m^2-1}{m(m+n)}} \ , \tag{28}$$

where v_J and v_j are the velocities of the excitations in the $U(1)$ and the $SU(m)_n$ K-M algebras respectively. In some sense the quasiparticles described by (27) are fundamental. All other quasiparticles can be constructed as bound states of the quasiparticles in (27).

In real samples we do not have the $SU(m)$ symmetry and different excitations in the $SU(m)$ K-M algebra do not move with the same velocity v_j. In this case we can still say some thing about the quasiparticle propagator. First we notice that the asymptotic form of the equal-time correlation can be obtained from the operator product expansion

$$\begin{aligned} &\Psi_q^{(l)\dagger}(x_1,t)\Psi_q^{(l)}(x_2,t) \propto (x_1 - x_2)^{-\nu(l-\frac{1}{m})^2-\frac{m^2-1}{m(m+n)}} = (x_1 - x_2)^{-g}, \\ &g = \frac{m}{m+n} + \frac{1-p}{(pm+n)(m+n)} + \frac{l^2 m - 2l}{pm+n} \ , \end{aligned} \tag{29}$$

and is independent of the Hamiltonian. If the edge excitations still have linear dispersion relation (possibly with many different velocities), then the "equal-space" correlation will has the form

$$\Psi_q^{(l)\dagger}(x,t_1)\Psi_q^{(l)}(x,t_2) \propto (t_1 - t_2)^{-g}. \tag{30}$$

The exponent in (30) can be directly measured in edge tunneling experiments (see Ref. [25]). The quasiparticle with $l = 0$ has the smallest exponent and may dominate the edge tunneling at low energies.

3. The Edge Excitations of Paired Non-Abelian FQH States

In this section we are going to study the edge excitations of another type of non-abelian FQH states, which involve electron pairing. The simplest paired non-abelian state is the p-wave paired state is proposed in Ref. [20] and described by the pfaffian wave function

$$\Psi_{pf}(\{z_i\}) = \mathcal{A}(\frac{1}{z_1 - z_2}\frac{1}{z_3 - z_4}\ldots)\prod_{i<j}(z_i - z_j)^2 e^{-\frac{1}{4}\sum_i |z_i|^2} \ , \tag{31}$$

where $\mathcal{A}$ is the antisymmetrization operator. Ψ_{pf} is an exact ground state of the following three-body potential[27]

$$V_{3bd} = -Uc^\dagger(z_1)c^\dagger(z_2)c^\dagger(z_3)\partial_{z_1^*}\delta(z_1 - z_2)\partial_{z_1}\partial^2_{z_3^*}\delta(z_3 - z_2)\partial^2_{z_3}c(z_3)c(z_2)c(z_1) \tag{32}$$

with zero energy. All other zero energy states have higher angular momenta and correspond to edge excitations. Thus To study the edge excitations of the pfaffian state we simply need to study the zero energy sector of (32).

We have calculated the spectrum of the Hamiltonian (32) on a plane with 8 electrons in the first 20 orbits (*i.e.*, $0 \le m \le 19$). Here we have chosen the symmetric gauge and m is the angular momentum of single-electron states. The numbers of states (NOS) of the low lying edge excitations at total angular momenta $M_0 + l$ (where M_0 is the angular momentum of the ground state) are found to be

$$\begin{array}{cccccccccc} l: & 0 & 1 & 2 & 3 & 4 & 5 & 6 & 7 & 8 \\ & 1 & 1 & 3 & 5 & 10 & 15 & 25 & 35 & 52 \end{array} . \tag{33}$$

The edge states (the zero-energy states) and the bulk states are clearly separated by a finite gap. The numerical results for $N_e = 4, 6, 8$ further indicate that the NOS for $l \le [\frac{N_e}{2}]$ have reached their thermodynamical value (*i.e.*, the NOS is unchanged as we increase N_e). Thus the edge excitations of the pfaffian state cannot be described by one branch of edge waves which has the following spectrum

$$\begin{array}{cccccccccc} l: & 0 & 1 & 2 & 3 & 4 & 5 & 6 & 7 & 8 \\ & 1 & 1 & 2 & 3 & 5 & 7 & 11 & 15 & 22 \end{array} \tag{34}$$

or two branches of edge waves which has

$$\begin{array}{cccccccccc} l: & 0 & 1 & 2 & 3 & 4 & 5 & 6 & 7 & 8 \\ & 1 & 2 & 5 & 10 & 20 & 36 & 65 & 110 & 180 \end{array} . \tag{35}$$

The pfaffian state contains more then one branch but less than two branches of edge excitations. The specific heat (per unit length) of edge excitations for the pfaffian state turns out to be 3/2 times of $\frac{\pi}{6}\frac{T}{v}$ – the specific heat of one branch of edge excitations. In this sense, the pfaffian state contains one and a half branches of edge excitations.

In the following we will show that the edge excitations of the pfaffian state are described by a $U(1)$ K-M algebra plus a chiral Majorana fermion theory. A system of a chiral Majorana fermion is described by the Lagrangian[28]

$$\mathcal{L} = i\lambda(\partial_t - v\partial_x)\lambda \ , \tag{36}$$

where λ is a *real* fermion field: $\lambda^\dagger = \lambda$ and v is the edge velocity. We will put the system on a circle $x \in [0, 2\pi)$. Because λ is real, it can only have two different boundary conditions $\lambda(0,t) = \pm\lambda(2\pi,t)$ (which will be called $\pm$ boundary condition). After quantization, (36) is described by the following Hamiltonian systems:

$$\begin{aligned} H &= v\sum_{k>0}^{+\infty} k a_k a_{-k} \ , \\ \{a_k, a_{k'}\} &= \delta_{k+k'} \ , \\ \lambda(x) &= (2\pi)^{-\frac{1}{2}} \sum_{k=-\infty}^{+\infty} a_k e^{ikx} \ , \end{aligned} \tag{37}$$

where k =integer (or $k = \frac{1}{2}$+integer) for the + (or −) boundary condition. The zero-momentum component a_0 is represented as $a_0 = \frac{1}{\sqrt{2}}(\alpha + \alpha^\dagger)$, where $\{\alpha, \alpha^\dagger\} = 1$. (37) is a free fermion system and is exactly soluble. Because the system (36) has a symmetry $\lambda \to -\lambda$, the quantity $(-)^{N_\lambda}$ is conserved, while the number of the fermions $N_\lambda = \sum_{k\ge 0} a_k a_{-k}$ is not conserved. The theory (36) thus contains four sectors with ±

boundary conditions and even/odd numbers of fermions, which are denoted by (+,even), (−,odd), *etc.* We can only create excitations within the same sector. It is straight forward to work out the numbers of the states at each total momentum k in each sector:

$$\begin{array}{cccccccccccc} \Delta k: & 0 & 1 & 2 & 3 & 4 & 5 & 6 & 7 & 8 & \text{sector} \\ & 1 & 0 & 1 & 1 & 2 & 2 & 3 & 3 & 5 & (-,\text{even}) \\ & 1 & 1 & 1 & 1 & 2 & 2 & 3 & 4 & 5 & (-,\text{odd}) \\ & 1 & 1 & 1 & 2 & 2 & 3 & 4 & 5 & 6 & (+,\text{odd}) \\ & 1 & 1 & 1 & 2 & 2 & 3 & 4 & 5 & 6 & (+,\text{even}) \end{array} , \tag{38}$$

where $\Delta k = k - k_0$ and k_0 is the momentum of the ground state in each sector. Because the Majorana fermion describe the critical point of the Ising model, we will call the system described by (36) an Ising system.[28] The system of the $U(1)$ K-M algebra is described by the following Hamiltonian:

$$\begin{aligned} H &= v\nu^{-1} \sum_{k=1}^{+\infty} \rho_k \rho_{-k} , \\ [\rho_k, \rho_{k'}] &= -\nu k \delta_{k+k'} , \end{aligned} \tag{39}$$

where ν is the filling fraction $\nu = 1/2$, and $\rho(x) = (2\pi)^{-\frac{1}{2}} \sum_k \rho_k e^{ikx}$ is the (1D) charge density on the edge.[29,16] (39) describes a collection of harmonic oscillators and is again exactly soluble. The number of the states at each total momentum k is given by (34). (Note the momentum along the circle Δk is equal to the angular momentum $l = L - M_0$.) After we put the $U(1)$ K-M algebra and the *Ising* system together (denoted by $U(1) \times Ising$), we find that the total numbers of the states in each sector at angular momenta ΔM are given by

$$\begin{array}{cccccccccccc} l: & 0 & 1 & 2 & 3 & 4 & 5 & 6 & 7 & 8 & \text{sector} \\ & 1 & 1 & 3 & 5 & 10 & 16 & 28 & 43 & 70 & (-,\text{even}) \\ & 1 & 2 & 4 & 7 & 13 & 21 & 35 & 55 & 86 & (-,\text{odd}) \\ & 1 & 2 & 4 & 8 & 14 & 24 & 40 & 64 & 100 & (+,\text{odd}) \\ & 1 & 2 & 4 & 8 & 14 & 24 & 40 & 64 & 100 & (+,\text{even}) \end{array} . \tag{40}$$

Comparing (33) and (40), we see that, for those that have reached the thermodynamical limit (*i.e.*, for $l \leq 4$), the NOS of the edge excitations in the 8-electron pfaffian state exactly match those in the (−,even) sector of the $U(1) \times Ising$ system. We also calculate the spectrum of (32) with 9 electrons in the first 21 orbits. The numbers of the low lying edge excitations are found to be

$$\begin{array}{cccccc} l: & 0 & 1 & 2 & 3 & 4 \\ & 1 & 2 & 4 & 7 & 13 \end{array} . \tag{41}$$

The above NOS are shown to have reached their thermodynamical values and exactly match the (−,odd) sector of the $U(1) \times Ising$ system.

From the above discussion, we see that the edge excitations of the pfaffian state are described by the $U(1) \times Ising$ system. This is closely related to the fact that the pfaffian wave function can be constructed from correlation functions in the $U(1) \times Ising$ conformal field theory.[20] The (−,even) [(−,odd)] sector describes the edge excitations for even (odd) numbers of electrons.

The dynamics of the edge excitations is governed by (36) and (39), which is a conformal field theory. To calculate the propagator we need to first identify the electron and the quasiparticle operators. The electron operator ψ_e must carry unit electric charge and be a fermionic operator. The only operator that satisfy the above conditions is $\lambda(x)e^{2i\phi(x)}$ which will be identified as the electron operator $\psi_e(x)$. Here $\frac{1}{2\pi}\partial_x \phi = \rho$.

From (36) and (37), we find that $\langle\psi_e^\dagger(x,t)\psi_e(0)\rangle \propto (x-vt)^{-3}$. This result can be confirmed through a calculation of the electron occupation number n_m in each angular momentum states. Here m is the angular momentum. For the pfaffian state of 10 electrons, we find $n_{m_0} : n_{m_0-1} : ... = 1 : 3.06 : 5.86 : 8.63 : ...$ which agrees with the theoretical prediction for $(x-vt)^{-3}$ propagator:[16] $1 : 3 : 6 : 10 : ....$ Here m_0 is the angular momentum of the last occupied orbit ($m_0 = 17$ for 10 electrons).

Next let us consider the quasiparticle operator. The $U(1) \times Ising$ system contains the following local operators: $e^{i\alpha\phi}$, λ and σ. Here σ is the disorder operator in the *Ising* system which changes the boundary condition of the fermion λ. Thus σ connects the + sector and the − sector. Not all the above operators are physical, *i.e.*, create an allowed excitation in the electron system. A physical quasiparticle operator must have a single-valued correlation function with the electron operator. This condition is closely related to the single valueness of the electron wave function in presence of the quasiparticle. The condition can be expressed through the operator product expansion: $\psi_e(w_1)\psi_q(w_2) \propto (w_1-w_2)^\gamma \hat{O}(w_2)$ where we require γ to be an integer. Here w is given by $x + vi\tau$ and τ is the imaginary time $\tau = it$. From the operator product expansion in the $U(1) \times Ising$ system:

$$\begin{aligned} e^{i\alpha\phi(w)}e^{i\beta\phi(0)} &\propto w^{\frac{1}{4}[(\alpha+\beta)^2-\alpha^2-\beta^2]}e^{i(\alpha+\beta)\phi(0)} , \\ \lambda(w)\sigma(0) &\propto w^{\frac{1}{2}}\mu(0) , \end{aligned} \tag{42}$$

we find that the following operator can be identified as quasiparticle operator: $\psi_q(x) = e^{i\frac{1}{2}\phi(x)}\sigma(x)$. ψ_q carries charge $\frac{1}{4}$ and corresponds to the non-abelian quasiparticle discussed in Ref. [20]. From the conformal theory result $\langle\sigma(w)\sigma(0)\rangle \propto w^{1/8}$, we find $\langle\psi_q(x,t)\psi_q(0)\rangle \propto (x-vt)^{-1/4}$.

From the above discussions of abelian and non-abelian FQH states we see that the FQH states contain very rich internal structures. Those rich electron correlations in the bulk are reflected in the rich structures of the edge excitations. We also see a close connection between the edge states and conformal field theories. The conformal field theory that reproduce the bulk wave function is exactly the conformal field theory that generate the edge spectra. However a more complete theory for the non-abelian FQH states is yet to be developed.

This work is supported by the National Science Foundation under Grant No. DMR91-14553.

References

1. D.C. Tsui, H.L. Stormer, and A.C. Gossard, *Phys. Rev. Lett.* **48**, 1559 (1982).
2. R.B. Laughlin, *Phys. Rev. Lett.* **50**, 1395 (1983).
3. F.D.M. Haldane, *Phys. Rev. Lett.* **51**, 605 (1983); B. I. Halperin, *Phys. Rev. Lett.* **52**, 1583 (1984). S. Girvin, *Phys. Rev. B* **29**, 6012 (1984); A.H. MacDonald and D.B. Murray, *Phys. Rev. B* **32**, 2707 (1985); M.P.H. Fisher and D.H. Lee, *Phys. Rev. Lett.* **63**, 903 (1989); J.K. Jain, *Phys. Rev. Lett.* **63**, 199 (1989); *Phys. Rev.* **B41**, 7653 (1991).
4. B.I. Halperin, *Phys. Rev.* **B25**, 2185 (1982).
5. Q. Niu and D.J. Thouless, *Phys. Rev.* **B35**, 2188 (1987).

6. S.A. Trugman, *Phys. Rev.* **B27**, 7539 (1983); A.H. MacDonald, and P. Streda, *Phys. Rev.* **B29**, 1616 (1984); P. Streda, J. Kucera and A.H. MacDonald, *Phys. Rev. Lett.* **59**, 1973 (1987); M. Buttiker, *Phys. Rev.* **B38**, 9375 (1988); J.K. Jain and S.A. Kivelson, *Phys. Rev.* **B37**, 4276 (1988); *Phys. Rev. Lett.* **60**, 1542 (1988).

7. P.L. McEuen, *et al.*, *Phys. Rev. Lett.* **64**, 2062 (1990); A.M. Chang and J.E. Cunningham, *Solid State Comm.* **72**, 651 (1989); L.P. Kouwenhoven *et al.*, *Phys. Rev. Lett.* **64**, 69 (1990); J.K. Wang and V.J. Goldman, *Phys. Rev. Lett.* **67**, 749 (1991).

8. X.G. Wen, *Int. J. Mod. Phys. B* **2**, 239 (1990); *Phys. Rev.* **B40**, 7387 (1989).

9. X.G. Wen and Q. Niu, *Phys. Rev.* **B41**, 9377 (1990).

10. J.K. Jain, *Phys. Rev.* **B40**, 8079 (1989).

11. B. Blok and X.G. Wen, *Phys. Rev.* **B42**, 8133 (1990); *Phys. Rev.* **B42**, 8145 (1990).

12. N. Read, *Phys. Rev. Lett.* **65**, 1502 (1990).

13. J. Fröhlich and A. Zee, *Nucl. Phys.* **B364**, 517 (1991).

14. X.G. Wen and A. Zee, *Phys. Rev.* **B46**, 2290 (1992).

15. C.W.J. Beenakker, *Phys. Rev. Lett.* **64**, 216 (1990); A.H. MacDonald, *Phys. Rev. Lett.* **64**, 220 (1990); F.D.M. Haldane, *Bulletin of APS* **35**, 254 (1990); M. Stone, *Annals of Physics* **207**, 38 (1991); *Int. J. Mod. Phys. B* **5**, 509 (1991); J. Fröhlich and T. Kerler, *Nucl. Phys.* **B254**, 369 (1991).

16. For a review see X.G. Wen, *Int. J. Mod. Phys.* **B6**, 1711 (1992).

17. X.G. Wen, *Mod. Phys. Lett. B* **5**, 39 (1991).

18. See, for example, P. Goddard and D. Olive, "Workshop on Unified String Theories, 1985", eds. M. Green and D. Gross (World Scientific, Singapore), p. 214; *Inter. J. Mod. Phys.* **1**, 303 (1986); V.G. Kac, "Infinite dimensional Lie algebra", (Birkhauser, Boston, 1983).

19. X.G. Wen, *Phys. Rev.* **B41**, 12838 (1990).

20. G. Moore and N. Read, *Nucl. Phys.* **B360**, 362 (1991).

21. X.G. Wen, *Phys. Rev. Lett.* **66**, 802 (1991).

22. B. Blok and X.G. Wen, *Nucl. Phys.* **B374**, 615 (1992).

23. I. Affleck, *Nucl. Phys.* **B265**, 409 (1986).

24. J. Fuchs and D. Gepner, *Nucl. Phys.* **B294**, 33 (1987).

25. X.G. Wen, *Phys. Rev.* **B44**, 5708 (1991).

26. V. G. Knishnik and A. B. Zamolodchikov, *Nucl. Phys.* **B247**, 83 (1984); D. Gepner and E. Witten, *Nucl. Phys.* **B278**, 493 (1986); A. Schellekens and S. Yankielowicz, *Nucl. Phys.* **B327**, 673 (1989).

27. M. Greiter, X.G. Wen and F. Wilczek, *Phys. Rev. Lett.* **66**, 3205 (1991); *Nucl. Phys.* **B374**, 567 (1992); M. Greiter and F. Wilczek, *Nucl. Phys.* **B370**, 577 (1992).

28. See for example, P. Ginsparg, in Les Houches lecture 1988 "Field theory, string and critical phenomena" (North-Holland, Amsterdam, 1990)

29. X.G. Wen, *Phys. Rev.* **B43**, 11025 (1991); *Phys. Rev. Lett.* **64**, 2206 (1990).

Field Theory of Fermions in the Lowest Landau Level

B. Sakita

Department of Physics, City College of the City University of New York, New York, NY 10031, USA

Abstract. A pedagogical presentation is given of the field theoretical description of a system of planar charged fermions in a strong magnetic field. The discussions include the kinematics of the lowest Landau level (LLL), symmetries of the fractional quantum Hall (FQH) system, W_∞ gauge transformations and electromagnetic interactions, and an electromagnetic effective Lagrangian for a circular droplet.

1. Introduction

The many body problem of planar charged fermions in a strong magnetic field is a fascinating subject, not only because it is relevant to the quantum Hall phenomena [1, 2] but also because of its importance to $c = 1$ string theory. The relation to $c = 1$ string theory is an unexpected one [3, 4], and there is intense activitiy in this area [5]. In this article, however, I do not discuss this subject. Rather, I describe the general field theory of fermions in the lowest Landau level (LLL), mostly based on the articles [3, 4, 6] I have written with Iso, Karabali and Ray, supplemented by my own notes.

2. Kinematics of the Lowest Landau Level [7, 2, 6]

Let us consider a single-particle Hamiltonian, H_0, for a planar electron in a magnetic field normal to the plane:

$$H_0 = \frac{1}{2m}\left[(\hat{\Pi}^x)^2 + (\hat{\Pi}^y)^2\right] , \qquad \hat{\Pi}^x \equiv \hat{p}_x - A^x(\hat{x},\hat{y}), \qquad \vec{A} \equiv \frac{1}{2}B(y,-x) + \vec{\nabla}\Lambda(x,y) \; . \quad (1)$$

We use the convention $\hbar = c = e = 1$ and assume that the constant magnetic field is in the negative $\hat{z}$ direction. Although we can proceed with the discussion in a gauge independent fashion, for simplicity we choose the symmetric gauge, i. e. $\Lambda = 0$.

Define dimensionless $\hat{\pi}$ and $\hat{\pi}^\dagger$

$$\hat{\pi} \equiv \frac{1}{\sqrt{2B}}(\hat{\Pi}^x - i\hat{\Pi}^y) , \qquad \hat{\pi}^\dagger \equiv \frac{1}{\sqrt{2B}}(\hat{\Pi}^x + i\hat{\Pi}^y) , \qquad [\hat{\pi},\hat{\pi}^\dagger] = 1 \; . \quad (2)$$

The Hamiltonian is written as

$$H_0 = \omega(\hat{\pi}^\dagger\hat{\pi} + \frac{1}{2}), \quad \omega = \frac{B}{m} \; . \quad (3)$$

Further, we define the guiding center coordinate operators [8]

$$\hat{X} \equiv \hat{x} - \frac{1}{B}\hat{\Pi}^y, \quad \hat{Y} \equiv \hat{y} + \frac{1}{B}\hat{\Pi}^x, \quad [\hat{X},\hat{Y}] = \frac{i}{B} \; , \quad (4)$$

Springer Series in Solid-State Sciences, Vol. 118
Correlation Effects in Low-Dimensional Electron Systems
Editors: A. Okiji and N. Kawakami

and their dimensionless holomorphic form

$$\hat{a} \equiv \sqrt{\frac{B}{2}}(\hat{X} + i\hat{Y}), \qquad \hat{a}^\dagger \equiv \sqrt{\frac{B}{2}}(\hat{X} - i\hat{Y}), \qquad [\hat{a}, \hat{a}^\dagger] = 1 . \tag{5}$$

Further,

$$[\hat{a}, \hat{\pi}] = [\hat{a}, \hat{\pi}^\dagger] = [\hat{a}^\dagger, \hat{\pi}] = [\hat{a}^\dagger, \hat{\pi}^\dagger] = 0 . \tag{6}$$

Thus we have two sets of independent raising and lowering operators.

There are two pairs of independent canonical operators defined in this Hilbert space, namely $\hat{x}, \hat{p}_x, \hat{y}, \hat{p}_y$. The operators $\hat{\pi}, \hat{\pi}^\dagger, \hat{a}, \hat{a}^\dagger$ are of course expressed in terms of $\hat{x}, \hat{p}_x, \hat{y}, \hat{p}_y$ through Eqs. (1), (4), (5). Useful relations are

$$\sqrt{\frac{B}{2}}(\hat{x} + i\hat{y}) \equiv \hat{z} = \hat{a} - i\hat{\pi}^\dagger, \quad \sqrt{\frac{B}{2}}(\hat{x} - i\hat{y}) \equiv \hat{\bar{z}} = \hat{a}^\dagger + i\hat{\pi} . \tag{7}$$

We define the following representations:

$$\text{(coordinate representation): } \hat{x}|\vec{x}\rangle = x|\vec{x}\rangle \ , \ \hat{y}|\vec{x}\rangle = y|\vec{x}\rangle, \ \langle\vec{x}|\vec{x}'\rangle = \delta(\vec{x} - \vec{x}') \tag{8}$$

and

$$\text{(Landau level representation): } \hat{\pi}^\dagger\hat{\pi}|n\rangle = n|n\rangle, \ \langle n|m\rangle = \delta_{n,m}, \tag{9}$$

$$\text{(coherent state representation): } \hat{a}|\zeta\rangle = \zeta|\zeta\rangle, \ \langle\zeta|\zeta'\rangle = e^{\bar{\zeta}\zeta'} , \tag{10}$$

where $|\zeta\rangle = e^{\hat{a}^\dagger\zeta}|0\rangle$ and $\int e^{-|\zeta|^2} d^2\zeta|\zeta\rangle\langle\zeta| = 1$, $d^2\zeta = d\mathrm{Re}\zeta d\mathrm{Im}\zeta/\pi$. We specify the coordinates of Hilbert space by $|\vec{x}\rangle$ or $|n\rangle \otimes |\zeta\rangle$. The transformation between them is given by

$$\langle\vec{x}|(|n\rangle \otimes |\zeta\rangle) = \sqrt{\frac{B}{2\pi}} \frac{i^n(z-\zeta)^n}{\sqrt{n!}} e^{-\frac{1}{2}|z|^2 + \bar{z}\zeta}, \quad z = \sqrt{\frac{B}{2}}(x + iy) . \tag{11}$$

Using the Landau level representation one can decompose a Schrödinger state as

$$|\psi\rangle = \sum_n |n\rangle\langle n|\psi\rangle = \sum_n |\psi_n\rangle , \tag{12}$$

where $|\psi_n\rangle$ stands for an n-th Landau level fermion state. From now on we consider only the lowest Landau level (LLL) states and denote $|\psi_0\rangle$ simply by $|\psi\rangle$.

When we decompose the field into independent components, it is convenient to use the angular momentum representation:

$$\hat{a}^\dagger\hat{a}|l\rangle = l|l\rangle, \quad \langle l|l'\rangle = \delta_{l,l'} , \quad \langle l|\zeta\rangle = \frac{\zeta^l}{\sqrt{l!}} . \tag{13}$$

Using (11) we obtain the LLL Schrödinger wave field

$$\psi(\vec{x},t) \equiv \langle\vec{x}|\psi(t)\rangle = \int \langle\vec{x}|(|0\rangle \otimes |\zeta\rangle) d^2\zeta e^{-|\zeta|^2} (\langle\zeta| \otimes \langle 0|)|\psi(t)\rangle = \sqrt{\frac{B}{2\pi}} e^{-\frac{1}{2}|z|^2} \psi(\bar{z},t) ,$$

where

$$(\langle z| \otimes \langle 0|)|\psi(t)\rangle \equiv \psi(\bar{z},t) = \sum_l \frac{\bar{z}^l}{\sqrt{l!}} C_l(t), \quad C_l(t) \equiv (\langle l| \otimes \langle 0|)|\psi(t)\rangle . \tag{14}$$

The quantization conditions are

$$\{\psi(\bar{z}), \psi^\dagger(z')\} = e^{\bar{z}z'}, \qquad \{C_l, C_m^\dagger\} = \delta_{l,m} . \tag{15}$$

3. Symmetries of the Fractional Quantum Hall System [4]

The fractional quantum Hall effect occurs in a system of electrons in two dimensions under a strong magnetic field. Using the second quantized Schödinger wave field for electrons we describe the Hamiltonian of the system as

$$\begin{aligned} H = & \int d\vec{x}\,[\tfrac{1}{2m}|(\vec{p}-\tilde{A}(\vec{x})-\vec{A}(\vec{x},t))\psi(\vec{x})|^2+(V(\vec{x})+A_0(\vec{x},t))\psi^\dagger(\vec{x})\psi(\vec{x})]+ \\ & +\tfrac{1}{2}\int\int d\vec{x}d\vec{x}'v(\vec{x}-\vec{x}'):\psi^\dagger(\vec{x})\psi(\vec{x})\psi^\dagger(\vec{x}')\psi(\vec{x}'):, \end{aligned} \tag{16}$$

where $\psi(\vec{x})$ is an unconstrained Schrödinger wave field, which obeys the quantization condition

$$\{\psi(\vec{x}),\psi^\dagger(\vec{x}')\}=\delta(\vec{x}-\vec{x}')\ . \tag{17}$$

We assume that the external vector potential consists of two parts. One, denoted by $\tilde{A}$, is that of the previous section, which gives rise to a large uniform magnetic field. The other, $A^\mu(\vec{x},t)$ is a small time dependent potential. $V(\vec{x})$ is a confining potential (a strong static potential provided by other materials outside the system). The last term in (16) represents a Coulomb interaction term.

Initially we neglect A^μ and project the field onto the LLL component and express (16) in terms of the constrained field (14). We obtain

$$H=\frac{1}{2}\omega N+\int d^2ze^{-|z|^2}V(z,\bar{z})\psi^\dagger(z)\psi(\bar{z})+H_C\,, \tag{18}$$

where H_C is a Coulomb interaction term:

$$H_C=\frac{1}{2}\int\int d^2ze^{-|z|^2}d^2z'e^{-|z'|^2}v(\sqrt{\frac{2}{B}}|z-z'|):\psi^\dagger(z)\psi(\bar{z})\psi^\dagger(z')\psi(\bar{z}'): \tag{19}$$

and

$$V(z,\bar{z})=V(\vec{x})|_{x=\sqrt{\frac{1}{2B}}(z+\bar{z}),y=\frac{1}{i}\sqrt{\frac{1}{2B}}(z-\bar{z})}\,,\qquad \psi(\bar{z})=\sum_l\frac{\bar{z}^l}{\sqrt{l!}}C_l\ . \tag{20}$$

N is the electron number operator and d^2z is the integration measure:

$$N=\int d^2ze^{-|z|^2}\psi^\dagger(z)\psi(\bar{z})\,,\qquad d^2z=\frac{\mathrm{dRe}z\mathrm{dIm}z}{\pi}\ . \tag{21}$$

It is obvious that the Hamiltonian (18) preserves the electron number. The operator N defined by Eq. (21) is a symmetry generator. Less obvious are the remnants of the translation and rotation symmetry of the original Hamiltonian (16) when the confining potential V is omitted. It is straightforward to show that the following operators commute with the Hamiltonian (18) when $V=0$:

$$L=\int d^2ze^{-|z|^2}\bar{z}z\psi^\dagger(z)\psi(\bar{z})\,,\quad a=\int d^2ze^{-|z|^2}\bar{z}\psi^\dagger(z)\psi(\bar{z})\,,\quad a^\dagger=\int d^2ze^{-|z|^2}z\psi^\dagger(z)\psi(\bar{z}) \tag{22}$$

It is likeweise easy to check that the operators L, a, $a^\dagger$ and N satisfy the following oscillator group Lie algebra commutation relations:

$$[L,a]=-a,\quad [L,a^\dagger]=a^\dagger,\quad [a,a^\dagger]=N,\quad [N,\mathrm{all}]=0\ . \tag{23}$$

These are the symmetry generators of the Hamiltonian (18) without V and are the remnants of the rotation and translation symmetry. Even in the case of $V=0$ the ground state can not be a singlet of this symmetry group, since N is necessarily non-zero. Namely the symmetry is broken and it implies the existence of zero energy modes.

If we further disregard the Coulomb term in (18), the Hamiltonian is just N. All the operators, which commute with N, generate a larger symmetry. In particular, the unitary transformation

$$C_n \to C'_n = u_{nm} C_m = \langle n|\hat{u}|m\rangle C_m \tag{24}$$

is a symmetry [4, 9]. An infinitesimal transformation of $\hat{u}(\approx 1 - i\hat{\xi})$ is generated by a hermitian operator which can be written as $\ddagger\xi(\hat{a}, \hat{a}^\dagger)\ddagger$. Here $\ddagger \cdots \ddagger$ denotes the anti-normal ordering symbol and ξ stands for a real function of $\hat{a}$ and $\hat{a}^\dagger$ in the case when $\hat{a}$ and $\hat{a}^\dagger$ are replaced by z and $\bar{z}$ respectively.

$$\delta C_l = -i\sum_m \langle l|\ddagger\xi(\hat{a}, \hat{a}^\dagger)\ddagger|m\rangle C_m = -i\int \langle l|z\rangle d^2z e^{-|z|^2}\xi(z,\bar{z})\sum_m \langle z|m\rangle C_m \ . \tag{25}$$

On multiplication by $\langle z|l\rangle$ and summation over l one obtains

$$\begin{aligned} \delta\psi(\bar{z}) &= -i\int e^{\bar{z}z'} d^2z' e^{-|z'|^2}\xi(z',\bar{z}')\psi(\bar{z}') = -i\ddagger\xi(\partial_{\bar{z}},\bar{z})\ddagger\psi(\bar{z}) & (26)\\ &= -i[\psi(\bar{z}), \int d^2z' e^{-|z'|^2}\xi(z',\bar{z}')\psi^\dagger(z)\psi(\bar{z}')] \equiv -i[\psi(\bar{z}),\ \rho[\xi]] \ . & (27)\end{aligned}$$

Using (26) (27) and the definition of $\rho[\xi]$ in (27) we find that the algebra satisfied by the generators $\rho[\xi]$ is given by

$$[\hat{\rho}[\xi_1], \hat{\rho}[\xi_2]] = \frac{i}{B}\hat{\rho}[\{\!\{\xi_1, \xi_2\}\!\}] \ , \tag{28}$$

where

$$\{\!\{\xi_1, \xi_2\}\!\} = iB\sum_{n=1}^{\infty}\frac{(-)^n}{n!}\left(\partial_z^n\xi_1\partial_{\bar{z}}^n\xi_2 - \partial_{\bar{z}}^n\xi_1\partial_z^n\xi_2\right) \tag{29}$$

is the Moyal bracket[10]. The Lie algebra (28) is the W_∞ algebra [11], which in this case is the algebra of $U(\infty)$.

4. W_∞ Gauge Transformations and Electromagnetic Interactions [12]

In the previous section we have found the symmetries of the system in the case when A^μ is neglected. In this section we restore A^μ and discuss the response of LLL electrons to the electromagnetic interaction. For simplicity we ignore the Coulomb interaction.

Let $\psi(\vec{x},t)$ be an unconstrained field, i.e. $\psi(\vec{x},t) = \sum_n \psi_n(\vec{x},t)$. The Lagrangian is invariant by gauge transformations

$$\delta\psi(\vec{x},t) = -i\Lambda(\vec{x},t)\psi(\vec{x},t), \qquad \delta A^\mu(\vec{x},t) = \partial^\mu\Lambda(\vec{x},t) \ . \tag{30}$$

Since

$$\Lambda(\vec{x},t)\psi(\vec{x},t) = \langle\vec{x}|\Lambda(\hat{a} - i\hat{\pi}^\dagger, \hat{a}^\dagger + i\hat{\pi}, t)|\psi(t)\rangle,$$

the gauge transformations cause the inter Landau level transformations. Nevertheless we demand that the effective electromagnetic interaction Lagrangian of LLL electrons be gauge invariant. This follows from the fact that the effective Lagrangian can in principle be obtained from the original Lagrangian (16) by integrating out the components of electron field, which correspond to the higher Landau levels [6]:

$$Z[A]=Z[A+\partial\Lambda]=\int\cdots\int\prod_n \mathcal{D}\psi_n\mathcal{D}\bar{\psi}_n e^{i\int\mathcal{L}(\psi,A)}=\int\cdots\int\mathcal{D}\psi_0\mathcal{D}\bar{\psi}_0 e^{i\int\mathcal{L}'(\psi,A)}\ .$$

Let us first consider the following projected gauge transformation:

$$\delta\psi(\bar{z},t)=-i\ddagger\Lambda(\partial_{\bar{z}},\bar{z},t)\ddagger\psi(\bar{z},t)\ .$$

The change in $L=\int d^2ze^{-|z|^2}\psi^\dagger(z,t)(i\partial_t-V(z,\bar{z}))\psi(\bar{z},t)$ is given by

$$\delta L=\int d^2ze^{-|z|^2}\psi^\dagger\psi(\partial_t\Lambda+\frac{1}{B}\{\!\{\Lambda,V\}\!\})\ .$$

Since in the large B limit, $\frac{1}{B}\{\!\{\Lambda,V\}\!\}\approx n_i\partial^i\Lambda,\ \ n_i=\frac{1}{B}\epsilon_{0ij}\partial^jV$, we add $-(A_0+n_iA^i)\psi^\dagger\psi$ to the Lagrangian to compensate for the change. Due to this new term we obtain a new change in L, which is linear in A^μ. We then add an additional term which is quadratic in A^μ to compensate it, and so on.

A smart way to achieve this task is to consider the following W_∞ gauge invariant Lagrangian [13]

$$L=\int d^2ze^{-|z|^2}\psi^\dagger(z,t)(i\partial_t-V(z,\bar{z})-\mathcal{A}(z,\bar{z},t))\psi(\bar{z},t) \tag{31}$$

and W_∞ gauge transformations

$$\delta\psi(\bar{z},t)=-i\ddagger\xi(\partial_{\bar{z}},\bar{z},t)\ddagger\psi(\bar{z},t),\qquad \delta\mathcal{A}=\partial_t\xi+\frac{1}{B}\{\!\{\xi,V\}\!\}+\frac{1}{B}\{\!\{\xi,\mathcal{A}\}\!\}. \tag{32}$$

Since the Moyal bracket is given by the Poisson bracket in the large B limit:

$$\{\!\{\xi_1,\xi_2\}\!\}\ \underset{B\to\infty}{\Longrightarrow}\ \{\xi_1,\xi_2\}_{P.B.}=\epsilon_{0ij}\partial^i\xi_1\partial^j\xi_2\ ,$$

the W_∞ gauge transformatiion (32) can be approximated by

$$\delta\mathcal{A}(\vec{x},t)\approx n_\mu\partial^\mu\xi+\frac{1}{B}\epsilon_{0ij}\partial^i\xi\partial^j\mathcal{A},\qquad n_0=1,\ \ n_i=\frac{1}{B}\epsilon_{0ij}\partial^jV\ . \tag{33}$$

We then demand (32) and (33) to be generated by $\delta A^\mu=\partial^\mu\Lambda$. We set

$$\mathcal{A}\ =\ a_\mu A^\mu+\frac{1}{B}a_{\mu\nu}A^\mu A^\nu+\frac{1}{B}a_{\mu\nu\sigma}A^\mu\partial^\nu A^\sigma+\cdots\ , \tag{34}$$

$$\xi\ =\ \Lambda+\frac{1}{B}\xi_{\mu\nu}\partial^\mu\Lambda A^\nu+\cdots\ . \tag{35}$$

On substituting back into (33) we obtain the following expressions for a set of minimal solutions:

$$\xi_{ij}=-\frac{1}{2}\epsilon_{0ij},\quad a_\mu=n_\mu,\quad a_{\mu\nu}=\frac{1}{2}\epsilon_{0\mu j}\partial^jn_\nu,$$
$$a_{ij0}=\epsilon_{0ij},\quad a_{i0j}=-\frac{1}{2}\epsilon_{0ij},\quad a_{ijk}=\epsilon_{0ij}n_k-\frac{1}{2}\epsilon_{0ik}n_j,\quad \text{all others are zero}\ . \tag{36}$$

$$\mathcal{A}_{\min}=n_\mu A^\mu+\frac{1}{2B}\epsilon_{\mu\nu\rho}A^\mu\partial^\nu A^\rho+\frac{1}{2B}\epsilon_{0ij}\partial^j(A^in_\mu A^\mu)+\cdots \tag{37}$$

The above leads to the following effective Lagrangian [14]:

$$\begin{aligned}L\ =\ &\int d^2ze^{-|z|^2}\psi^\dagger(z,t)(i\partial_t-V(z,\bar{z}))\psi(\bar{z},t)-\\ &-\int d^2ze^{-|z|^2}\psi^\dagger\psi(n_\mu A^\mu+\frac{1}{2B}\epsilon_{\mu\nu\rho}A^\mu\partial^\nu A^\rho+\frac{1}{2B}\epsilon_{0ij}\partial^j(A^in_\mu A^\mu))\ .\end{aligned} \tag{38}$$

As an application we briefly consider edge fermions in a circular droplet (a dot in a strong magnetic field). Let $V(z,\bar{z})$ be a confining potential:

$$V(z,\bar{z}) = \frac{1}{2}\alpha[(x^2+y^2) - R^2] = \frac{\alpha}{B}[\bar{z}z - BR^2/2] \,. \tag{39}$$

Neglecting the Coulomb interaction one obtains the total energy of the system

$$H = \int d^2z e^{-|z|^2}\psi^\dagger(z,t)V(z,\bar{z})\psi(\bar{z},t) = \frac{\alpha}{B}\sum_n (n+1-BR^2/2)C_n^\dagger C_n \,. \tag{40}$$

The ground state is constructed by filling the negative energy states:

$$|G\rangle = \prod_{n=0}^{N-1} C_n^\dagger|0\rangle \,, \qquad N = \frac{BR^2}{2} - \frac{1}{2} \,, \tag{41}$$

where N is the total number of sea electrons. We keep only those electron operators near the Fermi level. We define

$$C_{N+n-1/2} = b_n. \tag{42}$$

With respect to the new vacuum (41), b_n with n positive acts as an annihilation operator while for n negative acts as a hole creation operator. The Hamiltonian (40) is then

$$H = \frac{\alpha}{B}\sum_n (n + \frac{1}{2} - N)C_n^\dagger C_n = \ v_\theta \sum_{n=\text{half integer}} n\, b_n^\dagger b_n \,, \qquad v_\theta = \alpha/B \,. \tag{43}$$

We restrict the range of the sum to be $n \ll N$, although eventually after taking the large N limit we set $-\infty < n < \infty$.

We normal-order the density operator with respect to the new vacuum. This leads to

$$e^{-|z|^2}\psi^\dagger(z)\psi(\bar{z}) = e^{-|z|^2}\sum_{n=0}^{N-1}\frac{|z|^2}{n!} + e^{-|z|^2} : \psi^\dagger(z)\psi(\bar{z}) :\approx \theta(R-r) + \frac{2\pi}{BR}\delta(r-R) : \chi^\dagger(\theta)\chi(\theta) : \tag{44}$$

in the limit of large B and N with finite R. Here $\theta(r)$ is the standard step function and

$$\chi(\theta) = \frac{1}{\sqrt{2\pi}}\sum_n e^{-in\theta} b_n \,, \qquad z = re^{i\theta} \,. \tag{45}$$

Using (43) we can express the Hamiltonian as

$$H = \int d^2z e^{-|z|^2}\psi^\dagger(z,t)V(z,\bar{z})\psi(\bar{z},t) = \int d\theta : \chi^\dagger(\theta)(-iv_\theta\partial_\theta)\chi(\theta) : \tag{46}$$

up to an uninteresting additive constant.

Now it is straightforward to compute the effective Lagrangian (38). We obtain

$$L = L_{\text{bulk}} + L_{\text{boundary}} \,, \tag{47}$$

$$L_{\text{bulk}} = \int_{r<R} dxdy(-A^0 + \frac{1}{2\pi}b + \mathcal{L}_{C-S}) \,, \tag{48}$$

$$L_{\text{boundary}} = \oint d\theta\left[\chi^\dagger((i\partial_t - A_0) - v_\theta(i\partial_\theta - A_\theta))\chi - \frac{v_\theta}{4\pi}A_\theta^2\right] \,, \tag{49}$$

where

$$\mathcal{L}_{C-S} = -\frac{1}{4\pi}\epsilon_{0ij}\left(A^i\partial^0 A^j - 2A^i\partial^j A^0\right) \tag{50}$$

and

$$b = \epsilon_{0ij}\partial^j A^i, \quad A_\theta = \epsilon_{0ij}x^j A^i|_{r=R} \,. \tag{51}$$

We see that L_{bulk} contains a Chern-Simons term. Since the bulk is a compact domain with a boundary, L_{bulk} is not gauge invariant. The boundary Lagrangian (49) is the Lagrangian of chiral fermions interacting with an external electromagnetic field. The anomalies associated with this system and its corresponding bosonization has been discussed by many authors, for example in [15]. Taking these anomalies into account one proves that the Lagrangian (47) is gauge invariant [16, 17, 18].

Acknowledgements

This article is based on the work done with S. Iso, D. Karabali and R. Ray. I thank them for their generous collaboration. This work was supported by the NSF grant PHY90-20495 and the Professional Staff Congress Board of Higher Education of the City University of New York under grant no. 6-63351.

References

1. R. E. Prange amd S. M. Girvin eds. *"The Quantum Hall Effect"*, Springer, New York, 1990.
2. M. Stone ed. *"Quantum Hall Effect"*, World Scientific, Singapore, 1993
3. S. Iso, D. Karabali and B. Sakita, *Nucl. Phys.* **B 388** (1992) 700.
4. S. Iso, D. Karabali and B. Sakita, *Phys. Lett.* **B296** (1992) 143.
5. E. Brézin, V.A. Kazakov and A.B. Zamolodchikov, *Nucl.Phys.* **B338** (1990) 673; G. Parisi, *Phys.Lett.* **B238** (1990) 209; D. Gross and N. Miljković, *Phys.Lett.* **B238** (1990) 217; P. Ginsparg and J. Zinn-Zustin, *Phys.Lett.* **B240** (1990) 333; S. R. Das, A. Dhar, A. Sengupta and S. R. Wadia, *Mod. Phys. Lett* A5 (1990) 891; S. R. Das and A. Jevicki,*Mod. Phys. Lett* A5 (1990) 1639); J. Avan and A. Jevicki, *Phys. Lett.* **B266** (1991) 35; J. Polchinski, *Nucl.Phys.* **B346** (1990) 253; S. R. Das, A. Dhar, G. Mandal and S. R. Wadia; *Int. J. Mod. Phys.* **A7** (1992) 5165, Mod. Phys. Lett. **A7** (1992) 937.
6. R. Ray and B. Sakita *"The Electromagnetic Interactions of Electrons in the Lowest Landau Level" Ann. Phys.* (N.Y.) (in press).
7. S. M. Girvin and T. Jach, *Phys. Rev.* **B29** (1983) 5617.
8. R. Kubo, J. S. Miyake and N. Hashizume, *Solid State Physics* Vol.**17**, 269, eds. F. Seitz and D. Turnbull, Academic Press 1965
9. A. Capelli, C. Trugenberger and G. Zemba, *Nucl. Phys.* **B396** (1993) 465.
10. J. E. Moyal, *Proc. Camb. Phyl. Soc.* **45** (1949) 99
11. J. Hoppe, MIT Ph. D. Thesis (1982); J. Hoppe and P. Schaller, *Phys. Lett.* **B237** (1990) 407; C.N. Pope, L.J. Romans and X. Shen *"A Brief History of W_∞"* in Strings 90, ed. R. Arnowitt et al (World Scientific 1991) and references therein.
12. B. Sakita, *Phys. Lett.* **B315** (1993) 124.
13. This kind of gauge transformation was considered and used by Das, Dhar, Mandal and Wadia in the context of $c = 1$ matrix model: S. R. Das, A. Dhar, G. Mandal and S. R. Wadia; *Int. J. Mod. Phys.* **A7** (1992) 5165, Mod. Phys. Lett. **A7** (1992) 937.

14. I was informed by Shizuya that he also reached a similar electromagnetic interaction Lagrangian by extending his formulation: K. Shizuya, *Phys. Rev.* **B45** (1992) 11143.
15. J. Sonnenschein, *Nucl Phys.* **B 309** (1988) 752.
16. X.G. Wen, *Phys. Rev.* **B 43**, 11025, (1990)
17. M. Stone, *Ann. Phys.* (N.Y.), **207** (1991) 38; Int. J. Mod. Phys. **B5** (1991) 509.
18. J. Fröhlich & T. Kerler, *Nucl. Phys.* **B354** (1991) 365.

Matrix-Model Approach to Electron Correlations in a Strong Magnetic Field

S. Hikami

Department of Pure and Applied Sciences, University of Tokyo,
Meguro-ku, Komaba, Tokyo 153, Japan

Abstract. The lowest Landau level of the two dimensional electron in a strong magnetic field is investigated by the random matrix model technique. We discuss the universality of the electron correlation of the edge state for the arbitrary polynomial measure form of the eigenvalue, based upon the large N limit of the matrix formulations. The edge state is described by the $c = 1$ conformal field theory. The relation of the matrix model to the 2D quantum gravity is discussed. It is shown that a matrix model which describes the lowest Landau level state with an appropriate interaction becomes a new model of the 2D quantum gravity coupled to matter field with the central charge $c = 1$.

1. Introduction

The two dimensional electron system in a strong magnetic field is known experimentally to exhibit the quantum Hall effect. The edge state or the boundary state shows interesting one dimensional behavior while the bulk state is described by an incompressible liquid state. These bulk and edge state are explained by Laughlin's wave function [1]. The Laughlin's wave function becomes 2D coulomb gas partition function by the exponentiation and it is also related to some random matrix models with Gaussian eigenvalue distribution [2]. The position of electron is given by the two dimensional coordinate in the complex plane and then the related random matrix model has complex eigenvalues with a Gaussian distribution [3]. This correspondence between the 2D lowest Landau level electron system and a complex matrix model have been noted by several authors [4-6]. Indeed, the advantage of the reformulation by a matrix model is noticeable. It explains the incompressible liquid state of quantum Hall effect and it gives the scheme of the evaluation of electron correlations in the edge state. The edge state with interesting excitations and correlations is described by the $c = 1$ conformal field theory (CFT) [7-9]. The relation between the lowest Landau level and the $c = 1$ CFT theory have been also discussed recently by several authors [10,11].

The aim of the present paper is to discuss further the complex matrix model which is relevant to the lowest Landau level and to investigate the universalities of the correlation functions. The universality of the Hermitian matrix model, which has real eigenvalues, has been recently studied [12]. Our present matrix model has a simple structure and we prove the universality of correlation functions for the distribution deviated from Gaussian form.

The matrix model has been studied intensively as a model of 2D quantum gravity theory. This correspondence is realized at the critical value of the coupling constant, where the density of state shows a new singularity at the edge point. We investigate this relationship for our models and we find that a matrix model which describes the lowest Landau level model becomes a new model of the 2D quantum gravity coupled to matter field with the central charge $c = 1$.

Springer Series in Solid-State Sciences, Vol. 118
Correlation Effects in Low-Dimensional Electron Systems
Editors: A. Okiji and N. Kawakami

2. Correlation Functions in the Lowest Landau Level

Laughlin's wave function is given by

$$|\phi_m|^2 = \Pi_{i<j}|z_i - z_j|^{2m} \exp(-\sum_{i=1}^{N} z_i^* z_i) , \tag{1}$$

where z_i represents the coordinate of two dimensional electrons ($i = 1, ..., N$) in a complex plane $z_i = x_i + iy_i$. We consider these complex values z_i as eigenvalues of a rank N complex matrix M. The probability distribution $P(M)\mu(dM)$ becomes

$$P(M)\mu(dM) = \Pi_{i,j} dM_{i,j} dM_{i,j}^* \exp(-\mathrm{Tr} M^* M) , \tag{2}$$

in which we represent M by a complex matrix X by $M = XEX^{-1}$, where the column of X is eigenvector of M. E is a diagonal matrix with $E = diag(z_1, ..., z_N)$. Then we have $tr(M^*M) = tr[(X^*X)^{-1}E^*(X^*X)E]$ and we integrate out the part of the matrix X [3]. The joint probability distribution P becomes

$$P(z_1, ..., z_N) = C\Pi_{i<j}|z_i - z_j|^2 \exp(-\sum |z_i|^2). \tag{3}$$

Thus P reduces to the $m = 1$ integer quantum Hall case as (1). The coefficient C is determined by the normalization condition as $C = (\pi^N \Pi j!)^{-1}$ [2,3]. The n-point correlation functions are given by

$$R_n(z_1, ..., z_n) = \frac{N!}{(N-n)!} \int P(z_1, ..., z_N) \Pi_{i=n+1}^{N} dx_i dy_i. \tag{4}$$

These quantities are easily evaluated by the Vandermonde determinant, and expressed by [2],

$$R_n(z_1, ..., z_n) = \pi^{-n} \exp(-\sum_{i=1}^{n} |z_i|^2) \det[K_N(z_i, z_j)]_{i,j=1,...,n} , \tag{5}$$

where we have

$$K_N(z_i, z_j) = \sum_{l=0}^{N-1} (z_i z_j^*)^l / l!. \tag{6}$$

The density of eigenvalue becomes simply

$$R_1(z) = \pi^{-1} \exp(-|z|^2) \sum_{l=0}^{N-1} \frac{|z|^{2l}}{l!} , \tag{7}$$

which gives the constant behavior for $|z| \ll \sqrt{N}$ and shows a rapid decrease at $|z| \approx \sqrt{N}$. This density of state explains the incompressibility of liquid state of quantum Hall effect and the existence of the edge state at $|z| \approx \sqrt{N}$ [7]. Two-point correlation function $R_2(z_1, z_2)$ is given by (5),

$$R_2(z_1, z_2) = R_1(z_1)R_1(z_2) - |\rho(z_1, z_2)|^2 , \tag{8}$$

$$\rho(z_1, z_2) = \frac{1}{\pi} \exp(-\frac{1}{2}|z_1|^2 - \frac{1}{2}|z_2|^2) \sum_{l=0}^{N-1} \frac{(z_1 z_2^*)^l}{l!}. \tag{9}$$

This $\rho(z_1, z_2)$ becomes same as noninteracting one-particle Green function for the lowest Landau level in the large N limit, which appears in various physical situations, including the localization [13] and the superconducting fluctuation [14] in the lowest Landau level.

Cappelli et al. [7] have discussed that the edge state correlation $\rho(z_1, z_2)$ becomes the order of $1/\sqrt{N}$ for $|z_1| = |z_2| = \sqrt{N}$ which is also predicted by the $c = 1$ CFT. The effective Hamiltonian of the edge state is given by the relativistic chiral fermion in 1+1 dimension,

$$H_R = \frac{1}{4m}\int_0^{2\pi R} dx\psi^\dagger(-i\partial_x - \frac{R}{l^2})\psi + c.c. \ , \tag{10}$$

where R is the radius of the disk and l is the cyclotron radius $l = \sqrt{2\hbar c/eB}$. Two-point correlation becomes with $z_1 = R\exp(i\theta_1)$ and $z_2 = R\exp(i\theta_2)$,

$$\langle\psi^\dagger(Re^{i\theta_1})\psi(Re^{i\theta_2})\rangle = \sqrt{\frac{2}{\pi l^2}}\frac{1}{2\pi R}e^{i(N-1)(\theta_2-\theta_1)}\frac{e^{i\theta_2}}{e^{i\theta_2}-e^{i\theta_1}}. \tag{11}$$

This shows the algebraic long-range order at the boundary. The primary fields h becomes $1/2$ and $1/R$ dependence appears according to CFT [7].

3. Universalities of Correlation Function

We generalize the Gaussian distribution in (3) by considering a potential $V(z_i)$ which consists of arbitrary polynomial of the eigenvalues up to order $|z|^{2K}$,

$$V(z_i) = |z_i|^2 + \sum_{l=2}^{K}\frac{g_{2l}}{N^{l-1}}|z_i|^{2l} \ , \tag{12}$$

$$P(z_i, ..., z_N) = C\Pi|z_i - z_j|^2\exp(-\sum_{i=1}^{N}V(z_i)). \tag{13}$$

It is possible to construct the orthogonal polynomial $P_l(z)$, since we have

$$\int\int e^{-V(z)}z^l z^{*m}dxdy = C_l\delta_{l,m} \ . \tag{14}$$

Note that our potential (12) is different from $\mathrm{tr}(M^*M)^l$ which appears in the usual multi-polynomial complex matrix model [15]. It is known that any complex matrix is decomposed into two Hermitian matrices $M = M_1 + iM_2$ and the Haar measure is written by the product of the measure of these two matrices. Our measure in (3) is different from this due to the representation of $M = XEX^{-1}$. Nevertheless we consider it is worthwhile to investigate this generalized distribution (13) as a first step to include the interaction term since the eigenvalues represent physically the two dimensional coordinate of electrons. A conventional complex matrix model with a gauge field will be discussed later.

The inclusion of the higher terms in $V(z_i)$ modifies the constant density of state. The density of state becomes

$$R_1 = \exp(-V(z))\sum_{l=0}^{N-1}\frac{|z|^{2l}}{C_l} \ , \tag{15}$$

where C_l is given by (14). We use the saddle point method for the large N limit. The coefficient C_{N-1} is evaluated for the large N,

$$C_{N-1} = \int_0^\infty \exp(-V(\lambda))\lambda^{2N-1}d\lambda d\theta = 2\pi\tilde{\lambda}^{2N-1}e^{-V(\tilde{\lambda})} \ , \tag{16}$$

where $\tilde{\lambda}$ is given by the saddle point equation,

$$-2\tilde{\lambda}-\sum 2l\frac{g_{2l}}{N^{l-1}}\tilde{\lambda}^{2l-1}+\frac{2N}{\tilde{\lambda}}=0. \tag{17}$$

Taking $|z|=\tilde{\lambda}$, we have

$$R_1(z)=\frac{1}{\tilde{\lambda}} \tag{18}$$

which is order of $1/\sqrt{N}$.

We find that the saddle point method is effective for obtaining the density of state in a finite N [16], although the Riemann-Hilbert integral equation in the large N limit can not be solved analytically as Hermitian case [17]. We evaluated $R_1(z)$ for a finite N , taking only g_4 term as an interaction term. We find that the obtained curve $R_1(z)$ by the saddle point method is consistent with the numerical value of (15).

Two-point correlation function on the edge state is derived similarly,

$$\rho(z_1,z_2)=\exp(-\frac{1}{2}V(z_1)-\frac{1}{2}V(z_2))\sum_{l=0}^{N-1}\frac{(z_1z_2^*)^l}{C_l}. \tag{19}$$

We put $z_1=\lambda e^{i\theta}$, $z_2=\lambda e^{i\theta_2}$ and using the expression of (16), we find the term in the sum of (19) becomes,

$$\frac{e^{-V(\lambda)}\lambda^{2l}e^{i(\theta_1-\theta_2)l}}{\int e^{-V(\lambda)}\lambda^{2l+1}d\lambda d\theta}=\frac{1}{2\pi}\frac{e^{il(\theta_1-\theta_2)}}{\tilde{\lambda}}, \tag{20}$$

where $\tilde{\lambda}$ is a saddle point for a fixed l, and we consider $\lambda=\tilde{\lambda}$ at the edge state. Taking the sum in (19), we have

$$\rho(z_1,z_2)=\frac{1}{2\pi\tilde{\lambda}}\frac{e^{i(N-1)(\theta_1-\theta_2)}}{1-e^{-i(\theta_1-\theta_2)}}. \tag{21}$$

When the potential is $V(\lambda)=\lambda^2+g_4\lambda^4/N$, then the saddle point value of $\tilde{\lambda}$ becomes

$$\tilde{\lambda}=\sqrt{N}\sqrt{\frac{\sqrt{1+8g_4}-1}{4g_4}}. \tag{22}$$

Our result (21) corresponds to the previous expression (11) for the Gaussian ensemble. Since the density of state is modified by the general potential, two-point correlation function is also changed but it is propotional to the density of state. Our result shows that apart from this density of state, it is universal. Furthermore, we note that n-point correlation function is also written by the combination of $K_N(z_i,z_j)\exp(-|V(z_i)|^2/2-|V(z_j)|^2/2)$, which is order $1/\sqrt{N}$ at the edge state. Therefore, the n-point correlation functions are determined by the one and two-point correlation functions, and this is a property of Gaussian statistics. The modification of the potential of $V(z)$ does not change the situation of the Gaussian statistics for the correlations. We find that the angular dependence of the correlation $\rho(z_1,z_2)$ of (21) is consistent with the Green function of the edge state derived by the macroscopic theory [18].

4. 2D Gravity Coupled to Matter Field with Central Charge c

Matrix model with an interaction term as $(g_4/N)\mathrm{tr}M^4$ shows a new sigularity at a critical value g_{4c} in the density of state or in the free energy. This critical behavior describes the closed string behavior. The string embeded in d-dimensional space is described by a matrix model, which realizes 2D quantum gravity coupled to matter field with a central charge c ($c = d$) [19]. Our lowest Landau level model becomes interesting with respect to this critical behavior, since it has g_4 term in (12). We put simply $g_l = 0$, for $l > 4$. To avoid the divergence in the integral, we need an infinitesimal positive coefficient g_6 for example.

We take the partition function as the integral of $P(z_1, ..., z_N)$ in (13) and define the free energy F by the logarithm of it. Using the Vandermonde determinant [2], we write

$$F = -\ln \int \Pi dx_i dy_i \exp[-\sum V(z_i)] \Pi_{l=2}^{N} |z_l|^{2(l-1)}. \tag{23}$$

In the large N limit, by the help of saddle point method, we obtain $\lambda_l = z_l$,

$$F = \sum \lambda_l^2 + \frac{g_4}{N} \sum \lambda_l^4 - \sum (l-1) \ln \lambda_l^2 \ , \tag{24}$$

where λ_l is determined as $\lambda_l^2 = N(-1 + \sqrt{1 + 8g_4(l-1)/N})/4g_4$, and the free energy becomes in the large N limit,

$$F = N^2 \int_0^1 dt[-\frac{1}{4g}(-1+\sqrt{1+8gt}) - \frac{1}{16g}(-1+\sqrt{1+8gt})^2 + t\ln(N(1-\sqrt{1+8gt})/-4g)]. \tag{25}$$

It is interesting to note that (25) is same as free energy of the branched polymer [20] if we do not take the integral. But after taking the integration, we find the string susceptibility γ_{st} becomes $-1/2$, which coincides with the pure gravity value. The string susceptibility γ_{st} is defined by

$$F \approx (g_4 - g_{4c})^{2-\gamma_{st}}. \tag{26}$$

Since we are considering a bulk free energy F, the result $\gamma_{st} = -1/2$, which is same as a complex matrix model [14], is reasonable.

To obtain the model with $\gamma_{st} = 0$ as expected for 2D quantum gravity coupled to matter field $c = 1$, we investigate another model for the lowest Landau level, which is gauge invariant. As same as a superconducting fluctuation model which is described by the Ginzburg Landau model restricted to the lowest Landau level [13], we consider the Ginzburg Landau model with the generalization of the complex order parameter ϕ to a rank N complex matrix $\tilde{\phi}_{i,j}$. The free energy of this new matrix model is given by

$$F(\tilde{\phi}) = \frac{1}{2m} \mathrm{Tr} |(-i\nabla_\mu - eA_\mu)\tilde{\phi}|^2 + \alpha \mathrm{Tr} |\tilde{\phi}|^2 + \frac{\beta}{2N} \mathrm{Tr} |\tilde{\phi}|^4 \ , \tag{27}$$

where $\mu = x, y$. Subjecting on the lowest Landau level in a strong magnetic field with a gauge choice $A = (0, xB, 0)$, the order parameter is expanded by the harmonic oscillator form in x coordinate [14],

$$\tilde{\phi}_{i,j} = \sum_q a_{i,j}(q)(L_y)^{-1/2}(eB/\pi)^{1/4} \exp(iqy) \exp[-\frac{eB}{2}(x - \frac{q}{eB})^2] \ , \tag{28}$$

where $a_{i,j}(q)$ is a rank N complex matrix and L_y is the length of the system. We consider two dimensional system. The free energy is given by the integration of two dimensional space coordinates,

$$\begin{aligned} F \ &= \ \alpha_B \sum_q \mathrm{Tr} |a(q)|^2 + \sum_{q_i} \frac{\beta}{2L_y N} (\frac{eB}{2\pi})^{1/2} \exp[-\frac{1}{2eB}(\sum q_i^2 - \frac{1}{4}(\sum q_i)^2)] \\ &\times \ \delta_{q_1+q_2, q_3+q_4} \mathrm{Tr}[a(q_1)^* a(q_3) a(q_2)^* a(q_4)] \ , \end{aligned} \tag{29}$$

where $\alpha_B = \alpha + (e/4m)B$. We find a nonlocal interaction in this one dimensional momentum representation. We denote $eB\beta/4\pi\alpha_B^2$ by g. The perturbation of this coupling constant g reduces to the Gaussian integral, and we have a factor $1/TG$, where T is a number of Euler path on each Feynman diagram and G is a symmetry factor. The perturbation of the coupling constant g for the free energy in the large N limit is easily obtained by the evaluation of these factors. The diagrams become planar in the large N limit. It is expanded in the powers of g up to order g^8 as

$$\begin{aligned} \frac{F}{N^2} &= 2g - 17g^2 + 248.888g^3 - 4687.466g^4 + 102344.3g^5 - 2464055.7g^6 \\ &+ 63603713.1g^7 - 1729741366.0g^8. \end{aligned} \tag{30}$$

The numerical estimate for γ_{st} is obtained by the the ratio method employed in [21,22]. We have no inverse logarithmic singularity, which is observed in the usual $d = 1$ Hermitian matrix model [17]. The analysis is free from the logarithmic singularity in the ratio method, and it gives precisely $\gamma_{st} = 0$ with $g_c = -1/40.0$. Thus 2D quantum gravity coupled to matter field with $c = 1$ is realized by this complex matrix model with a gauge field in the lowest Landau level. The central charge $c = 1$ may be understood by the one dimensional representation of (29) with the gauge choice $A = (0, xB, 0)$. Although we used this gauge, the physical correlation function should be gauge independent. Our result $\gamma_{st} = 0$ means the lowest Landau level described by a rank N matrix model indeed belongs to the class with the central charge $c = 1$. We have also obtained the series expansion of order $1/N^2$ of the correction to the free energy (30), and found that it satisfies the scaling law.

5. Discussion

In this article, we discussed the matrix model representation of the lowest Landau level N electrons. The space positions of N electrons are expressed by the N eigenvalues of a complex matrix. We have investigated the interaction terms and their effects on the electron correlations. In the large N limit, the correlation becomes universal as discussed on the edge state. This is related to the universality in the matrix model theory in the large N limit. We have argued that the saddle point method for the eigenvalues becomes a strong tool to understand these universality on the correlations. The forms of interactions, which appear in 2D electron system, are numerous, but from the matrix model point of view, there exist universality classes in the large N limit.

We have also investigated a matrix model of the lowest Landau level electrons with Ginzburg- Landau type interaction in the large N limit, which manifests the critical behavior of 2D quantum gravity coupled to matter field with central charge $c = 1$. This model is closely related to 2D superconductor fluctuation model and the vortex lattice melting. It is easy to extend this model to $c > 1$ case. For example, $c = 2$ case is obtained by introducing the degree of z coordinate by considering three dimensional Ginzburg-Landau model with a strong magnetic field. Or we consider coupled two matrices similar to matrix model of [21,22,23]. These matrix models are important since they have new fixed point at the critical coupling constant. For the discussion of the universality, we have to know the renormalization group flow which goes to a stable fixed point. We hope the matrix model approach becomes a strong tool for understanding the electron correlation. The details will be presented elsewhere.

References

[1] R. B. Laughlin, Phys. Rev. Lett. **50**, 1395 (1983).

[2] M. L. Mehta, " Random Matrices and the Statistical Theory of Energy Levels", Academic Press 1967, New York.

[3] J. Ginibre, J. Math. Phys. **6**, 440 (1965).

[4] D. J. Callaway, Phys. Rev. **B43** 8641 (1991).

[5] I. Kogan, A. M. Perelomov and G. W. Semenoff, Phys. Rev. **B45**, 12084 (1992).

[6] A. Cappelli, C. A. Trugenberger and G. R. Zemba, Phys. Lett. **B306**, 100 (1993).

[7] A. Cappelli, G. V. Dunne, C. A. Trugenberger and G. R. Zemba, Nucl. Phys. **B398**, 531 (1993).

[8] X. G. Wen, Phys. Rev. **B41**, 12838 (1990).

[9] M. Stone, "Quantum Hall Effect", World Scientific, 1992, Singapore, and references therein.

[10] G. Cristofano, G. Maiella, R. Musto and F. Nicodemi, "Theoretical Aspects of Quantum Hall Effect and Two Dimensional CFT", preprint 1991.

[11] S. Iso, D. Karabali and B. Sakita, Nucl. Phys. **B388**, 700 (1992).

[12] E. Brézin and A. Zee, Nucl. Phys. **B402**, 613 (1993).

[13] S. Hikami, Phys. Rev. **B29**, 3726 (1984).

[14] E. Brézin, A. Fujita and S. Hikami, Phys. Rev. Lett. **65**, 1949 (1990).

[15] T. R. Morris, Nucl. Phys. **B356**, 703 (1991).

[16] S. Hikami, Physica **A** in press, (preprint hep-th.9309005).

[17] E. Brézin, C. Itzykson, G. Parisi and J. B. Zuber, Commun. Math. Phys. **59**, 35 (1978).

[18] D.H. Lee and X.G. Wen, Phys. Rev. Lett. **66**, 1765 (1991).

[19] V. A. Kazakov, " fields, strings and critical phenomena", Les Houches 1988, edited by Brézin and J. Zinn-justin, P.369. North-Holland, Amsterdam (1990).

[20] S. Hikami and E. Brézin, J. Phys. **A12** 759 (1979).

[21] E. Brézin and S. Hikami, Phys. Lett. **B283** 203 (1992).

[22] S. Hikami and E. Brézin, Phys. Lett. **B295** 209 (1992).

[23] S. Hikami, Phys. Lett. **B305** 327 (1993).

Narrow-Channel Fractional Quantum Hall States and Their Projection into One Dimension – Non-Tomonaga-Luttinger-Liquid Behavior

D. Yoshioka and M. Ogata

Institute of Physics, University of Tokyo, Komaba,
Meguro-ku, Tokyo 153, Japan

Abstract. Two-dimensional electron system confined into a narrow channel by a parabolic potential is considered. Laughlin state is realized in the presence of a strong magnetic field even in such a confined system. Conditions for realization of the Laughlin state is clarified numerically. The present system can be considered as a one-dimensional system. Laughlin state is investigated as an asymptotically exact ground state of a model Hamiltonian. It is shown that Laughlin state is neither a Fermi liquid nor Tomonaga-Luttinger-liquid for certain choice of a parameter.

1. Introduction

Recently it has become possible to fabricate good quality quantum wire with sub-micron width. In the first part of this paper we review our results on the ground state of two-dimensional electron system confined in such a wire in a strong magnetic field [1]. In the absence of the confining potential electrons are known to form the fractional quantum Hall (FQH) state at Landau level filling factor $\nu = p/q$, where p and q are integers with q being odd [2]. It is interesting to know if the FQH state is realized in a quantum wire, and if at all, under what condition it will be. For this purpose we have considered two-dimensional electrons which is confined in one direction by a parabolic potential well. This model is convenient since single electron state is obtained analytically. We diagonalize Hamiltonian of a small system to obtain the ground state. Overlap of this ground state with Laughlin wave function [3] is calculated to search for the condition for realization of the FQH state.

Another notable aspect of our model is that it is possible to consider it as a strongly correlated one-dimensional system. The single-particle part of our Hamiltonian has the same parabolic dispersion as the one-dimensional electrons in a continuous space. The only difference from the genuine 1-d system is the interaction part of the Hamiltonian. Our system has a slightly complicated form, although the difference may not be very important. Thus this is a good model to study one-dimensional highly correlated electron systems. In the second part of this paper we review our results on this kind of 1-d system [4]. Namely, we can construct a Hamiltonian of one-dimensional electron system where Laughlin state is the asymptotic ground state. When projected into one-dimension, Laughlin state has a parameter, which is the Larmor radius ℓ in the original two-dimensional system. When considered as a ground state of one-dimensional system, the wave function changes character as this parameter is changed: It changes continuously from typical Tomonaga-Luttinger liquid at $\ell = 0$ to non-Tomonaga-Luttinger liquid when ℓ exceeds some critical value. Thus we have pointed out that a state which is neither a Fermi liquid nor a Tomonaga-Luttinger liquid is possible as a ground state of one-dimensional strongly correlated system.

Springer Series in Solid-State Sciences, Vol. 118
Correlation Effects in Low-Dimensional Electron Systems
Editors: A. Okiji and N. Kawakami

2. Model

We consider two-dimensional electrons with mass m_e on the xy-plane in an external confining potential. The potential is flat in the x-direction, but parabolic in the y-direction. The length of the system in the x-direction is L_x, and we impose a periodic or antiperiodic boundary condition in this direction. A strong magnetic field, B, is applied in the z-direction. Thus the single-particle part of the Hamiltonian is given by

$$\mathcal{H}_0 = \frac{1}{2m_e}[(p_x+eBy)^2+p_y^2]+\frac{1}{2}m_e\omega_0^2y^2. \tag{1}$$

Here ω_0 gives the strength of the confining potential. The eigenstate of this Hamiltonian is easily obtained as

$$\psi_{k_x,n}(\mathbf{r}) = \exp(ikx - \frac{\tilde{y}^2}{2\ell^2})\mathrm{H}_n(\frac{\tilde{y}}{\ell}), \tag{2}$$

where $\ell = \sqrt{\hbar/m_e\Omega}$ is the effective Larmor radius, $\Omega = \sqrt{\omega_0^2+\omega_c^2}$ and $\omega_c = eB/m_e$. The wave function is localized in the y-direction around $\tilde{y} = 0$, where $\tilde{y} = y + (\hbar k\omega_c)/(m_e\Omega^2)$. The wave number in the x-direction is quantized to be $k = (2\pi/L_x)m$, with quantum number m being an integer (half-odd integer) under the periodic (antiperiodic) boundary condition. The momentum $\hbar k$ determines the center coordinate of the wave function in the y-direction. The function H_n is the Hermite polynomial.

This state has the energy, $E_{k,n} = (n+1/2)\hbar\Omega + (\hbar^2k^2\omega_0^2)/(2m_e\Omega^2)$. Hereafter we assume that the magnetic field is strong enough that we are allowed to consider only the states with $n = 0$, and spin of electrons is completely polarized. Thus our system has the same dispersion as free 1-d electrons with effective mass m_e^* which can be quite heavy: $m_e^* = (\Omega^2/\omega_0^2)m_e$. We introduce a creation (annihilation) operator $a_k^\dagger(a_k)$ corresponding to the wave function $\psi_{k,0}(\mathbf{r})$. Then the second quantized single-electron Hamiltonian is written as follows with the exclusion of a constant term:

$$\mathcal{H}_0 = \sum_m \frac{\hbar^2k^2}{2m_e^*}a_k^\dagger a_k. \tag{3}$$

Without interaction, our system is equivalent to that of free, spinless fermions. When there are n_e electrons, they occupy states with $|k| \le k_\mathrm{F}$, where $k_\mathrm{F} = \pi n_e/L_x$. In real space it means that electrons condense into a strip along the x-axis. Inside of the strip every state is singly occupied, namely the filling factor ν is unity there. The width of the strip is about $2k_\mathrm{F}\ell^2$.

Now we introduce interaction between electrons $V(\mathbf{r})$. Considering the boundary condition in the x-direction, we truncate the potential such that $V(x,y) = 0$ for $|x| > L_x/2$. Then the interaction part of the Hamiltonian is given as

$$\mathcal{H}_\mathrm{int} = \frac{1}{2}\sum_k\sum_p\sum_q f(p,q)a_{k+q}^\dagger a_{k-p}^\dagger a_{k-p+q}a_k, \tag{4}$$

where

$$f(p,q) = \frac{1}{\sqrt{2\pi}\ell L_x}\int_{-L_x/2}^{L_x/2}\mathrm{d}x\int_{-\infty}^{\infty}\mathrm{d}yV(x,y)\exp[-iqx-\frac{1}{2\ell^2}(y+\eta\ell^2p)^2-\frac{\eta^2\ell^2q^2}{2}], \tag{5}$$

where $\eta = \omega_c/\Omega$. Thus our Hamiltonian $\mathcal{H} = \mathcal{H}_0 + \mathcal{H}_\mathrm{int}$ looks like that of an one-dimensional system with slightly complicated interaction part. The ground state of this model depends on the interaction $V(\mathbf{r})$ and on a dimensionless parameter $\lambda = 2k_\mathrm{F}\ell$, which is proportional to linear density in the x-direction.

3. Coulomb Interaction

In this section we investigate the ground state of a system with Coulomb interaction $V(\mathbf{r}) = e^2/4\pi\epsilon r$ between electrons. Here we have three dimensionless parameters: $\lambda = 2k_F\ell$, $\eta = \omega_c/\Omega$, and V_c/E_0, where $V_c = e^2/4\pi\epsilon\ell$ gives the measure of the Coulomb interaction and $E_0 = \hbar^2/2m^*\ell^2$ is chosen as the unit of energy. Among these parameters we will set $\eta = 1$ in the following calculation, assuming $\omega_c >> \omega_0$.

When $V_c << E_0$, electrons condense into $\nu = 1$ state at the bottom of the confining potential. As we increase V_c/E_0 at a fixed λ, the region with finite electron density expands laterally, and the filling factor decreases from unity. In the course of expansion various ground states are realized. Laughlin's FQH states at $\nu = 2/3$ and $1/3$ are possible candidates for the ground states. Here we focus our attention on where these FQH states are realized. We calculate the ground state wave function numerically for six electrons system, and compare it with Laughlin's wave function.

The results of such calculation are shown in Fig. 1. The calculation has been done for ten choices of λ. The ranges of V_c/E_0 where the ground state has overlap more than 90% with the $\nu = 1$, $2/3$, and $1/3$ FQH states are indicated by vertical lines. There are two cases for the termination of the ranges. One case is indicated by a filled circle, which means that there is a cross-over of the ground state: the new ground state has little overlap with the FQH state. The other case is indicated by an open circle, where the overlap with the FQH state decreases to 90% before the cross-over occurs. In the former case the boundary is definite and the ground state within the range has overlap more than 94%. There are successive first order phase transitions as the Coulomb potential is changed. We notice such a case is realized when the system can be considered as two-dimensional: The width of the occupied region becomes larger than L_x, when λ becomes larger than 4.76 and 3.75 for state at $\nu = 2/3$ and for that at $\nu = 1/3$, respectively. Since the overlap is sufficiently large, we know that the Laughlin state is realized even in the presence of the confining potential, and that there is no remarkable edge reconstruction in these states.

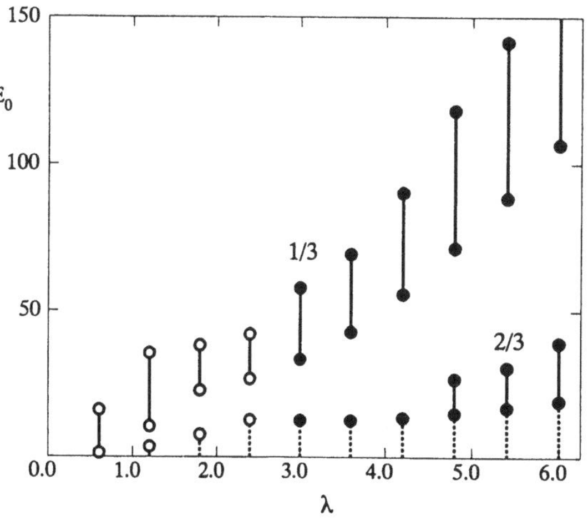

Fig.1 Ranges of V_c/E_0 where the ground state has more than 90% overlap with Laughlin's wave function are shown by solid lines for $\nu = 1/3$ and $2/3$ and by dashed lines for $\nu = 1$.

As λ become smaller, the 2/3-state vanishes, and the boundary of the 1/3-state becomes obscure. This is because the system becomes one-dimensional. In the next section we see that even the 1/3-Laughlin state has density distribution similar to that of $\nu = 1$-state.

4. Model One-dimensional System

We can consider the Hamiltonian $\mathcal{H} = \mathcal{H}_0 + \mathcal{H}_{\rm int}$ as that for one-dimensional spin-less electrons in the absence of the magnetic field. On the other hand, we know that if we choose model pseudo-potential the Laughlin state is the exact ground state of the interaction part of the Hamiltonian [5]. Thus in the limit of $m_e^* \to \infty$ our Hamiltonian with model pseudo-potential has an exact ground state, which is Laughlin state projected into one-dimension. In this case λ is no longer related to the line density of electrons, and it becomes a free parameter.

For model pseudo-potential, $f(p,q)$ in $\mathcal{H}_{\rm int}$ has the following form,

$$f(p,q) = \exp[-\frac{1}{2}(p^2+q^2)\ell^2]\sum_{n=0}^{\infty} V_n \int_{-\infty}^{\infty} \frac{dx}{2\pi} \mathrm{L}_n(x^2+\ell^2 q^2)\exp[-\frac{1}{2}(x+\mathrm{i}\ell p)^2], \tag{6}$$

where L_n is the Laguerre polynomial. We can verify that if $V_i > 0$ for $i \leq m-1$, and $V_i = 0$ for $i \geq m$, the following n_e-electron wave function $\Psi_m(\lambda)$ is the ground state of $\mathcal{H}_{\rm int}$ with eigenvalue zero:

$$\Psi_m(\lambda) = \sum_{\{k_i\}} A(k_1, k_2, ..k_{n_e}) \exp(\frac{1}{8}\lambda^2 \sum_{i=1}^{n_e} \frac{k_i^2}{k_{\rm F}^2}) \prod_{i=1}^{n_e} a_{k_i}^\dagger |0\rangle, \tag{7}$$

where $|0\rangle$ is the vacuum. The coefficients A's are determined such that $\Psi_m(0)$ coincides with the second quantized form of the n_e-electron wave function

$$\Phi_m(x_1, x_2, ..., x_{n_e}) = \prod_{i>j} \{\exp[\frac{2\pi \mathrm{i}}{L}(x_i - X)] - \exp[\frac{2\pi \mathrm{i}}{L}(x_j - X)]\}^m, \tag{8}$$

where x_i's are electron coordinates and $X = \sum_i x_i / n_e$ is the center of mass. The function Ψ_m is nothing but the Laughlin wave function [6] projected into one-dimension. Without $\mathcal{H}_0$ the ground state is degenerated. For example, we can increase the momentum of every electron uniformly, and still get the ground state. The degeneracy is removed by $\mathcal{H}_0$. Although our wave function eq.(7) is not the eigenstate of $\mathcal{H}_0$, it is characterized as the state where the expectation value of $\mathcal{H}_0$ becomes minimum among the ground states of $\mathcal{H}_{\rm int}$, as far as λ is not so large. Thus in the limit that $V_i >> k_{\rm F}^2/2m_e^*$ for $i \leq m-1$, our wave function asymptotically becomes the exact unique ground state of the total Hamiltonian.

The wave function eq.(8) is the same as the wave function for $1/r^2$ potential that is considered by Sutherland [7]. This state shows typical TL-liquid behavior [8]. The momentum distribution shows a power law singularity at $k_{\rm F}$:

$$n(k) \simeq n_{k_{\rm F}} - C[\log|1-k/k_{\rm F}|]^\beta |1-k/k_{\rm F}|^\alpha \mathrm{sgn}(k-k_{\rm F}), \tag{9}$$

and the structure factor diverges at $2k_{\rm F}$:

$$S(k) \simeq S_0 + a_\pm [\log|1-(k/2k_{\rm F})|]^\gamma |1-(k/2k_{\rm F})|^{\eta-1}, \quad (k \simeq 2k_{\rm F}), \tag{10a}$$

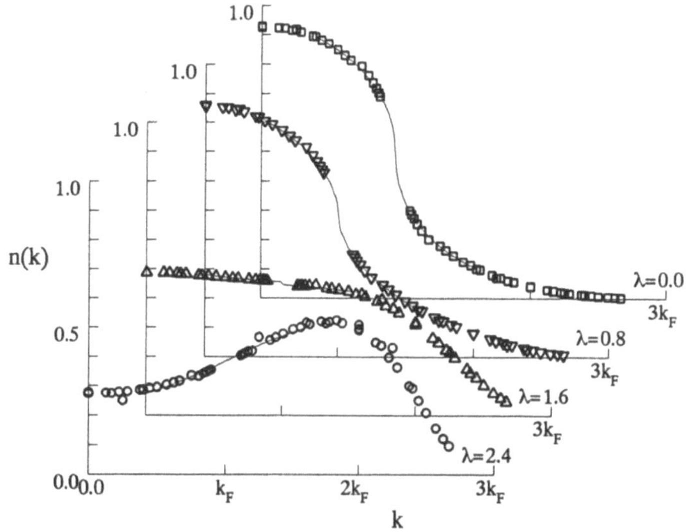

Fig.2 The momentum distribution function $n(k)$ for $\lambda = 0.$, 0.8, 1.6, and 2.4. The data for four to nine electron systems are shown. The thin lines through the data are the result of least square fitting by eq.(9).

or

$$S(2k_F) \simeq S_0 + a[\log n_e]^\gamma n_e^{\eta-1}, \quad (k = 2k_F). \tag{10b}$$

The exponent α and η are given by the TL-liquid parameter K to be $\alpha = (K+1/K-2)/2$ and $\eta = 2K$, while K is calculated by asymptotic Bethe ansatz [9] to be $K = 1/m$. We confirmed these relations numerically by calculating $n(k)$ and $S(k)$ for finite size systems up to 12 electrons at $m = 3$. The least square fitting gives $\alpha \simeq \eta \simeq 2/3$. Although $\gamma = -0.002$ indicates absence of the log correction in $S(k)$, $\beta \simeq 0.5$ is not negligible. This may indicate inadequacy of the system size to investigate the delicate log correction in $n(k)$.

Now we consider the properties of the wave function for $\lambda > 0$. For finite λ there is no asymptotic Bethe ansatz, so we must calculate the exponents numerically. The wave function for finite number of electrons ($n_e < 10$) is obtained by diagonalization of the Hamiltonian rather than manipulating the wave function itself, since the former method is easier. We investigated wave functions with $m = 3$ and $m = 5$. Since the results are qualitatively the same, we concentrate on the case of $m = 3$ here.

For large enough λ ($\lambda > 10$) the interaction becomes short ranged in the momentum space, and the momentum distribution ceases to be continuous: for fixed number of electrons every m-th momentum is occupied by an electron. However, for small λ ($\lambda < 3$) the momentum distribution lies on a single smooth curve for all the choices of electron number at each λ. Figure 2 shows as examples the momentum distribution functions for several choices of λ. Thus we calculate the exponents and coefficients in eq.(9) and eq.(10b) by fitting the numerical data. It is evident from Fig.2 that the singularity at k_F becomes smaller as λ becomes larger, and the distribution eventually becomes smooth at k_F. We find that the exponent α remains almost constant for $\lambda < 1.6$, but the coefficient

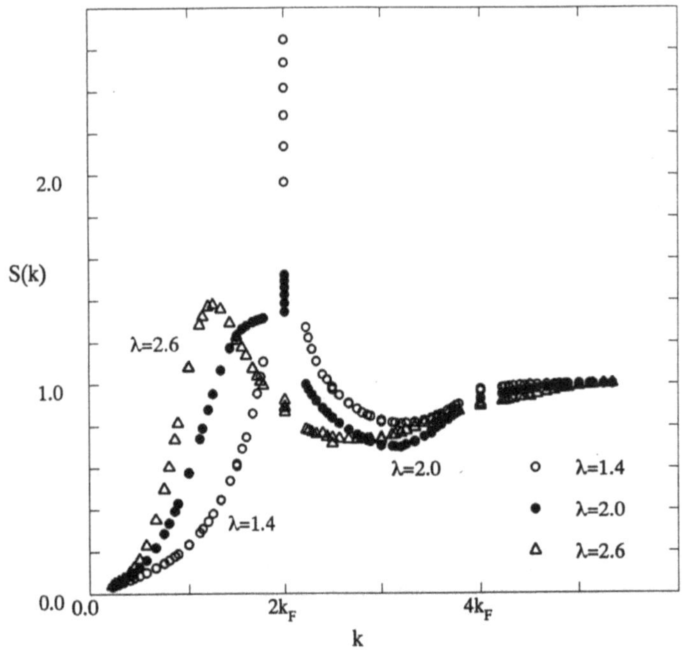

Fig.3 The structure factor $S(k)$ for the choice of $\lambda = 1.4$, 2.0, and 2.6 are shown by open circles, closed circles and triangles, respectively.

C vanishes linearly around $\lambda = 1.6$. For $\lambda > 1/6$ there seems to be no singularity at k_F, so $n(k)$ at $\lambda > 1.7$ is quite different from the TL-liquid behavior.

Corresponding to this qualitative difference in $n(k)$ above and below $\lambda \simeq 1.6$, the structure factor $S(k)$ changes its shape. Figure 3 shows $S(k)$ at several choices of λ. The exponent η remains to be 2/3 as far as $\lambda = 2.0$, while the exponent γ remains to be small ($|\gamma| < 0.01$). However, fitting becomes difficult for $\lambda > 2.0$, where a shoulder develops in the low momentum side of the $2k_F$ peak. The shoulder eventually becomes a peak around $k = 1.25k_F$, while the singularity at $2k_F$ almost vanishes. This behavior of $S(k)$ is again quite different from that of the TL-liquid.

5. Conclusion

I this paper we investigated two-dimensional electron system in a strong magnetic field confined into a narrow channel. In the first part of this paper we examined the ground state, and find that FQH state expressed by Laughlin's wave function is realized for suitable choice of parameters. It seems that even in the presence of the confining potential Laughlin wave function gives good approximation to the true ground state. However, it is possible that slight but qualitative difference exists at the edges. Detailed investigation around the edges is left for future investigation.

In the latter half of this paper we regard the present system as one-dimensional system in zero magnetic field. Knowledge of the original two-dimensional system allowed us to construct a model Hamiltonian for which there is an asymptotically exact ground state wave function. We have seen that behavior quite different from that of the TL-liquid is observed for suitable choice of the dimensionless parameter λ. It should be emphasized that the transition from the TL-liquid to the non-TL-liquid is continuous as can be seen from eq(7), which shows that the wave function has the common form for any value of λ. Further investigation of the ground state properties in the non-TL-liquid region will give us better understanding of the one-dimensional strongly correlated systems.

Acknowledgements

This work is supported by Grant-in-Aid for Scientific Research on Priority Areas "Computational Physics as a New Frontier in Condensed Matter Research" (04231105) from the Ministry of Education, Science and Culture. DY thanks Aspen center for Physics where part of this work was done.

References

[1] D. Yoshioka, J. Phys. Soc. Jpn. **62**, 839 (1993).

[2] For a review see *The Quantum Hall Effect,* edited by R.E. Prange and S.M. Girvin (Springer-Verlag, New York, 1990).

[3] R.B. Laughlin, Phys. Rev. Lett. **50**, 1395 (1983).

[4] D. Yoshioka and M. Ogata, J. Phys. Soc. Jpn. **62**, 2990 (1993).

[5] F.D.M. Haldane, in Ref.[2] Chap.8.

[6] D.J. Thouless, Surf. Sci. **142**, 147 (1984).

[7] B. Sutherland: Phys. Rev. A**4**, 2019 (1971), *ibid.* A**5**, 1372 (1972).

[8] J. Sólyom: Adv. Phy. **28**, 201 (1979).

[9] N. Kawakami and S.-K. Yang: Phys. Rev. Lett. **67**, 2493 (1991).

Chern-Simons Ginzburg-Landau Theory of the Fractional Quantum Hall System with Edges

N. Nagaosa[1] *and M. Kohmoto*[2]

[1]Department of Applied Physics, University of Tokyo, Tokyo 113, Japan
[2]Institute for Solid State Physics, University of Tokyo, Tokyo 106, Japan

Abstract. The superfluidity analogy of the fractional quantum Hall effect via the Chern-Simons Ginzburg-Landau (CSGL) theory is extended to the system with edges. The advantage of this method is, in addition to its simplicity without employing any advanced ideas, that the bulk and edge properties can be discussed in a unified fashion. The electromagnetic response of the system is discussed by deriving the effective action for the external vector potential A_μ. The Green's function of the physical electron is also studied, and the results by Wen is rederived.

1. Introduction

The seach for the order parameter or broken symmetry characterizing the incompressible quantum Hall Liquid has been accomplished reasonably well by the Chern-Simons Ginzburg-Landau (CSGL) theory where the original electrons (fermions) are transmuted into hard core bosons by introducing CS gauge field [1]. The CS gauge flux cancels the external magnetic field on average when the filling fraction $\nu = \frac{1}{2k+1}$ (k:integer). Hence the bosons are expected to show the superfluidity, which is translated into the incompressibility and $\sigma_{xy} = \nu\frac{e^2}{h}$. On the other hand, the important roles played by the edge states have been stressed for finite size sample [2-6]. Some authors discussed that the integer quantum Hall effect can be explained only by the currents flowing in the edge channels [4]. However one must be careful to extract only the one dimensional edge channel because its existence is closely linked to the bulk properties. In the case of fractional quantum Hall liquid, the edge wave can be interpreted as the surface wave of the incompressible liquid, like water, where the shape of the boundary is vibrating. Another analogous phenomenon is the supercurrent of a superconductor. Although the Meissner effect is a bulk effect, it is sustained by the supercurrent which is localized near the surface within the penetration depth. Hence the physics at the edge can not be separated naively from the bulk properties, and it is desirable to develope a theoretical framework to treat both the bulk and edge on an equal footing.

2. Model

We start with the following Lagrangian density L[1]

$$L = L_a + L_\phi \quad , \tag{1}$$

where

$$L_a = \frac{1}{4\eta}\varepsilon^{\mu\nu\lambda} a_\mu \partial_\nu a_\lambda, \tag{2}$$

Springer Series in Solid-State Sciences, Vol. 118
Correlation Effects in Low-Dimensional Electron Systems
Editors: A. Okiji and N. Kawakami

$$L_\phi = \phi^\dagger(i\partial_t + A_0 + a_0 + U(x) + \frac{1}{2m}(-i\nabla - \vec{A} - \vec{a})^2)\phi - \frac{1}{2}\int d^2x'\delta\rho(x)V(x-x')\delta\rho(x'). \quad (3)$$

The role of CS term L_a is to attach the CS gauge flux $b = \nabla \times \vec{a}$ to the particle as

$$\frac{1}{2\eta}\nabla \times \vec{a} = \rho = \phi^\dagger\phi \ , \quad (4)$$

which can be easily obtained from $\delta_{a_0}S = 0$. When the statistical angle η is $(2k+1)\pi$ (k:integer), the boson-gauge flux composite particle represents the fermion through the Aharonov-Bohm effect. The interaction $V(x-x')$ between the bosons is assumed to be short ranged, i.e., $V(x-x') = V_0\delta(x-x')$, for simplicity. The action S is given by

$$S = \int dt \int_{\text{whole space}} d^2\vec{r}\ L, \quad (5)$$

where the spatial integral extends over the whole space. It can be easily seen that the action S in eq.(5) is gauge invariant w.r.t. the following "internal" gauge transformation,

$$\phi \to \phi e^{i\varphi} \qquad (\phi^\dagger \to \phi^\dagger e^{-i\varphi}), \quad (6a)$$

$$a_\mu \to a_\mu + \partial_\mu\varphi. \quad (6b)$$

We now assume that the electrons are confined to the finite region S due to the potential $U(x)$. Then our action is

$$S = S_a + S_\phi = \int dt\Big\{\int_{\text{wholespace}} d^2\vec{r}\ L_a + \int_S d^2\vec{r}\ L_\phi\Big\} \ , \quad (7)$$

where $U(x) = 0$ inside the sample S. It is noted that the CS gauge field a_μ is defined even outside the sample, which guarantees the gauge invariance of S_a. Now let us devide A_μ as

$$A_\mu = \bar{A}_\mu + \delta A_\mu, \quad (8)$$

where $\bar{A}_\mu$ corresponds to the external magnetic field while δA_μ is the test field to study the response of the system. Correspondingly the CS gauge field is decomposed as

$$a_\mu = \bar{a}_\mu + \delta a_\mu, \quad (9)$$

where $\bar{a}_\mu$ is the mean field part and δa_μ is the fluctuating part. For the filling fraction $\nu = \frac{1}{2k+1} = \frac{\pi}{\eta}$

$$B = \nabla \times \vec{\bar{A}} = -\nabla \times \vec{\bar{a}} = -2\eta\bar{\rho} \ , \quad (10)$$

where $\bar{\rho}$ is the average density. Hence $\bar{A}_\mu + \bar{a}_\mu = 0$ at the mean field level, and ϕ field will show the superfluidity. Expressing ϕ in terms of the amplitude A and the phase θ as $\phi = Ae^{i\theta}$ ($\phi^\dagger = Ae^{-i\theta}$) and integrating over the small amplitude fluctuation δA, we obtain the effective Lagrangian for the phase field θ in the Meissner phase as

$$L_\phi = L_\theta = \frac{1}{V_0}(\partial_t\theta - A_0 - a_0)^2 - \frac{\rho_s}{2m}(\nabla\theta - \vec{A} - \vec{a})^2, \quad (11)$$

where ρ_s is the superfluidity density. Equations (2),(7) and (11) constitute our starting point for the following analysis. Before going into the explicit calculation, let us demonstrate the usefulness of the superfluidity analogy by showing that the Meissner effect actually guarantees $\sigma_{xy} = \nu\frac{e^2}{h}(= \frac{\nu}{2\pi}$ in the present dimensionless unit $e = c = \hbar = 1)$ for the geometry with edges. The sample we consider is the annular film with inner ($r = r_1$) and outer ($r = r_2$) limbs [2,3]. Besides the uniform magnetic field B, the additional magnetic flux Φ is confined to region $r < r_1$. Consider the line integrals $I_A(C) = \oint_C \vec{A} \cdot d\vec{\ell}$, $I_a(C) = \oint_C \vec{a} \cdot d\vec{\ell}$ along the closed loop C inside the sample. The bose field ϕ is coupled with the sum $A_\mu + a_\mu$, and the corresponding line integral is

$$I_{A+a}(C) = I_A(C) + I_a(C) = \oint_C (\vec{A} + \vec{a}) \cdot d\vec{\ell}.$$

When the bose field is in the Meissner phase, it shows the flux quantization $I_{A+a}(C) = n\phi_0 = 2\pi n$ (n:integer) and the valleys with different n's are separated by the potential barrier of the macroscopic size. Now let us increase Φ adiabatically [2,3]. The voltage drop along the contour C is given by

$$\Delta V = \oint_C \vec{E} \cdot d\vec{\ell} = \int (\nabla \times \vec{E})_z \cdot dS = -\int \frac{\partial B}{\partial t} dS = -\frac{d\Phi}{dt} \tag{12}$$

which is independent of the choice of C. The current J across the film is obtained as follows. Because $I_{A+a}(C)$ is fixed,

$$\frac{dI_{A+a}(C)}{dt} = \frac{dI_A(C)}{dt} + \frac{dI_a(C)}{dt} = \frac{d\Phi}{dt} + \frac{dI_a(C)}{dt} = 0. \tag{13}$$

On the other hand, from eq.(4) $\frac{dI_a(C)}{dt}$ is 2η times the total charge flowing across the contour C per unit time, i.e., $\frac{dI_a(C)}{dt} = 2\eta J$. Then

$$J = \frac{1}{2\eta}\frac{dI_a(C)}{dt} = -\frac{1}{2\eta}\frac{d\Phi}{dt} = \frac{1}{2\eta}\Delta V. \tag{14}$$

The off-diagonal conductance σ_{xy} is

$$\sigma_{xy} = \frac{J}{\Delta V} = \frac{1}{2\eta} = \frac{\nu}{2\pi} = \nu\frac{e^2}{h}\quad, \tag{15}$$

where e and $\hbar$ is recovered in the last equation. The above argument is robust against disorder or imperfections, i.e., it assumes only the long ranged phase coherence and the rigidity of the bose system.

3. Boundary Condition and Edge Mode

We now present the simplest discussion of the edge excitation. It should be noted that eqs.(2) and (11) and hence (7) are alredy quadratic w.r.t. θ, a_μ and A_μ, and the equations of motion derived from $\delta S = 0$ are linear. Explicitly $\delta_a S = 0$ gives

$$\frac{1}{2\eta}(\partial_x a_y - \partial_y a_0) = \frac{2}{V_0}(\partial_t - a_0 - A_0) \equiv j_0, \tag{16a}$$

$$\frac{1}{2\eta}(\partial_y a_0 - \partial_t a_y) = -\frac{\rho_s}{m}(\partial_x\theta - a_x - A_x) \equiv -j_x, \tag{16b}$$

$$\frac{1}{2\eta}(\partial_x a_0 - \partial_t a_x) = \frac{\rho_s}{m}(\partial_y\theta - a_y - A_y) \equiv j_y, \tag{16c}$$

while $\delta_\theta S = 0$ gives

$$\partial_t j_0 + \partial_x j_x + \partial_y j_y = 0, \tag{17a}$$

$$\vec{j} \cdot \vec{n}|_{\partial S} = 0, \tag{17b}$$

where $\vec{n}$ is the unit vector normal to the boundary ∂S, and this boundary condition eq.(17b) expresses that the current can not flow to the outside of the sample. It is noted that $\delta_a S = 0$ gives no boundary condition because S_a is the integral over the whole space and S_ϕ does not include the derivative of a_μ. The continuity equation (17a) is derived from eqs.(16), and hence we have three independent equations for four quantities θ, a_0, a_x, and a_y. This uncertainty is related to the gauge degrees of freedom, and we now impose the Coulomb gauge condition

$$\partial_x a_x + \partial_y a_y = 0, \tag{18}$$

besides eqs.(16). It is easy to derive the following equations from eqs.(16) and (18),

$$\frac{\rho_s}{m}\nabla^2\theta = \frac{1}{2\eta}\partial_t b, \tag{19a}$$

$$\left[\partial_t^2 - \frac{V_0\rho_s}{2m}\nabla^2 + \left(\frac{2\eta\rho_s}{m}\right)^2\right] b = 0, \tag{19b}$$

where $b = \partial_x a_y - \partial_y a_x$ and A_μ is put to be zero (no external field). Now let us look for the solution localized near the edge. For simplicity we assume that the sample is the semi-infinite film extending for $x < 0$ with the boundary ∂S being $x = 0$. Putting $b(\vec{r}, t) = Ae^{\alpha x}e^{-\mathrm{i}\omega t + \mathrm{i}k_y y}$ into eq.(19b) gives

$$\left[-\omega^2 + \frac{V_0\rho_s}{2m}(k_y^2 - \alpha^2) + \left(\frac{2\eta\rho_s}{m}\right)^2\right] A = 0, \tag{20}$$

which determines α(= the inverse of the penetration depth λ). In the case of small ω and k_y, α is given by

$$\alpha^2 = \frac{1}{\lambda^2} \cong \frac{(4\eta)^2\rho_s}{2mV_0}, \tag{21}$$

and $\theta(\vec{r}, t) = -\frac{m}{2\eta\rho_s\alpha^2}\mathrm{i}\omega Ae^{\alpha x}e^{-\mathrm{i}\omega t + \mathrm{i}k_y y}$ which is also localized near the edge. In the Coulomb gauge, the spatial component $\vec{a}$ of the gauge field is determined from b and given by $a_x(\vec{r}, t) = -\frac{\mathrm{i}k_y}{\alpha^2}Ae^{\alpha x}e^{-\mathrm{i}\omega t + \mathrm{i}k_y y}$, $a_y(\vec{r}, t) = -\frac{1}{\alpha}Ae^{\alpha x}e^{-\mathrm{i}\omega t + \mathrm{i}k_y y}$. Then the boundary condition eq.(17b) becomes

$$-\frac{m}{2\eta\rho_s\alpha}\mathrm{i}\omega + \frac{1}{\alpha^2}\mathrm{i}k_y = 0, \tag{22}$$

which determines the acoustic dispersion relation

$$\omega = ck_y, \tag{23}$$

with the sound velocity $c = \frac{2\eta\rho_s}{m\alpha}$. Using eq.(21) one can easily obtain $c^2 = \frac{\rho_s V_0}{2m}$ which is the sound velocity of the Bogoliubov mode (Goldstone mode) of the bose system. In summary we have obtained the phonon-like sound mode localized near the edge. The dispersion relation takes only one possibility $\omega = ck_y$ from the usual expression $\omega^2 = c^2k_y^2$, and is called "chiral" [5,6]. It is the collective mode of the charge density (gauge flux b) and the phase of the condensed bose field. The penetration depth λ in eq.(21) includes the boson-boson interaction V_0 and is different from the magnetic penetration depth λ_m of the charged bose system. Assuming $mV_0 \sim 1$, which is reasonable for the hard core bosons, λ is of the order of magnitude of the magnetic length $\ell = \sqrt{\frac{\hbar}{eB}}$.

4. Effective Action for A_μ

The effective action $S[A_\mu]$ for the external elctromegnetic field A_μ is obtained by integrating over the phase field θ and the CS gauge field a_μ as

$$Z[A_\mu] = Z[A_\mu = 0] e^{\mathrm{i}S[A_\mu]} = \int Da_\mu D\theta e^{\mathrm{i}S[\theta, a_\mu, A_\mu]}. \tag{24}$$

The integral over a_μ contains the infinite gauge volume, and we have to fix the gauge. Here we take the temporal gauge ($a_0 = 0$) because S_a does not contain the spatial derivative of $\vec{a}$ in this gauge, and $\vec{a}$ can be integrated out for each $\vec{r}$ separately. The condition $a_0 = 0$ does not eliminate the gauge degrees of freedom completely because it allows the time-independent gauge transformation. However the divergence coming from the corresponding gauge volume is absorbed into $Z[A_\mu = 0]$ and the gauge invariant quantity $S[A_\mu]$ is finite. After integrating over $\vec{a}$, one obtains

$$S[\theta, A_\mu] = \int \frac{d\omega}{2\pi} \int_S d^2\vec{r} \Big\{ \sum_{\alpha,\beta=x,y} J_\alpha(\vec{r}, -\omega) \Gamma_{\alpha,\beta}(\omega) J_\beta(\vec{r}, \omega) + \frac{1}{V_0} J_0(\vec{r}, -\omega) J_0(\vec{r}, \omega) \Big\}, \tag{25}$$

where $J_\mu = \partial_\mu \theta - A_\mu$ and the matrix $\Gamma(\omega)$ is given by

$$\Gamma(\omega) = \frac{(\frac{\rho_s}{2m})^2}{(\frac{\rho_s}{2m})^2 - (\frac{\omega}{2\eta})^2} \begin{bmatrix} \frac{2m}{(4\eta)^2 \rho_s}\omega^2, & \frac{1}{4\eta}\mathrm{i}\omega \\ -\frac{1}{4\eta}\mathrm{i}\omega, & \frac{2m}{(4\eta)^2\rho_s}\omega^2 \end{bmatrix} \equiv \begin{bmatrix} \alpha(\omega), & \beta(\omega) \\ -\beta(\omega), & \alpha(\omega) \end{bmatrix}. \tag{26}$$

Next we integrate over θ. This is achieved by solving $\delta_\theta S[\theta, A_\mu] = 0$, and putting the solution $\theta(A_\mu)$ back into $S[\theta, A_\mu]$ to obtain $S[A_\mu] = S[\theta(A_\mu), A_\mu]$. We now specify the sample to be a semi-infinite film extending for $x < 0$ as in section 3. The equation $\delta_\theta S[\theta, A_\mu] = 0$ is explicitly given as

$$\alpha(\omega)\nabla^2\theta - \frac{\omega^2}{V_0}\theta = \alpha(\omega)\nabla \cdot \vec{A} + \beta(\omega)\nabla \times \vec{A} - \frac{i\omega}{V_0}A_0 \equiv f \;, \tag{27a}$$

with the boundary condition

$$\alpha(\omega)(\partial_x\theta - A_x) + \beta(\omega)(\partial_y\theta - A_y) = 0 \quad \text{on } \partial S. \tag{27b}$$

Equation (27b) is the same condition as eq.(17b). Because our system is translationally invariant along the y-axis, we introduce the Fourier transformation for y. Then eqs.(27) become

$$\Big[\alpha(\omega)(\frac{d^2}{dx^2} - k_y^2) - \frac{\omega^2}{V_0}\Big]\theta = f, \tag{28a}$$

$$(\alpha(\omega)\frac{d}{dx} + ik_y\beta(\omega))\theta|_{x=0} = \alpha(\omega)A_x + \beta(\omega)A_y|_{x=0}. \tag{28b}$$

This boundary value problem can be solves by an elementary method, and we finally obtain the action $S_{\mathrm{edge}}[A_\mu]$ localized near the edge as

$$S_{\mathrm{edge}}[A_\mu] = \frac{c}{4}\int \frac{d\omega}{2\pi} \int \frac{dk_y}{2\pi} \frac{\nu}{2\pi} \frac{\omega + ck_y}{\omega - ck_y} (\bar{A}_y(-k_y, -\omega) + \frac{1}{c}\bar{A}_0(-k_y, -\omega))(\bar{A}_y(k_y, \omega) + \frac{1}{c}\bar{A}_0(k_y, \omega)) \tag{29}$$

where

$$\bar{A}_\mu(k_y, \omega) = \frac{1}{\lambda}\int_{-\infty}^{0} e^{x/\lambda} A_\mu(x, k_y, \omega) dx \tag{30}$$

is the average of A_μ near the edge. Equation (29) is equivalent to the form predicted by Wen, and describes the current-current correlation of the one-dimensional edge excitation [5,6].

5. Green's Function of the Electron near the Edge

The statistical transmutation of the field operator is written as [7]

$$\psi(\vec{r}) = \exp\left[-i\frac{\eta}{\pi}\int d\vec{r}'\varphi(\vec{r}-\vec{r}')\rho(\vec{r})\right]\phi(\vec{r}) \ , \tag{31}$$

where $\psi(\vec{r})(\phi(\vec{r}))$ is the annihilation operator of the electron (boson), $\rho(\vec{r})$ is the density, i.e., $\rho(\vec{r}) = \psi^\dagger(\vec{r})\psi(\vec{r}) = \phi^\dagger(\vec{r})\phi(\vec{r})$, and $\varphi(\vec{r}-\vec{r}')$ is the angle of the vector $\vec{r}-\vec{r}'$ measured from the positive x-axis. It is easy to show $\psi(\vec{r})$ obeys the anti-commutation relations of the fermions. Assuming the bose condensation eq.(31) becomes

$$\psi(\vec{r}) \simeq \phi_o \exp\left[i\theta(\vec{r}) - i\frac{\eta}{\pi}\int d\vec{r}'\varphi(\vec{r}-\vec{r}')\rho(\vec{r}')\right] \equiv \phi_o \exp[i\Phi(\vec{r})], \tag{32}$$

where only the phase $\theta(\vec{r})$ of the bose field is retained. By taking the gradient of $\phi(\vec{r})$ we obtain

$$\nabla_{\vec{r}}\Phi(\vec{r}) = \nabla_{\vec{r}}\theta(\vec{r}) - \frac{\eta}{\pi}\int d\vec{r}'\nabla_{\vec{r}}\varphi(\vec{r}-\vec{r}')\rho(\vec{r}') = \nabla_{\vec{r}}\theta(\vec{r}) - \vec{a}(\vec{r}) = \frac{m}{\rho_s}\vec{j}(\vec{r}) \ , \tag{33}$$

where the Φ is related to the current $\vec{j}(\vec{r})$. It is important to note that $j_\mu(\vec{r})$ in eq.(33) is the 2D current density, and is related to the 1D current density $j_\mu^{(1D)}$ by

$$j_\mu(x, k_y, \omega) = \frac{1}{\lambda}e^{x/\lambda}j_\mu^{(1D)}(k_y, \omega). \tag{34}$$

When we consider the electron on the very edge ($x = 0$) of the sample, $j_\mu(0, k_y, \omega) = \frac{1}{\lambda}j_\mu^{(1D)}(k_y, \omega)$. Now eq.(31) represents the correlation function of this 1D current density $j_\mu^{(1D)}$. From this Wen expressed $j_\mu^{(1D)}$ in terms of chiral boson operator ϕ_{chiral} as [5]

$$j_y^{(1D)} = \frac{\sqrt{\nu}}{2\pi}\frac{\partial}{\partial t}\phi_{\text{chiral}} = \frac{\sqrt{\nu}}{2\pi}c\frac{\partial}{\partial y}\phi_{\text{chiral}}, \tag{35a}$$

$$j_0^{(1D)} = \frac{\sqrt{\nu}}{2\pi}\frac{\partial}{\partial y}\phi_{\text{chiral}}. \tag{35b}$$

From eqs.(33) and (35a),

$$\frac{\partial}{\partial y}\Phi = \frac{m}{\rho_s}\frac{1}{\lambda}j_y^{(1D)} = \frac{m}{\rho_s}\frac{c}{\lambda}\frac{\sqrt{\nu}}{2\pi}\frac{\partial}{\partial y}\phi_{\text{chiral}} = \frac{1}{\sqrt{\nu}}\frac{\partial}{\partial y}\phi_{\text{chiral}}, \tag{36}$$

where eq.(21) and $c = \frac{2\eta\rho_s}{m\alpha}$ have been used.
Then

$$\psi(\vec{r}) \sim e^{i\Phi} = e^{i\frac{1}{\sqrt{\nu}}\phi_{\text{chiral}}}, \tag{37}$$

which is exactly the result obtained by Wen, and it is straightforward to see [5,6]

$$\langle\psi^\dagger(t,y)\psi(0,0)\rangle \sim (ct+y)^{-(2k+1)}. \tag{38}$$

As discussed above, the phase of the electron is related to the current-density, and its Green's function is determined by the 1D chiral boson near the edge. It is an interesting problem to study the crossover of the Green's function from the edge to the bulk behavior, which is left for future investigations.

6. Conclusions

We have discussed CSGL theory with edges. The bulk and edge properties can be studied on an equal footing in this method. As an application the edge excitation has been studied, and the chiral acoustic mode is obtained explicitly which shows all the properties predicted by more sophisticated method [5,6]. This method can be applied to more general situations. Actually we can investigate the electromagnetic response of the system with arbitrary shape by solving the boundary value problem for the partial differential equation, i.e., eqs.(27a), (27b). The interaction between edges can be also studied. When the sample is a strip confined in $-L < x < 0$, i.e., quantum wire, we have found that the electromagnetic response always contains both $\omega + ck_y$ and $\omega - ck_y$ as the denominators. This means that the system is always compressible with the right and left going chiral bosons. The detailed discussion of this highly correlated 1D system will be published elsewhere.
The authors thank Y.S.Wu, X.G.Wen and Y.Hatsugai for fruitful discussions.

References

[1] For a review see S.C.Zhang, Int. J. Mod. Phys. B6, 25 (1992).

[2] B.I.Halperin, Phys. Rev. B **25**, 2185 (1982).

[3] D.J.Thouless, in " The Quantum Hall Effect ", ed. R.E.Prange and S.M.Girvin (Springer-Verlag, New York, 1987) Chap.4.

[4] M.Buttiker, Phys. Rev. Lett. **57**, 57 (1986).

[5] X.G.Wen, Phys. Rev. B **41**, 12838 (1990).

[6] X.G.Wen, Phys. Rev. B **43**, 11025 (1991).

[7] G.Semenoff, Phys. Rev. Lett. **61**, 517 (1988).

Part V

2D Correlated Electrons

$Sr_{n-1}Cu_{n+1}O_{2n}$: From One Dimension to Two Dimensions via Trellis Lattices

T.M. Rice[1], *S. Gopalan*[1], *M. Sigrist*[2], *and F.C. Zhang*[3]

[1]Theoretische Physik, ETH-Hönggerberg, CH-8093 Zürich, Switzerland
[2]Paul Scherrer Institut, CH-5232 Villigen PSI, Switzerland
[3]Physics Department (UCTP), ML11, University of Cincinnati, Cincinnati, OH 45221, USA

Abstract. The key elements in all known cuprate superconductors are lightly doped CuO_2-planes. Recently a new homologous series of compounds $Sr_{n-1}Cu_{n+1}O_{2n}$ have been reported in which the planes contain a parallel array of line defects which form a trellis lattice with ladder-segments of the square lattice weakly coupled through triangular line defects. The magnetic properties of undoped compounds will be dominated by the properties of the ladders. Heisenberg $s = 1/2$ ladders can have a spin liquid groundstate with a spin gap if the number of rungs is odd so that a short range RVB groundstate is predicted for such trellis lattices. Using a t-J model to describe the doped material leads to the prediction of a d-wave RVB superconducting groundstate with a large spin gap.

1. Introduction

The simplest model to describe the doping of holes in a $s = 1/2$ antiferromagnet is the so called t-J model. As such it plays a central role in the attempts to construct a microscopic theory for the high-T_c cuprate superconductors. In all the cuprate superconductors the doped CuO_2-planes are the key element and the competition between kinetic energy of the doped carriers and the local antiferromagnetic order of the copper spins is the essential feature captured in the t-J model [1]. It has proved very difficult to reliably analyse the doped t-J model on a square lattice, partly because there are several possible phases in competition namely, the d-wave paired superconducting state, states with incommensurate antiferromagnetic order of spiral or other type and flux states characterized by orbital currents. The case of one dimension (i.e. a t-J chain) has been extensively analysed [2] and much of this meeting is devoted to the special case of one-dimensional models. The connection between the well understood case of one dimension and the poorly understood case of a two-dimensional square lattice is the main theme of this talk.

Recently we have pointed out that a new class of cuprates of the general form $Sr_{n-1}Cu_{n+1}O_{2n}$ which have been reported [3] should have properties intermediate between one and two dimensions [4]. Experiments on the magnetic and electronic properties of these materials would be most interesting to test the ideas discussed here.

2. $Sr_{n-1}Cu_{n+1}O_{2n}$ Compounds

The key feature of these new cuprates is the presence of a parallel array of domain walls, which act to break up the CuO_2 planes into segments in the form of ladders [3]. In a CuO_2 plane the O-atoms form a square lattice and the Cu-atoms sit in the centers of one half of the squares. A domain wall separates two domains with different choices for the Cu sites as illustrated in *Fig. 1.* The domain wall itself has a CuO-structure of a double chain with edge-sharing of the CuO_4 squares. As a result the superexchange across the edge is much

Springer Series in Solid-State Sciences, Vol. 118
Correlation Effects in Low-Dimensional Electron Systems
Editors: A. Okiji and N. Kawakami

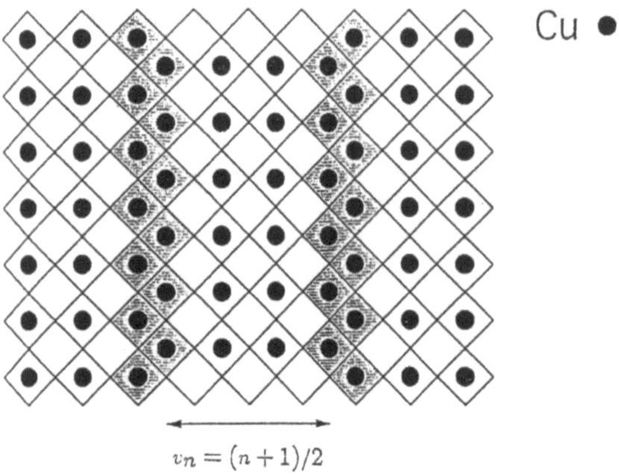

Figure 1. The structure of the Cu-O planes in $Sr_{n-1}Cu_{n+1}O_{2n}$ for n=7 following Hiroi et al. [3]. The shaded region denotes the edge-sharing CuO double chains parallel to the y-axis. An O-atom sits at the corner of each square. The Cu-atoms are shown as large dots.

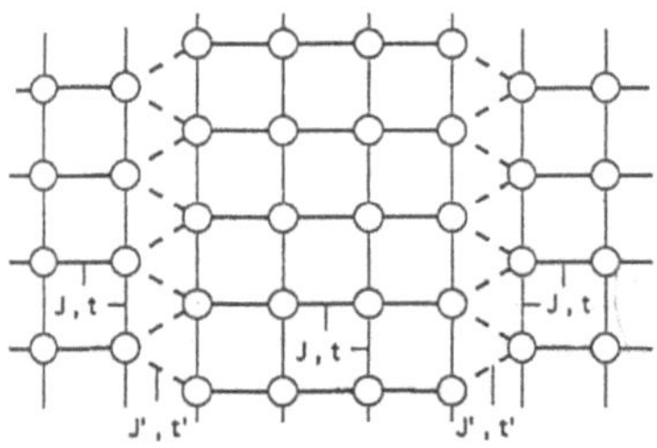

Figure 2. The model Hamiltonian with t-J ladders (solid lines) and interladder t' and J' couplings (dashed lines).

weaker than the strong antiferromagnetic exchange coupling of the 180° Cu-O-Cu bond in the corner-sharing CuO_2-planes. In addition there is a relative displacement of the square lattices of the Cu-ions.

Fig. 2 shows the form of the exchange couplings in the Heisenberg model [4] for a cuprate plane in the $n = 7$ compound ($Sr_6Cu_8O_{14}$). Inside the ladder segments are strong AF-coupling constants J. The ladder segments are staggered with respect to each other and are joined by triangular segments. The coupling constants that connect the ladders, J' are much weaker and ferromagnetic in sign ($| J' | /J \sim 0.1$). The result is an unusual magnetic lattice with a mixture of segments of the usual square lattice in the form of ladders coupled to each other by triangular segments. The whole makes a trellis lattice. The width of the ladder segments depends on the parameter, n, which controls the number of legs in the ladder segments. This number we denote by v_n and $v_n = (n+1)/2$. It can be even or odd and we shall see that there are important differences between the series of compounds with even values of v_n (i.e. $n = 3, 7, 11 \cdots$ corresponding to $v_n = 2, 4, 6 \cdots$) and those with odd values of v_n (i.e. $n = 1, 5, 9 \cdots$ corresponding to $v_n = 1, 3, 5, \cdots$).

The electronic band structure parallels the magnetic structure. In our earlier work [4] we reported on tight-binding calculations of the antibonding band composed of hybridized $3d_{x^2-y^2}$-orbitals on the Cu-ions and $2p_{x(y)}$-orbitals on the O-ions. In the infinite CuO_2-plane this band forms a broad two-dimensional band with predominantly n.n. tight-binding character. The stoichiometric compound has this band half-filled. Hole doping lowers the electron Fermi energy in this band and leads to the one band t-J model in the strong correlation description. The presence of domain walls is a strong perturbation on the band structure. It breaks the single antibonding band into a set of subbands. Again the coupling across the edge sharing CuO_4-squares is much weaker and occurs only through a $2p_\pi$-overlap between neighboring O-atoms. This band structure can be parameterized with a standard hopping matrix element, t, inside the ladder segments and a much weaker matrix element t' ($t'/t \sim 0.1$) on the triangular links between the ladders. The single two-dimensional band is essentially replaced by the ladders which act as a set of quantum wells. The number of subbands is again v_n – the number of legs in the ladder segment.

The simplest description then of the $Sr_{n-1}Cu_{n+1}O_{2n}$-compounds is as a generalized t-J model on the trellis lattice with strong and uniform couplings (t,J) inside the ladders and much weaker values (t', J') on the triangluar links between the ladders.

3. Magnetic Properties: Spin Gaps in Trellis Lattices

The trellis lattices discussed above have ladders whose width or number of rungs depends on the parameter, n, with strong antiferromagnetic interactions within the ladders and weak ferromagnetic interactions within the triangular segments. Further a glance at *Fig. 2* shows that these interactions connecting the ladders through the triangular segments are frustrated. Therefore to a first approximation we can ignore these interactions and consider the magnetic properties as determined by the properties of isolated ladders.

The properties of $s = 1/2$ Heisenberg ladders have been studied recently by several groups [5]-[7]. A single Heisenberg chain has a long range powerlaw fall off of the magnetic correlation $\langle \vec{S}_i \cdot \vec{S}_{i+r} \rangle \propto (-1)^r r^{-1} ln\, r$. However a ladder with two legs and a single rung behaves very differently. The excitation spectrum above the singlet groundstate at low energies consists of a single triplet mode with dispersion $\omega(k)$ with k a wavevector along the ladder [6]. The excitation energy $\omega(k)$ remains positive for all k and has a minimum at $k = \pi$. This minimum determines the spin-gap ω_g $(= \omega(\pi))$ and the finite correlation lengths ξ for the spins along the ladder $\langle \vec{S}_i \cdot \vec{S}_{i+r} \rangle \propto (-1)^r\ e^{-r/\xi}$. The result is a spin liquid groundstate with a spin gap and purely shortrange correlations.

The parameters of the spin liquid are functions of the ratio $J_\perp/J$ between the coupling across the rungs $(J_\perp)$ and along the legs J. However it appears that there is not a critical value of the ratio $J_\perp/J$ but that for all values there is a finite gap. Note the case $J_\perp/J < 0$ scales as a $s = 1$ chain also has a spin gap – the well known Haldane gap. The limit $(J_\perp/J) \to \infty$ clearly corresponds to singlets on each rung and in this case the dispersion $\omega(k) = J_\perp + J\cos(k)$ and $\omega_g = J_\perp - J$ [6]. As the ratio $J_\perp/J$ decreases show that the spingap reduces but remains finite while the dispersion near $k = \pi$ becomes more linear [6]. Only in the limit $J_\perp/J \to 0$ does $\omega_g \to 0$ and the single chain result with a linear dispersion near $k = \pi$ is recovered.

In the present case we are interested in the isotropic limit $J_\perp = J$. In this case an extrapolation of the reuslts of a Lanczos calculation on ladders up to 2×12 in length led Barnes et al. [6] to a value $\omega_g = 0.5J$. The dispersion in $\omega(k)$ has a almost linear variation near $k = \pi$.

Several years ago Sachdev and Bhatt [8] introduced a meanfield scheme to treat such spin liquids. In the present case of a single rang ladder, a representation of the $s = 1/2$ operators in terms of rung singlet, s_i, and triplet, $t_{i\alpha}$, is used [9]. The spin liquid groundstate corresponds to a coherent condensation of singlets $\langle s \rangle$ and triplets $\langle t \rangle$ and the ratio depends on $J_\perp/J$ with $\langle t \rangle \to 0$ as $J_\perp/J \to \infty$. The triplet excitation $\omega(k)$ evolves from the strong coupling form for $J_\perp/J \gg 1$ to a more linear dispersion near $k = \pi$ as $J_\perp/J$ decreases, but the value of the spin gap ω_g is too small in the isotropic limit $J_\perp = J$. This can be improved by introducing a term to correct for the wrong treatment of short range effects in the mean field approximation.

The behavior of ladders with an odd number of legs (i.e. an even number of rungs) is again gapless as is the case of single chain. There is an odd-even effect in the leg number reminiscent of the case half-integer and integer spins in the Haldane chains. The spin gap, ω_g, for even leg number is a strong function of the ladder width. At present there are no numerical determinations of ω_g for wider ladders. Gopalan et al. used the mean field treatment and found a value $\omega_g = 0.07J$ in the isotropic limit for the case of a 3-rung (4-leg) ladder. They show that there is a strong variation of ω_g with the ladder width. Note in the Sachdev-Bhatt scheme the transition to long range AF correlations is signaled by $\omega_g \rightarrow 0$.

Gopalan et al. [9] also considered the trellis lattice within the mean field scheme. As experted the weak and frustrated ferromagnetic interactions, J', on the triangular segments connecting ladders have only a small effect on ω_g. In fact a small increase in the value of ω_g results. This leads to the prediction that the trellis lattices with odd number of rungs in the ladders will be $s = 1/2$ spin liquids and as such realizations of a short range RVB (resonant valence bond) groundstate of the type proposed some years ago by Anderson [1].

4. Hole Doping of Trellis Lattices

We turn now to a discussion of hole doping described in the t-J model. In the case of a single t-J chain, the properties are well studied [2]. In the region of interest, $J/t \sim 1/3$, a Luttinger liquid with predominantly incommensurate magnetic correlations results. Pairing correlations which could lead to superconductivity are suppressed at all hole doping in this parameter range. On the other hand several studies [10],[11] of modified models which, unlike the single Heisenberg chain have a spin gap when undoped, led to more promising results. They found that the spin gap remained for a finite range of doping and evidence for singlet pairing correlations in this region.

Single rung ladders and other ladders with an even number of legs have spin gaps in the undoped state so that we can expect that they are more promising models to examine. In addition a drawback of single chain models is that there can be no analog of the d-wave RVB state, which is not the case here. For these reasons it is very interesting to examine the effect of doping on the ladders.

Sigrist, Rice and Zhang [12] have carried out mean field calculations for the single rung ladders using a fermion scheme. The difficult local constraint has been approximated by separate Gutzwiller renormalization factors, g_t, for the kinetic and, g_J magnetic parts [13]. This scheme is known to reproduce quite well results of variational Monte Carlo calculation which treat the local constraint exactly. Starting with the undoped or half-filled case, it is well known that the most favored state without long range AF-order is a state which can be written in various forms as a projected fermion wavefunction, namely as a Affleck-Marston flux phase [14], or a d-wave RVB state etc. All these fermion wavefunctions when projected to satisfy the local constraint are equivalent through local SU(2)-gauge transformations (e.g. see [13]) to the identical spin wavefunction. The lack of long range AF order in this spin wavefunction means that it is not a good representation for the infinite plane which has AF long range order. In the case of the single rung ladder, the AF order is strictly short range and so this type of mean field is much more promising. Sigrist et al. [12] have carried out an analogous mean field calculation for the single rung ladder. In this case the quasiparticle spectrum $E^2_{\vec{k}} = \varepsilon(\vec{k})^2 + \Delta(\vec{k})^2$ has an energy gap and is doubly degenerate. The wave vector is $\vec{k}(= k_x, k_y)$ with continuous values for k_x, $-\pi \leq k_x \leq \pi$ and two values for k_y namely $k_y = 0, \pi$. The degeneracy in the quasiparticle spectrum is between the values $k_y = 0, \pi$ corresponding to even (bonding) or odd (antibonding)

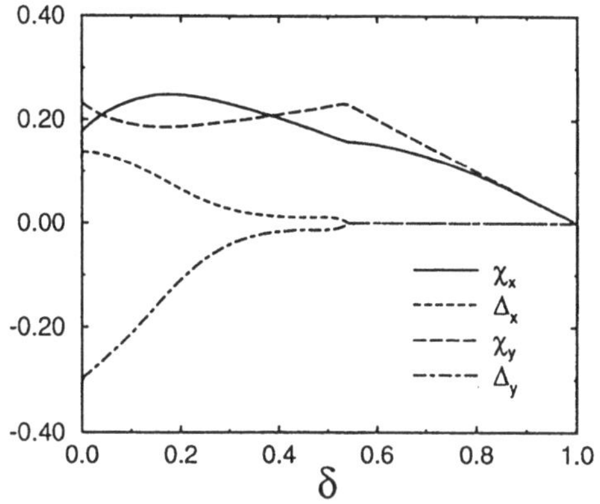

Figure 3. The mean fields on the legs (x) and rungs (y) of a single rung ladder as a function of the doping concentration for $(J)J_\perp = J = 0.3t$.

combinations on a rung. The functions $\varepsilon(\vec{k})$ and $\Delta_{\vec{k}}$ are determined by two types of meanfields defined on legs and rungs, $\chi_x = \langle c^+_{i+1,\alpha,s} c_{i,\alpha,s} \rangle_0$, $\chi_y = \langle c^+_{i,l,s} c_{i,r,s} \rangle_0$ and $\Delta_x = \langle c_{i+1,\alpha,s} c_{i,\alpha,-s} \rangle_0$, $\Delta_y = \langle c_{i,l,s} c_{i,r,-s} \rangle_0$ in the unprojected fermion wavefunction. The SU(2) gauge transformation alters this mean field but not the quasiparticle spectrum. The quasiparticle spectrum for the ladder has a gap and the lowest excitations have wavevectors $\vec{q} = (0,0), (0,\pi), (\pi,0)$ or (π,π). The presence of the gap is a consequence of reduced Fermi surface here relative to the two-dimensional plane. Note if we identify the Fermi wavevectors $\{\vec{k}_F\}$ as those where $E^2_{\vec{k}}$ is minimal then $\vec{k}_F = (\pm\pi/2, 0)$ and $(\pm\pi/2, \pi)$.

The gap in the quasiparticle spectrum leads to a spin gap but the value is much too big ($\omega_g = 2J$) for the isotropic case ($J_\perp = J$). The fault lies in the nature of the quasiparticles which are free spinons. The low energy triplet excitation discussed earlier corresponds to a local triplet on a single rung, when $J_\perp \gg J$ which is delocalized along the ladder. Such an excitation is a bound state or triplet excitation in the spinon gap. Sigrist et al. [12] examined the residual interactions between the quasiparticles with different k_y-values, i.e. between those with $k_y = 0$ and $k_y = \pi$. This attraction leads to a triplet excitation well below the spinon or quasiparticle continuum which lowers the spin gap or threshold for $s = 1$ excitations from $2J$ to $\omega_g = 0.15J$ at a value $\vec{q} = (\pi,\pi)$. The absolute value of ω_g is too small compared with the exact value $\omega = 0.5J$ quoted earlier but the qualitative feature of a low lying triplet excitation with minimum at $\vec{q} = (\pi,\pi)$ is correct. This fermion mean field description based on an Afleck-Marston or d-wave RVB state is a reasonable and qualitatively correct description of the half-filled single rung ladder.

Upon doping with holes this degeneracy is lifted and the flux states and d-wave RVB states are different physically. It is known that hole kinetic energy favors the d-wave RVB state at least for the 2-dim. plane, so it is natural to follow the evolution of the d-wave, RVB state with doping. Sigrist et al. [12] found that the gap remained but the degeneracy between bonding (or $k_y = 0$) and antibonding ($k_y = \pi$) quasiparticles is lifted as the hole doping, δ increases. The quasiparticle gap drops rapidly with doping due to a reduction in the pairing meanfield amplitudes $\Delta_{x(y)}$. This is illustrated in *Fig. 3* which show the evolution in the meanfield amplitudes $\chi_{x(y)}$ and $\Delta_{x(y)}$. The pairing amplitudes $\Delta_{x(y)}$ reach small values at $\delta = 0.3$ and vanish at $\delta = 0.5$. Of especial interest is the evolution of the triplet excitation spectrum with doping. This is shown in *Fig. 4* where we see an initial increase in the gap $\omega_g(\delta)$. This is due to the drastic reduction in the binding energy relative to the continuum, so that at values of $\delta \gtrsim 0.13$ the excitation has essentially merged into the continuum. Note that the existence of a positive ω_g in this description corresponds to a stable spin liquid and a negative ω_g would imply an instability to a spin state with

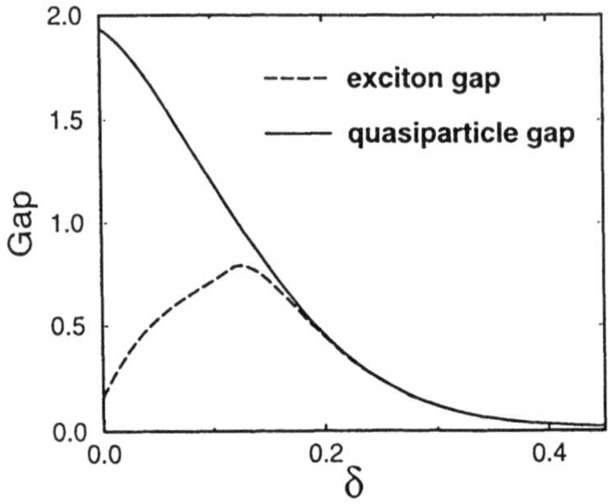

Figure 4. The spin gap as a function of the doping concentration for $J = J_{\perp} = 0.3t$. The solid line denotes the bottom of the quasiparticle spectrum and the dashed line the gap of the exciton (or bound state of two quasiparticles).

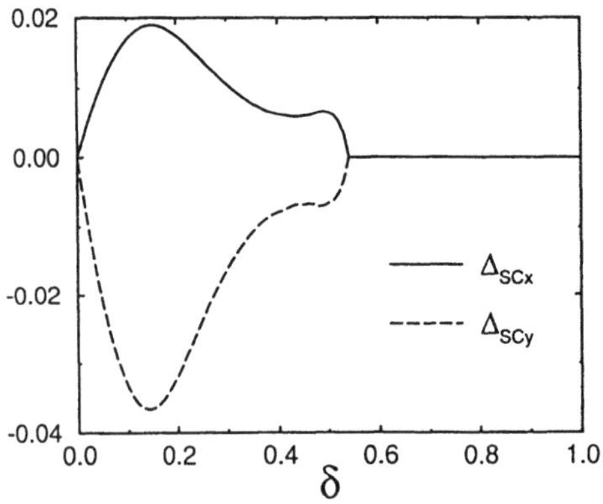

Figure 5. The superconductivity order parameter versus the doping concentration for $(J)J_{\perp} = J = 0.3t$.

long range AF order. This tendency to increase ω_g with doping, δ, points to a suppression of AF-fluctuations and a persistence of the spin gap. The spin gap can vanish for two distinct reasons, namely through a triplet excitation becoming unstable at small doping or through a collapse of the quasiparticle gap at large doping. In the present case this latter transition will be a transition to Luttinger liquid state with gapless spin excitations.

Upon doping, the mean field amplitudes become finite also for the projected wavefunctions and acquire a physical meaning. In particular the pairing amplitudes $\Delta_{x(y)}$ when evaluated for the projected wavefunction, grow linearly with δ initially, go through a maximum before decreasing as the quaisparticle gap vanishes. These pairing amplitudes, plotted in *Fig. 5* are now superconducting order parameters and the opposite signs of Δ_x and Δ_y correspond to a modified *d*-wave symmetry. Because of the inequivalence of the x and y directions in the ladder, the symmetry will not be pure *d*-wave. Similarly the restricted Fermi surface of the ladders means that the *d*-wave state can have a full gap.

To date we have only carried out mean field calculations but this problem of t-J ladders is amenable to more powerful numerical treatment which can test the predictions of the mean field theory. In particular it will be very interesting to verify the persistence of the spin gap to finite doping and the emergence of modified d-wave superconducting fluctuations on a separate and much lower energy scale than the spin gap.

There is another way to look at the d-wave RVB state. If we assume that the system scales to the large $J_\perp$ limit then we can examine that limit to gain insight into the correct physical picture. This limit is dominated by singlet formation on each rung. In terms of the bonding and antibonding state on a rung $\left[c^+_{i,\pm,\sigma} = \frac{1}{\sqrt{2}}(c^+_{i,l,\sigma} \pm c^+_{i,r,\sigma})\right]$ such a singlet has the form $\frac{1}{\sqrt{2}}(c^+_{i+\uparrow}\, c^+_{i+\downarrow} - c^+_{i-\uparrow}\, c^+_{i-\downarrow})$, i.e. a coherent superposition of two electrons in bonding and in antibonding orbitals. Further if we identify the bonding with $k_y = 0$ and the antibonding orbitals with $k_y = \pi$, then this corresponds to a change of sign of the pairing amplitude between the Fermi points at $k_y = 0\left[(\pm\pi/2, 0)\right]$ and at $k_y = \pi\left[(\pm\pi/2, \pi)\right]$. Again we see the d-wave pattern, this time as a consequence of the formation of local singlets on a rung. Note the bonding and antibonding configurations are equally present in the groundstate at half-filling and this is confirmed by the degeneracy of the quasiparticle bands with $k_y = 0$, and π at half-filling. This is very different to the bare band structure and corresponds to a scaling of $t_\perp \to 0$ at half-filling.

5. Conclusions

One of the key problems at present in the theory of strongly correlated fermions is how to connect the deep knowledge and understanding of the 1-dim. systems to 2 dimensions. Trellis lattices offer an interesting set of systems whose properties interpolate between the one-dimensional chains and the two-dimensional square lattice. Since examples of such lattices have been reported in the homologous compound series $Sr_{n-1}Cu_{n+1}O_{2n}$ there is reason to hope that the predictions can be tested experimentally. If it is possible to make systematic experiments on the magnetic properties and even observe superconductivity in such compounds, then the link between one and two dimensions can be determined and the predictions of the theoretical models can be severely tested.

The trellis lattices are also of especial interest because they can be realizations of Anderson's original short range RVB idea. They offer the possibility of observing a $s = 1/2$ Heisenberg spin liquid system and testing the link between superconductivity and a spin gap. The factors controlling the competition between magnetic order and superconductivity upon doping can be examined. It is a curious fact that the trellis lattices with odd number of rungs (even number of legs) in the ladder segments, on the one hand frustrate the magnetism and on the other hand promote the superconducting state. This points to the close relationship between the two possible groundstates but also to the fact the departures from perfect periodicity, in this case by replacing the simple square lattice by a trellis lattice, can suppress magnetic order and enhance superconductivity and not vice versa as is often assumed.

Acknowledgments

We are grateful to R. Hirsch, H. Tsunetsugu and F. Mila for useful conversations. This work was supported by Schweizerischer Nationalfonds.

References

[1] P.W. Anderson, Science 235, 1196 (1987).
[2] M. Ogata, M.U. Luchini, S. Sorella and F.F. Assaad, Phys. Rev. Lett. 66, 2388 (1991).
[3] Z. Hiroi, M. Azuma, M. Takano and Y. Bando, J. Sol. St. Chem. 95, 230 (1991); M. Takano, Z. Hiroi, M. Azuma and Y. Takeda, Jap. J. of App. Phys. 7, 3 (1992); see also H. Müller-Buschbaum, Ang. Chem. 89, 704 (1977).
[4] T.M. Rice, S. Gopalan and M. Sigrist, Europhys. Lett. 23, 445 (1993).
[5] R. Hirsch, Diplomarbeit Uni. Köln, 1988 (unpublished).
[6] E. Dagotto, J. Riera and D. Scalapino, Phys. Rev. B 45, 5744 (1992); T. Barnes et al., Phys. Rev. B 47, 3196 (1993).
[7] S.P. Strong and A.J. Millis, Phys. Rev. Lett. 69, 2419 (1992).
[8] S. Sachdev and R.N. Bhatt, Phys. Rev. B 41, 9323 (1990).
[9] S. Gopalan et al., (to be published).
[10] M. Ogata, M.U. Luchini and T.M. Rice, Phys. Rev. B 44, 12083 (1991).
[11] M. Imada, Phys. Rev. B 48, 550 (1993) and references therein.
[12] M. Sigrist, T.M. Rice and F.C. Zhang, ETH-preprint TH/93-35.
[13] F.C. Zhang, C. Gros, T.M. Rice and H. Shiba, Super. Sci. and Tech. 1, 36 (1988).
[14] I. Affleck and J.B. Marston, Phys. Rev. B 37, 3774 (1988).

Spin-Charge Decoupling and the One-Hole Green's Function in a Quantum Antiferromagnet

S. Sorella and Q.F. Zhong

International School for Advanced Study, Via Beirut 4, I-34013 Trieste, Italy

Abstract. The one-hole spectral weight for the two dimensional lattice is studied numerically using a recently developed technique, based on the Lanczos iteration scheme. This method is applied to the $t - J_z$ model for $J_z \to 0$, directly in the infinite size lattice. The presence of closed loop paths inhibit spin charge decoupling in two spatial dimensions although our calculation cannot rule out the existence of spin charge decoupling at long distance where the numerical calculation becomes prohibitive. Instead in the Bethe lattice case, for arbitrary coordination number, spin charge decoupling is satisfied at any distance-as in one dimension-due to the absence of closed loop paths.

1. Introduction

Recently there has been a considerable amount of numerical work [1, 2] suggesting the occurrence of spin-charge decoupling in two spatial dimension in the $t - J$ model. Spin-charge decoupling is well known to occur in one dimension where a one electron excitation can be decomposed in a spinon excitation which carries spin and no charge and an holon excitation which carries charge and no spin. As a consequence the one electron Green's function in space-time can be written as a simple product of the two contributions:

$$G(R,t) = G^{spinon}(R,t)\ G^{holon}(R,t). \tag{1}$$

This expression is *exact* in the one dimensional $U \to \infty$ Hubbard model [3] or in the $J \to 0$ $t - J$ model. It is also asymptotically valid at large distance $R \to \infty$ for the finite U Hubbard model as well as for the finite J $t - J$ model in the Luttinger liquid region of its phase diagram [4].

2. The Model

In this work we consider a different but related model that can be conveniently studied numerically, but that still is capable to show spin-charge decoupling in the one electron Green's function.

In the following we mimic the properties of a quantum antiferromagnet by use of the simpler antiferromagnetic Ising hamiltonian whose *exact* ground state is the Néel state and we address the problem of the single hole Green's function in the $J_z \to 0$ limit of the $t - J_z$ model defined by:

$$H = -t \sum_{<i,j>,\sigma} (c^{\dagger}_{i\sigma} c_{j\sigma} + h.c.) + J_z \sum_{<i,j>} S^z_i S^z_j \tag{2}$$

Springer Series in Solid-State Sciences, Vol. 118
Correlation Effects in Low-Dimensional Electron Systems
Editors: A. Okiji and N. Kawakami

where the constraint of no double occupancy is understood. The two dimensional square lattice (2D) will be considered. The Nagaoka theorem [5] provides a lower bound e_F to the ground state energy of one hole in the $J_z \to 0$ limit: $e_F = -zt$, where z is the coordination number of the lattice ($z = 4$ for the 2D case).

If $J_z \to 0$ there is no dynamic for spinons and $G^{spinon}(R,t) = \Omega(R)$ is expected in Eq. (1) and the holon Green's function is free within the assumption of spin-charge decoupling. We get therefore that, as a relevant consequence of spin-charge separation, the one hole Green's function should be written in the following way:

$$G(R,t) = \Omega(R)\; G^{Free}(R,t). \tag{3}$$

The free electron Green's function is nothing but the free propagator in the Nagaoka limit:

$$G(k,\omega) = \frac{1}{\omega - \epsilon_k + i\delta \operatorname{sgn} \omega}\,, \tag{4}$$

where $\epsilon_k = 2t(\cos k_x + \cos k_y)$ is the energy of a free hole.

By Fourier transforming in time and by taking the imaginary part of the equation (3) we obtain the spectral weight as a function of the final position of the hole and the frequency ω:

$$A(\omega,R) = \Omega(R)\; A^{Free}(\omega,R). \tag{5}$$

The previous expression is exact in one dimension even for this simplified $t - J_z$ model where the spinon function is particularly simple $\Omega(R) = \delta_{R,O}$.

Equation (5) can be considered a direct and measurable consequence of spin-charge separation. In fact, by measuring the spectral weight for two different positions of the hole we should get that the ratio:

$$\frac{A(\omega,R')}{A(\omega,R)} = \frac{\Omega(R')}{\Omega(R)} = \text{independent of } \omega. \tag{6}$$

The above property can be verified even in two spatial dimensionality where we have used the so called " Lanczos spectra decoding" method [6] to evaluate directly the spectral weight in real space for the 2D $J_z \to 0$ case.

3. The Lanczos Spectra Decoding

In order to evaluate the spectral weight in real space we take, as a trial state ψ_T, the one with the hole at the origin surrounded by the Néel background. After N Lanczos steps we get eigenvalues and eigenvectors ϵ_i and ϕ_i that can be written in terms of the simpler non orthogonal basis of states $H^n|\psi_T>$ for $n = 0, 1, \ldots N$:

$$|\phi_i> = \sum_{j=0}^{N} Z_{i,j} H^j |\psi_T>\,, \tag{7}$$

where $Z_{i,j}$ is the matrix of the basis change.

By propagating ψ_T with powers of the hamiltonian the hole can move at position R, leaving the Néel background unchanged. We name ψ_T^R the corresponding one-hole state. Thus we can have contributions to the spectral weight $A(\omega, R)$ if n is large enough.

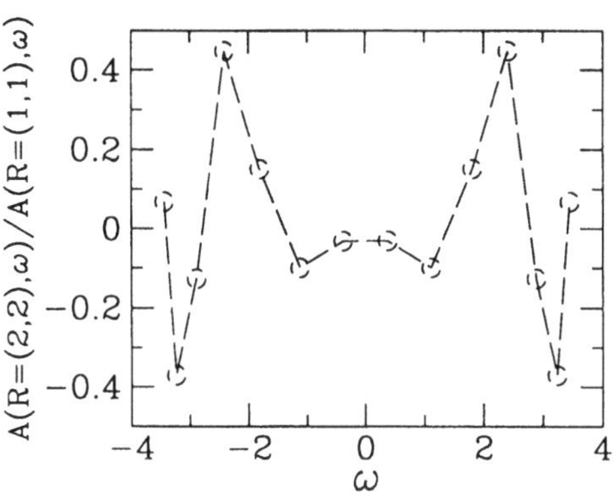

Figure 1: Ratio of the spectral function $A(R,\omega)$ between the first two sites along the diagonal direction. Empty dots are numerical estimates at given frequencies by the Lanczos Spectra decoding[6]. The dashed line is a guide to the eye. Frequencies are measured in unit of t.

In the restricted Lanczos basis ϕ_i the spectral weight can be written in the form:

$$A(\omega, R) = \sum_{i=0}^{N} Z_i^R \, \delta(\omega - \epsilon_i) \,, \tag{8}$$

where:

$$Z_i^R = <\phi_i|\psi_T><\psi_T^R|\phi_i> = \left(\sum_{j=0}^{N} Z_{i,j} <\psi_T|H^j|\psi_T>\right)\left(\sum_{j=0}^{N} Z_{i,j} <\psi_T^R|H^j|\psi_T>\right). \tag{9}$$

The procedure of coarse graining of the δ−functions, as described in [6], can be applied even in this case. The matrix elements $<\psi_T^R|H^j|\psi_T>$ can be independently evaluated since they represent the number of paths leaving unchanged the Néel order and moving the hole from the origin to the final position R in j steps. An example is given by the Trugman path [7], where the hole moves from the origin to the nearest site along the diagonal direction $R = (1,1)$ in six steps turning around the corresponding square plaquette. Since in two spatial dimension we can perform exactly 13 Lanczos steps, it is possible to evaluate with a reasonable accuracy $A(\omega, R)$ for $R = (1,1)$ and $R' = (2,2)$. We show in Fig.1 the ratio $\frac{A(\omega,R')}{A(\omega,R)}$ that, according to expression (6), should be independent of ω if spin-charge decoupling is satisfied. This is clearly not true in the present case as the occurrence of closed loop paths strongly renormalize the spectral weight as a function of the hole distance from the origin.

4. Bethe Lattice Case

In order to verify that the spin-charge decoupling can be a property of the lattice, we have studied the Green's function for $J_z \to 0$ (or $J \to 0$) in the Bethe lattice case with arbitrary coordination number in the $t - J_z$ or in the $t - J$ models. Roughly speaking the Bethe lattice has been introduced to mimic lattices with dimension higher than one, by introducing an arbitrary large coordination number z. However this Bethe lattice cannot contain closed loop paths since there is *only one path* connecting the hole at two different positions.

After a cumbersome calculation it turns out that expressions (3) and (5) are exact also in this special lattice, where G^{Free} and $\Omega(R)$ are given below.

$$G^{Free}(\omega, R_n) = K(\omega)^n \; G_{BR}(\omega) \quad , \tag{10}$$

where n is the number of hops necessary for the hole to reach the site R_n of the Bethe lattice starting from the origin and :

$$K(\omega) == \frac{1}{2(z-1)t}\left[\omega - sgn\omega\sqrt{\omega^2 - 4(z-1)t^2}\right] \quad , \tag{11}$$

whereas G_{BR} is just the Brinkman Rice expression for the local Green's function [8]:

$$G_{BR}(\omega) = \frac{\omega(z-2) - z\left[\omega^2 - 4(z-1)t^2\right]^{1/2}}{2(z^2t^2 - \omega^2)}. \tag{12}$$

The spinon contribution to the Green's function is $\Omega(R) = \delta_{R,O}$ in the Ising $J_z \to 0$ case while in the $t-J$ model for $J \to 0$ it is given by a more involved correlation function on the Heisenberg ground state Φ_H:

$$\Omega(\{R_l\}^n) = < \Phi_H | \chi_{R_n,R_{n-1}} \chi_{R_{n-1},R_{n-2}} \cdots \chi_{R_1,R_0} | \Phi_H > \quad , \tag{13}$$

where χ_{R_i,R_j} is the operator that interchanges the two spins at positions R_i and R_j:

$$\chi_{R_i,R_j} == \frac{1}{2} n_{R_i} n_{R_j} + 2\mathbf{S}_{R_i} \cdot \mathbf{S}_{R_j}. \tag{14}$$

Based on the numerical results and the previous analytical one we conclude that spin-charge decoupling can occur mainly in lattices where closed loops are forbidden by the geometry.

Closed loop paths are essential to prove the Nagaoka theorem [5]. In fact a fundamental property used in this theorem is a sort of ergodic behavior of the kinetic hamiltonian, i.e. *any* possible one-hole state -with the hole fixed at some position and the surrounding spins with definite S_z- can be obtained by a repeated application of the hamiltonian to a generic state in the given Hilbert space. Without closed loops the hole can move to another position in a *unique* way. Thus starting from a given state all the possible states generated by the Bethe lattice hamiltonian may be labelled only by the final position of the hole. It is clear therefore that many of the possible one hole state differing one another for the spin configurations cannot be generated.

For the above considerations the Nagaoka theorem cannot be applied in the Bethe lattice case where spin-charge separation is exactly fulfilled. We have thus found an interesting relation between the Nagaoka theorem, spin-charge decoupling and presence or absence of closed loop paths in a given lattice.

5. Conclusions

Recent numerical work by Putikka *et al.* [1] suggests that spin-charge decoupling can actually remain in the two dimensional $t-J$ model. This result is based on high temperature expansion for correlation functions evaluated up to the 10th order in β, the inverse temperature. This kind of calculation is equivalent to compute the trace of H^n up to $n = 10$. As it is clear for the one hole $t - J_z$ case the first closed loop path contributing to the trace (or to the density-density and spin-spin static correlation functions) is generated at 12th order in β [8]. Thus at least in the $t - J_z$ model, i.e. without the spin-flip term that can allow more complicated contributions, the spin-charge separation *cannot* be evidenced at this low order in β. In another numerical work Chen and Lee [2] applied the power method and the quantum Monte Carlo technique to the $t - J$ hamiltonian at low density. They compute powers of the hamiltonian $< \psi_T|H^n|\psi_T >$ on a given trial state for n up to sixteen. Although this is surely not a small number the calculation may be not well converged since it is limited by the use of the power method, instead of the more accurate Lanczos method used here. Therefore the relevance of closed loop paths may be still hidden in the numerical data at the present stage.

Based on the above considerations it would be therefore interesting to apply the high temperature expansion or the quantum Monte Carlo to the $t - J_z$ model or to the Bethe lattice case (where spin-charge decoupling may be exact at any filling). In this way it should be possible to verify if the results obtained in [1] and [2] crucially depend on the spin-flip term of the Heisenberg hamiltonian or on the presence or absence of closed loops in the lattice. Alternatively spin-charge decoupling may represent only a crossover effect- but still enough interesting- that can be detected only at high temperature where closed loop paths are not important.

We believe that further study is necessary to understand if spin-charge decoupling can exist at zero temperature in the two dimensional $t - J$ model. As far as the one hole problem is concerned we have shown that spin-charge decoupling breaks down at short distance for the two dimensional lattice. However preliminary results on the two chain model suggest that spin-charge separation is asymptotically satisfied within our numerical scheme [9].

Acknowledgements

We acknowledge useful discussions with A. Parola and E. Tosatti. One of us S.S. is grateful to M. Imada for his warm hospitality in the ISSP in Tokyo.

References

[1] W. O. Putikka, R. I. Glenister , R. R. Singh and H. Tsumetsugu, preprint NHMFL-93-988 , ETH-TH/93-33 , available on cond-mat@babbage.sissa.it by e-mail.

[2] Y. C. Chen and T. K. Lee preprint available on babbage by e-mail cond-mat@babbage.sissa.it.

[3] A. Parola and S. Sorella, *Phys. Rev. B* **45**, 13156 (1992); S. Sorella and A. Parola, *J. Phys. Cond. Mat.* **4** 3589 (1992).

[4] M. Ogata , M. Luchini, S. Sorella and F. Assaad , Phys. Rev. Lett. **66**, 2388 (1991).

[5] Y. Nagaoka, Phys. Rev. **147**, 392 (1966).

[6] S. Sorella Q. F. Zhong and A. Parola, submitted to Phys. Rev. B.

[7] S.A. Trugman, Phys. Rev. B **37**, 1597 (1988).

[8] W.F. Brinkman and T.M. Rice, Phys. Rev. B **2**, 1324 (1970).

[9] Q. F. Zhong and S. Sorella , in preparation.

Charge-Spin Separation and Pairing by Chiral Spin Fluctuations

C. Kübert and A. Muramatsu

Physikalisches Institut, Universität Würzburg, Am Hubland, D-97074 Würzburg, Germany

Abstract. An effective field-theory for holes moving in a slowly varying antiferromagnetic spin-background is constructed, where the spin degrees of freedom are integrated out within an expansion in generalized Berry-phases. By choosing a spin-quantization axis for the fermions that rotates with the antiferromagnetic order-parameter, a gauge-theory is obtained where the fermions are coupled to a vector gauge-field, whose fluctuations are controlled by the CP^1 model and fermionic contributions. As a consequence of the confining potential produced by the U(1) gauge-fields in (2+1) dimensions, bound states result corresponding to charge-spin separation and pairing. An alternative representation in the laboratory reference frame gives a coupling of spin- and fermionic currents that was first obtained by Shraiman and Siggia for the *t-J* model. It is also seen in this reference frame that the gauge-fields correspond to chiral fluctuations of the staggered magnetic order-parameter.

1. Introduction

A realistic description of the high-T_c superconductors is given by the three-band Hubbard model [1], that explicitly takes into account the fact that the doped holes mainly reside in σ-like bonded oxygen orbitals in the CuO_2 planes [2]. Quantum Monte Carlo (QMC) simulations of the three-band Hubbard model [3] show that a consistent description of several normal-state properties as well as of pairing correlation functions is obtained with a parameter-set situated in the strong-coupling region. In this limit, the initial model can be reduced to the spin-fermion Hamiltonian [1] (or two-band model), where the local magnetic moments on the Cu-sites persist from the insulating to the metallic phase in agreement with neutron scattering experiments [4]. A further reduction to a one-band model can be achieved, as was shown first by Zhang and Rice [5], by considering the strong exchange-interaction between the Cu-hole and a symmetric combination of O-orbitals around the Cu-site. This mapping leads then from the spin-fermion model to the *t-J* one, that attracts presently a great deal of attention. Although the latter model is particularly suited for exact diagonalization studies [6] due to the reduced Hilbert space, it seems to be until now rather intractable for analytic studies [7].

The present field-theoretic treatment centers on an expansion around the continuum limit of the spin-fermion model. This limit is appropriate when the correlation-length of the spin-background is large in comparison to the lattice constant. It is therefore restricted to the low doping region. Furthermore we restrict ourselves to antiferromagnetic (AF) short-range order most suited to YBCO compounds and do not take into account for the moment incommensurate configurations, which are observed in the case of $La_{2-x}Sr_xCuO_4$ [8]. The continuum limit also restricts the validity of the results to the low-energy region. However, it has the advantage that no approximations are necessary with respect to the strength of the coupling constants of the microscopic model.

Springer Series in Solid-State Sciences, Vol. 118
Correlation Effects in Low-Dimensional Electron Systems
Editors: A. Okiji and N. Kawakami

In looking for an effective description of the low-energy physics of the system, we integrate out the fast magnetic modes such that only the interaction of the fermions with the fluctuating AF order-parameter remains. This step is achieved by an expansion in generalized Berry-phases [9, 10] that takes into account the non-trivial algebra of the spins exactly.

Further insight is achieved by considering the system from two different reference-frames for the spin-quantization axis of the fermions. We choose on the one hand, a space- and time-dependent frame that rotates with the AF order-parameter. The rotation is performed using a CP^1 representation, such that the fluctuations of the magnetic order-parameter are described by bosonic degrees of freedom $\bar{Z} = (\bar{z}_1, \bar{z}_2)$. This leads to the appearance of a gauge-field that couples to the fermions. Thus, the system reduces to Z-bosons for the magnetic degrees of freedom and spin-$\frac{1}{2}$ fermions coupled via an $U(1)$ gauge-field. The dynamics of the gauge-field is obtained in a $1/N$ expansion after generalizing the magnetic degrees of freedom to a CP^{2N-1} model and the fermions to $SU(2)^{\otimes N}$.

In the course of the $1/N$ expansion, a mass for the Z-bosons is generated dynamically, such that a new energy scale comes into play. In this way, a "spin-gap" enters naturally in the description of the low-energy properties of the system. Moreover, the importance of this new energy scale is further stressed by the fact that it determines the strength of the propagator of the gauge-fields, and hence, the coupling of the gauge-fields to the Z-bosons and fermions. The low-energy behavior is given by the non-relativistic limit, that in two dimensions (2-D) leads to a logarithmic confining potential. The resulting bound-states correspond to charge-spin separation and pairing. It is seen therefore, that the existence of a gap in the spin-excitation spectrum directly determines the low-energy fermionic excitations.

By returning to a global reference-frame, a coupling of spin- and fermionic currents results, that was first obtained by Shraiman and Siggia [11] for the t-J model. These authors argued that the presence of such a term would lead to a spiral deformation of the spin-background around the doped mobile holes. Whether such a deformation can drive the system to an incommensurate phase is still not clear. However this question can be investigated in the present framework by studying the stability of the antiferromagnetic background as a function of the microscopic parameters. Although such a study is beyond the scope of the present paper, we would like to mention that it is presently being carried out.

Finally, by returning back to the global reference frame, the physical meaning of the gauge-fields becomes clear, since in this frame, the "magnetic" field associated to the vector potential describes chiral fluctuations of the AF order-parameter. After such an identification is made, a verification of the relevance of gauge-fields for the high-temperature cuprates can be made either numerically or experimentally.

2. Expansion in Generalized Berry-Phases

We describe the CuO_2 layer by the following spin-fermion Hamiltonian that is obtained in an expansion in the Cu-O hibridization up to fourth order [12]:

$$
\begin{aligned}
\hat{H} &= \bar{t} \sum_{\substack{<j,j'> \\ \sigma}} (-1)^{\beta_{jj'}} c^\dagger_{j,\sigma} c_{j',\sigma} + \tilde{t} \sum_{\substack{<j,j';i> \\ \sigma}} (-1)^{\alpha_{ij}+\alpha_{ij'}} c^\dagger_{j,\sigma} c_{j',\sigma} \\
&+ J_K \sum_i \vec{R}_i \cdot \vec{S}_i + J_H \sum_{<i,i'>} \vec{S}_i \cdot \vec{S}_{i'} \ . \qquad (1)
\end{aligned}
$$

The index i (j) runs over the Cu- (O) sites. The kinetic part in Eq. (1) includes both a direct O-O hopping ($\bar{t}$) as well as an effective hopping ($\tilde{t}$) mediated by a nearest-neighbor Cu-site i. The third term is a non-local Kondo-like interaction between the spin on the Cu-site and the holes on the surrounding O-sites, with

$$\vec{R}_i = \sum_{\substack{(j,j';i) \\ \alpha,\beta}} (-1)^{\alpha_{ij}+\alpha_{ij'}} c^\dagger_{j,\alpha} \, \vec{\sigma}_{\alpha\beta} \, c_{j',\beta} \ , \tag{2}$$

where $\vec{\sigma}$ are the Pauli matrices and $(j,j';i)$ means that a summation is performed on j and j' nearest-neighbors to i. The phase factors α_{ij} and $\beta_{jj'}$ take into account the d- and p-symmetry of Cu- and O-orbitals, respectively [5]

$$\alpha_{ij} = \begin{cases} 1, & \text{if } j = i + \frac{1}{2}\hat{x} \text{ or } j = i - \frac{1}{2}\hat{y} \\ 2, & \text{if } j = i - \frac{1}{2}\hat{x} \text{ or } j = i + \frac{1}{2}\hat{y} \end{cases} \tag{3}$$

and

$$\beta_{jj'} = \begin{cases} 1, & \text{if } j' = j - \frac{1}{2}\hat{x} - \frac{1}{2}\hat{y} \text{ or } j' = j + \frac{1}{2}\hat{x} + \frac{1}{2}\hat{y} \\ 2, & \text{if } j' = j + \frac{1}{2}\hat{x} - \frac{1}{2}\hat{y} \text{ or } j' = j - \frac{1}{2}\hat{x} + \frac{1}{2}\hat{y} \end{cases} . \tag{4}$$

In a path-integral description of the system, we use a Grassmann representation for the O-holes, while for the Cu-spins we use the method of generalized Berry-phases [9, 10]. Following Ref. [10] we decouple the Heisenberg interaction by introducing a field $\vec{\phi}_i(\tau)$ via a Hubbard-Stratonovich transformation. Now we have to deal with a spin on site i moving in the magnetic field $\vec{\Psi}_i(\tau)$, which is built up by the Hubbard-Stratonovich field $\vec{\phi}_i(\tau)$ and the operator $\vec{R}_i(\tau)$:

$$\vec{\Psi}_i(\tau) = 2J_H \sum_{(i';i)} \left(\vec{\phi}_i(\tau) - \vec{\phi}_{i'}(\tau)\right) + J_K \vec{R}_i(\tau) \, . \tag{5}$$

In the low-temperature limit $\beta \to \infty$ and for smooth paths of $\vec{\Psi}_i(\tau)$ in time, the trace over the spin-degrees of freedom can be performed by applying the method of generalized Berry-phases. It consists essentially in a systematic expansion around the adiabatic limit of the spin-system coupled to the time-dependent magnetic field $\vec{\Psi}$. The effective low-temperature action obtained after this expansion is given by [13]

$$\begin{aligned} S_{eff} &= \int_0^\beta d\tau \Big\{ \sum_{j,\sigma} c^\dagger_{j,\sigma}(\tau)\, \partial_\tau c_{j,\sigma}(\tau) + \bar{t} \sum_{\substack{<j,j'> \\ \sigma}} (-1)^{\beta_{jj'}} c^\dagger_{j,\sigma}(\tau)\, c_{j',\sigma}(\tau) \\ &+ \tilde{t} \sum_{\substack{<j,j';i> \\ \sigma}} (-1)^{\alpha_{ij}+\alpha_{ij'}} c^\dagger_{j,\sigma}(\tau)\, c_{j',\sigma}(\tau) + \frac{J_H}{2} \sum_{<i,i'>} \left(\vec{\phi}_i(\tau) - \vec{\phi}_{i'}(\tau)\right)^2 - s \sum_i |\vec{\Psi}_i(\tau)| \\ &+ is \sum_i A^\alpha_i[\vec{\Psi}_i(\tau)]\, \partial_\tau \Psi^\alpha_i(\tau) + s \sum_i M^{\alpha\beta}_i[\vec{\Psi}_i(\tau)]\, \partial_\tau \Psi^\alpha_i(\tau)\, \partial_\tau \Psi^\beta_i(\tau) + \ldots \Big\} , \end{aligned} \tag{6}$$

where $s = \frac{1}{2}$ in our case. The first three terms in the expression above determine the kinetic part for the fermions. The fourth term stems from the Hubbard-Stratonovich transformation. The contributions from the expansion in generalized Berry-phases are given by the remaining terms. The first of them corresponds to the sum of the instantaneous eigenvalues for the spin i in the slowly varying magnetic field $\vec{\Psi}_i(\tau)$ and is determined by the adiabatic theorem. Of special interest is the second term, which gives the Berry-phase of the spin at site i originating from the adiabatic evolution. All the other contributions are corrections that become important when one relaxes the hypothesis of adiabaticity.

$A_i^\alpha[\vec{\Psi}_i(\tau)]$ is the well-known Dirac monopole potential that fulfills in our case

$$\epsilon^{\alpha\beta\gamma}\frac{\delta A_i^\gamma[\vec{\Psi}_i(\tau)]}{\delta\Phi_i^\beta(\tau)}=\frac{\Psi_i^\alpha(\tau)}{|\vec{\Psi}_i(\tau)|^3}, \tag{7}$$

where

$$\vec{\Phi}=2J_H\sum_{(i';i)}\left(\vec{\phi}_i(\tau)-\vec{\phi}_{i'}(\tau)\right). \tag{8}$$

On the other hand, $M_i^{\alpha\beta}[\vec{\Psi}_i(\tau)]$ is proportional to the transverse projector in spin space:

$$M_i^{\alpha\beta}[\vec{\Psi}_i(\tau)]=\frac{1}{2|\vec{\Psi}_i(\tau)|^3}\left(\delta^{\alpha\beta}-e_i^\alpha(\tau)\,e_i^\beta(\tau)\right)\ ;\ e_i^\alpha=\frac{\Psi_i^\alpha(\tau)}{|\vec{\Psi}_i(\tau)|}. \tag{9}$$

From Eqs. (2),(5), and the expressions steming from the generalized phases, a complicated dependency on the fermionic variables seems to be induced. However, since we use a Grassmann representation for the holes, an expansion of the individual parts of S_{eff} in powers of $\vec{R}_i(\tau)$ gives only terms up to $\mathcal{O}(\vec{R}_i^2)$ which are non-vanishing. In order to make this point clear we define symmetric oxygen states with respect to the central copper site [5]:

$$\tilde{c}_{i,\sigma}(\tau)=\frac{1}{2}\sum_{(j;i)}(-1)^{\alpha_{ij}}\,c_{j,\sigma}(\tau). \tag{10}$$

In this representation the operator $\vec{R}_i(\tau)$ is given by:

$$\vec{R}_i(\tau)=4\tilde{c}_{i,\alpha}^\dagger(\tau)\,\vec{\sigma}_{\alpha,\beta}\,\tilde{c}_{i,\beta}(\tau). \tag{11}$$

Thus $|\vec{R}_i(\tau)|^2=16{J_K}^2(-6)\tilde{n}_{i,\uparrow}(\tau)\tilde{n}_{i,\downarrow}(\tau)$, where $\tilde{n}_{i,\sigma}(\tau)$ are the occupation numbers for the symmetric states. Remembering that $\tilde{c}_{i,\sigma}(\tau)$ are anticommuting Grassmann variables we see that all higher terms are identically zero. Here we only take into account the terms linear in $\vec{R}_i(\tau)$, since we are interested in the low doping regime of the model ($\delta \ll 1$). The resulting action is as follows [13]:

$$\begin{aligned}
S_{eff}=&\int_0^\beta d\tau\left\{\sum_{j,\sigma}c_{j,\sigma}^\dagger(\tau)\,\partial_\tau c_{j,\sigma}(\tau)+\bar{t}\sum_{\substack{<j,j'>\\ \sigma}}(-1)^{\beta_{jj'}}\,c_{j,\sigma}^\dagger(\tau)\,c_{j',\sigma}(\tau)\right.\\
&+\ \tilde{t}\sum_{\substack{<j,j';i>\\ \sigma}}(-1)^{\alpha_{ij}+\alpha_{ij'}}\,c_{j,\sigma}^\dagger(\tau)\,c_{j',\sigma}(\tau)+\frac{J_H}{2}\sum_{<i,i'>}\left(\vec{\phi}_i(\tau)-\vec{\phi}_{i'}(\tau)\right)^2\\
&-\ sJ_H\sum_i|\vec{\Phi}_i(\tau)|\left[1+\frac{J_K}{J_H}\frac{\left(\vec{\Phi}_i(\tau)\cdot\vec{R}_i(\tau)\right)}{|\vec{\Phi}_i(\tau)|^2}\right]+s\frac{1}{J_H}\sum_i\frac{1}{2|\vec{\Phi}_i(\tau)|^3}\left(\partial_\tau\vec{\Phi}_i(\tau)\cdot\partial_\tau\vec{\Phi}_i(\tau)\right)\\
&\left.+\ is\frac{1}{J_H}\sum_i\left(A_i^\alpha[\vec{\Phi}_i(\tau)]\,\partial_\tau\Phi_i^\alpha(\tau)-J_K\epsilon^{\alpha\beta\gamma}\frac{R_i^\alpha(\tau)\,\Phi_i^\beta(\tau)}{|\vec{\Phi}_i(\tau)|^2}\,\partial_\tau\Phi_i^\gamma(\tau)\right)\right\}.
\end{aligned} \tag{12}$$

3. Long-Wavelength Expansion

The long-wavelength expansion is performed around a staggered field $\vec{\phi}_i(\tau) = s(-1)^i \vec{n}_i(\tau)$. A gradient expansion of the local order-parameter $\vec{n}_i(\tau)$

$$\vec{n}_{i'}(\tau) = \vec{n}_i(\tau) + a^{\mu}_{i\to i'}\partial_\mu \,,\vec{n}_i(\tau) + \frac{1}{2} a^{\mu}_{i\to i'} a^{\nu}_{i\to i'} \partial_\mu \partial_\nu \vec{n}_i(\tau) + \cdots \; ; \; \mu = x, y \,, \tag{13}$$

where $\vec{a}_{i\to i'}$ is a vector joining the lattice sites (i,i'), leads to a continuum description of the magnetic excitations. The resulting fermionic action [13] describes the dynamics of the O holes in the presence of the slowly varying antiferromagnetic background of the Cu spins, represented by the O(3) non-linear σ-model,

$$S_{Nl\sigma} = \frac{1}{2g} \int_0^\beta d\tau \int d^2r \left(c\big(\partial_\mu \vec{n}(\vec{r},\tau)\big)^2 + \frac{1}{c}\big(\partial_\tau \vec{n}(\vec{r},\tau)\big)^2 \right) \tag{14}$$

with a coupling constant $g = a2\sqrt{2}/s$ and the spin wave velocity $c = 2\sqrt{2}aJ_H s$. Interestingly, the usual constraint on the length of $\vec{n}$ results automatically when taking the continuous limit since the following form appears in the generating functional:

$$\begin{aligned} \Delta &= \exp\Big(-\frac{8J_H s^2}{a^2} \sum_i a^2 \int_0^\beta d\tau \, \big(|\vec{n}_i(\tau)|^2 - 1\big)^2 \Big) \\ &\xrightarrow{a\to 0} \delta(|\vec{n}_i(\tau)|^2 - 1) \,. \end{aligned} \tag{15}$$

For the following development it should be pointed out that the Kondo-like interaction in the Hamiltonian Eq. (1) goes over to essentially the same form but with the order-parameter instead of the spin-operator:

$$- sJ_K \sum_i (-1)^i \left[\vec{n}_i(\tau) \cdot \vec{R}_i(\tau) \right] \,. \tag{16}$$

In order to derive the continuum limit of the lattice action in the fermionic variables, it is convenient to go over from the fixed laboratory reference frame to a spin reference frame defined by a local and time dependent spin $SU(2)$ transformation [11, 14] $U^\dagger(\vec{r},\tau)\, \vec{\sigma} \cdot \vec{n}(\vec{r},\tau)\, U(\vec{r},\tau) = \sigma^z$. We use a CP^1 representation [15] for the transformation U

$$U = \begin{pmatrix} z_1 & -\bar{z}_2 \\ z_2 & \bar{z}_1 \end{pmatrix} \tag{17}$$

with $\bar{Z}Z = 1$ and $\bar{Z} = (\bar{z}_1, \bar{z}_2)$. After such a rotation, Eq. (16) becomes a static staggered potential as it would appear in a mean-field treatment. Furthermore, the hopping terms for the fermions will be modified by the change of frame in such a way that for a slowly varying order-parameter gauge-fields appear. After a lengthy algebra [13], the effective action takes, in the continuum limit, the form

$$
\begin{aligned}
S &= \int_0^\beta d\tau \int d^2r \Big\{ p^{\dagger\lambda}(\vec r,\tau)\Big[\partial_\tau + i\,\sigma^z A_\tau - s\frac{J_K}{4J_H}\,K_\tau\Big]p^\lambda(\vec r,\tau) \\
&+ \frac{1}{2m_r}(\bar D_r - K_r)p^{\dagger\lambda}(\vec r,\tau)(D_r + K_r)p^\lambda(\vec r,\tau) \\
&- g_r\Big((\bar D_r - K_r)p^{\dagger\lambda}(\vec r,\tau)K_r p^\lambda(\vec r,\tau) - K_r p^{\dagger\lambda}(\vec r,\tau)(D_r + K_r)p^\lambda(\vec r,\tau)\Big) \\
&- g_r K_r p^{\dagger\lambda}(\vec r,\tau)K_r p^\lambda(\vec r,\tau) \\
&+ \frac{c}{2g}(\partial_\mu + iA_\mu)\bar Z\,(\partial_\mu - iA_\mu)Z\Big\}\ ;\ r = (\perp, \|)\,,
\end{aligned}
\tag{18}
$$

where we have defined a covariant derivative $D_\mu = \partial_\mu + i\,\sigma^z A_\mu$, $(\mu \in (\tau, \vec r))$ with $A_\mu = -i\bar Z\partial_\mu Z$. A non-diagonal contribution also appears:

$$
K_\mu = \begin{pmatrix} 0 & \bar z_2\partial_\mu \bar z_1 - \bar z_1\partial_\mu \bar z_2 \\ z_1\partial_\mu z_2 - z_2\partial_\mu z_1 & 0 \end{pmatrix}. \tag{19}
$$

Transverse and perpendicular masses m_r with $r = (\perp, \|)$ at $k_{min} = (\pm\frac{\pi}{2}, \frac{\pi}{2})$ were defined, where $\vec k_\perp = \frac{1}{\sqrt2}(\vec k_x \pm \vec k_y)$ and $\vec k_\| = \frac{1}{\sqrt2}(\vec k_x \mp \vec k_y)$ for the two independent valleys, which are perpendicular and parallel, respectively, to the boundary of the magnetic Brillouin-zone. Their values as a function of the original coupling constants is:

$$
m_\perp = \frac{J_K}{(J_K - \bar t + 2\tilde t)(J_K + \bar t - 2\tilde t)} \tag{20}
$$

and

$$
m_\| = \frac{2(\bar t - \tilde t) - J_K}{2\bar t(\bar t - 2\tilde t - J_K)}\,. \tag{21}
$$

These hole-pockets result from the diagonalization of the fermionic subsystem in the presence of the static staggered potential. The last term of Eq. (18) is the nonlinear σ-model in the CP^1 representation. The dependencies of the various coupling constants on the parameters of the spin-Fermion model are given by the following set of equations

$$
g_\perp = \frac{(\bar t - 2\tilde t)^2}{J_K} \tag{22}
$$

and

$$
g_\| = \frac{\bar t(2\bar t^2 + 4\tilde t^2 - 6\bar t\tilde t - 2J_K\bar t + 4J_K\tilde t + J_K^2)}{(2(\bar t - \tilde t) - J_K)^2}\,. \tag{23}
$$

Finally, we denoted by p the fermions in the rotated reference-frame after the above mentioned diagonalization.

The effective action Eq. (18) is invariant under the following transformations:

$$
\begin{aligned}
Z(\vec r,\tau) \to Z'(\vec r,\tau) &= e^{i\Lambda(\vec r,\tau)}Z(\vec r,\tau) \\
p_\sigma(\vec r,\tau) \to p'_\sigma(\vec r,\tau) &= e^{-i\sigma\Lambda(\vec r,\tau)}p_\sigma(\vec r,\tau),
\end{aligned}
\tag{24}
$$

where $\sigma = \pm$ corresponds to the up- and down-projections of the spin. Therefore, an U(1) local gauge-invariance is present and the action reduces to a gauge-theory for fermions coupled to a vector gauge-field. A hole with spin-up carries a charge $\tilde Q = -1$ and a hole with spin-down carries $\tilde{Q} = +1$, where $\tilde{Q}$ refers to the charge associated to the gauge-field introduced in Eq. (18) and not to charge that couples to the conventional electromagnetic field. At this point, the gauge-fields do not posses any dynamics. It will be generated by fluctuations of the CP^1 non-linear σ-model as well as by the fermions. The derivation of these contributions is performed in the next section.

It is possible to rotate back to the fixed uniform reference frame for the spins by means of the following transformations:

$$U(\vec{r},\tau)\, D_\mu\, U^\dagger(\vec{r},\tau) = \partial_\mu - i\vec{\sigma}\cdot(\vec{n}(\vec{r},\tau)\times\partial_\mu\vec{n}(\vec{r},\tau)) \tag{25}$$

and

$$U(\vec{r},\tau)\,(D_\mu + K_\mu)\,U^\dagger(\vec{r},\tau) = \partial_\mu\,. \tag{26}$$

These transformations lead to an alternative form for the action (18) that describes the low-energy dynamics of the spin-fermion system from a uniform reference frame:

$$\begin{aligned} S \;=\; & \int_0^\beta d\tau \int d^2r \left\{ \alpha^\dagger(\vec{r},\tau)\partial_\tau\alpha(\vec{r},\tau) + \frac{1}{2m_r}\,\partial_r\alpha^\dagger(\vec{r},\tau)\,\partial_r\alpha(\vec{r},\tau) \right. \\ - \;& g_r\,(\partial_r\vec{n})^2\,\alpha^\dagger(\vec{r},\tau)\alpha(\vec{r},\tau) + \frac{1}{2g}\left[(\partial_\tau\vec{n}(\vec{r},\tau))^2 + \frac{1}{c^2}\,(\partial_r\vec{n}(\vec{r},\tau))^2\right] \\ + \;& ig_r\,[\vec{n}(\vec{r},\tau)\times\partial_r\vec{n}(\vec{r},\tau)]\cdot\left[\partial_r\alpha^\dagger(\vec{r},\tau)\,\vec{\sigma}\,\alpha(\vec{r},\tau) - \alpha^\dagger(\vec{r},\tau)\,\vec{\sigma}\,\partial_r\alpha(\vec{r},\tau)\right] \\ + \;& \left. i\left(s\frac{J_k}{4J_H} - 1\right)[\vec{n}(\vec{r},\tau)\times\partial_\tau\vec{n}(\vec{r},\tau)]\left[\alpha^\dagger(\vec{r},\tau)\,\vec{\sigma}\,\alpha(\vec{r},\tau)\right]\right\}. \end{aligned} \tag{27}$$

The fermions in the uniform reference-frame are now denoted by α. These fermions are not the same as the original ones in Eqs. (1) and (2), since the rotation performed in Eq. (5) allowed to diagonalize the Kondo-like interaction. The fermions α correspond to this new basis after a rotation back to the uniform reference-frame [13].

Now we discuss the different contributions appearing in the field-theory (27). The first two terms determine the kinetic part (bare band-structure) for the fermions. The third term gives a renormalization of the spin-stiffness of the nonlinear σ-model (fourth term), whenever the corresponding site is occupied by a hole. The fifth term denotes the coupling between the magnetization current and the spin current of the holes found by Shraiman and Siggia for the t-J model [11]. The last term gives a coupling of the local spin-density to the background magnetization $\vec{m} \propto \vec{n}\times\partial_\tau\vec{n}$. Such a contribution stems from the Berry-phase for the holes induced by the adiabatic motion of the spin background.

Thus, by considering a local and a global reference-frame for the spin quantization axis, we obtain two equivalent representations for the continuum limit of the spin-fermion model: gauge-fields are present in the local reference-frame, whereas an interaction between magnetic currents appear in the global reference-frame.

4. Propagator of the Gauge-Fields

As already stated above, the gauge-fields present in Eq. (18) do not posses at that level any dynamics. However it will be generated by fluctuations of the CP^1 fields as well as the fermionic ones. In order to properly take into account this fact, a generalization of the model is convenient [16]. The CP^1 fields go over into CP^{2N-1} ones, where now $\bar{Z} = (\bar{z}_1, ..., \bar{z}_{2N})$, and the fermions are generalized to $SU(2)^{\otimes N}$, i.e. we introduce N copies of $p_\uparrow$ and $p_\downarrow$ fermions. In the following we will deal with the case $K_\mu = 0$. Wen [14] argued that since $Z^T i\sigma_2 \partial_\mu Z = (\partial_\mu \vec{n}) \cdot Z^T i\sigma_2 Z$, such terms will be suppressed by the non-linear σ-model that is quadratic in $\partial_\mu \vec{n}$. In any case, already with this restriction, we will be able to recognize the non-trivial structure of the low-energy theory.

After properly normalizing the action (18), the fermionic and Z-fields should be integrated out, leading to the following form for the action:

$$S = N\, Tr\, \ln \Delta_B - N\, Tr\, \ln \Delta_F + \int_0^\beta d\tau \int d^2 r\, \frac{i\sqrt{N}}{2g} \lambda(\vec{r}, \tau)\,, \tag{28}$$

where

$$\begin{aligned} \Delta_B &= -\frac{1}{2g} D_\mu^B D_\mu^B + M^2 - \frac{i\lambda}{\sqrt{N}}\,, \\ \Delta_F &= -\frac{1}{2m} D_r^F D_r^F + D_\tau^F\,, \\ D_\mu^B &= \partial_\mu + \frac{i}{\sqrt{N}} A_\mu\,, \\ D_\mu^F &= \partial_\mu + \frac{i}{\sqrt{N}} A_\mu\, \sigma^z\,. \end{aligned} \tag{29}$$

Here we have introduced a mass term M for the Z-bosons which, due to the constraint $\bar{Z}Z = 1$, corresponds to an irrelevant overall factor in the path-integral. Later we will see, how the value of M, which is arbitrary at this stage, will be fixed by a consistency consideration in the large N limit, giving rise to a univocal relationship between M and the renormalized mass for the Z-bosons. Furthermore, we included the constraint via a Lagrange-multiplier field $\lambda(\vec{r}, \tau)$ into the action. After these modifications an expansion in powers of $1/N$ of the determinants in Eq. (28) can be performed, yielding the following effective action:

$$S_{eff} = \sum_{\nu=1}^{\infty} N^{1-\nu/2}\, S^{(\nu)} + \text{const.}\,. \tag{30}$$

In the large N limit only the first two summands are relevant:

$$S^{(1)} = i\lambda(p=0)\left(\frac{1}{2g} - \frac{1}{\beta}\sum_{\omega_n}\int \frac{d^2 p}{(2\pi)^2} \frac{1}{\vec{p}^2 + \omega_n^2 + M^2}\right) \tag{31}$$

and

$$\begin{aligned} S^{(2)} = \frac{1}{2}\int \frac{d^2 p}{(2\pi)^2} \frac{1}{\beta} \sum_{\omega_n} \Big\{ & A_\mu(\vec{p}, \omega_n) \left(\Pi_{\mu\nu}^B(\vec{p}, \omega_n) + \Pi_{\mu\nu}^F(\vec{p}, \omega_n)\right) A_\nu(-\vec{p}, -\omega_n) \\ & - \lambda(\vec{p}, \omega_n)\, F(\vec{p}, \omega_n)\, \lambda(-\vec{p}, -\omega_n) \Big\}\,, \end{aligned} \tag{32}$$

where $\omega_n = 2\pi n/\beta$ are Matsubara frequencies. Here we have denoted the individual contributions of the fermions and Z-fields to the polarization tensor by $\Pi_{\mu\nu}^F$ and $\Pi_{\mu\nu}^B$, respectively. They are given by the following expressions:

$$\begin{aligned} \Pi_{\mu\nu}^B(\vec{p}, \omega_n) = {} & 2\delta_{\mu\nu} \frac{1}{\beta} \sum_{\nu_m} \int \frac{d^2 q}{(2\pi)^2} \frac{1}{\nu_m^2 + \vec{q}^2 + M^2} \\ & - \frac{1}{\beta} \sum_{\nu_m} \int \frac{d^2 q}{(2\pi)^2} \frac{(2q_\mu + p_\mu)(2q_\nu + p_\nu)}{\left[\vec{q}^2 + \nu_m^2 + M^2\right]\left[(\vec{q} + \vec{p})^2 + (\nu_m + \omega_n)^2 + M^2\right]} \end{aligned} \tag{33}$$

and

$$\begin{aligned}\Pi^F_{\mu\nu} &= \delta_{\mu r}\Big\{2\delta_{rs}\frac{1}{2m}\frac{1}{\beta}\sum_{\nu_m}\int\frac{d^2q}{(2\pi)^2}\,\frac{1}{i\nu_m-\frac{1}{2m}\vec{p}^2}\\ &- \Big(\frac{1}{2m}\Big)^2\frac{1}{\beta}\sum_{\nu_m}\int\frac{d^2q}{(2\pi)^2}\,\frac{(2p+q)_r\,(2p+q)_s}{[i\nu_m-\frac{1}{2m}\vec{p}^2][i(\nu_m+\omega_n)-\frac{1}{2m}(\vec{p}+\vec{q})^2]}\Big\}\delta_{s\nu}\\ &- \delta_{\mu 0}\Big\{\frac{1}{2\beta}\sum_{\nu_m}\int\frac{d^2q}{(2\pi)^2}\,\frac{1}{[i\nu_m-\frac{1}{2m}\vec{p}^2][i(\nu_m+\omega)-\frac{1}{2m}(\vec{p}+\vec{q})^2]}\Big\}\delta_{0\nu}\\ &- i\frac{1}{2}\Big(\frac{1}{2m}\Big)\delta_{\mu 0}\Big\{\frac{1}{\beta}\sum_{\nu_m}\int\frac{d^2q}{(2\pi)^2}\,\frac{(2p+q)_r}{[i\nu_m-\frac{1}{2m}\vec{p}^2][i(\nu_m+\omega_n)-\frac{1}{2m}(\vec{p}+\vec{q})^2]}\Big\}\delta_{r\nu},\end{aligned} \tag{34}$$

where ν_m is a Matsubara frequency corresponding to bosons for Eq. (33) and fermions for Eq. (34). The function $F(\vec{p},\omega_n)$ for the Lagrange multiplier field λ is given by:

$$F(\vec{p},\omega_n) = \frac{1}{\beta}\sum_{\nu_m}\int\frac{d^2q}{(2\pi)^2}\,\frac{1}{\vec{q}^2+\nu_m^2+M^2}\,\frac{1}{(\vec{q}+\vec{p})^2+(\nu_m+\omega_n)^2+M^2}\,. \tag{35}$$

The integral in Eq. (31) is ultraviolet divergent such that a renormalization procedure has to be introduced. This can be achieved by defining a renormalized coupling g_R and a renormalized mass μ such that:

$$\frac{1}{2g} = \frac{1}{2g_R} + I_1(\mu)\,, \tag{36}$$

where $I_1(\mu)$ is the same integral as in Eq. (31) but with M replaced by μ. It is interesting to notice that as in the one-dimensional case, since $I_1(\mu)$ diverges in the ultraviolet, the bare coupling constant has to go to zero, i.e. asymptotic freedom appears. This is not surprising since $S^{(1)}$ is fully determined by the non-linear σ-model, that shows such a phenomenon [16].

Since in the large N limit, the saddle-point condition $S^{(1)}=0$ has to be imposed [16], as can be seen from Eq. (30), a relation between the bare and the renormalized masses is established:

$$M = \frac{2}{\beta}\operatorname{arcsinh}\Big(\sinh\Big(\frac{\beta}{2}\mu\Big)\exp\Big(-\frac{\pi\beta}{g_R(\mu)}\Big)\Big)\,. \tag{37}$$

In this way, the mass M that initially had no physical meaning due to the constraint of the CP^{2N-1} model, gets a definite relationship to the renormalized mass μ. This is the same relationship obtained by Chakravarty, Halperin and Nelson [17] for the inverse correlation length, and the same discussion applies here. However, since the system we consider is a doped one, we will assume in the following that the mass M does not vanish, i.e. the doping is large enough as to be in the quantum disordered phase [17].

Using the explicit form of the function $F(\vec{p},\omega_n)$ given by Eq. (35) we obtain a massiv propagator for the λ field, where the corresponding mass $\sim M^{-1}$, and therefore, the constraint becomes ineffective as long as the correlation length $\xi \sim M^{-1}$ is large. From now on, we can omit the contribution related with the Lagrange multiplier.

As could have been expected on the basis of gauge invariance, the polarization tensor $\Pi_{\mu\nu} = \Pi^B_{\mu\nu} + \Pi^F_{\mu\nu}$ for the gauge field $A_\mu(\vec{p},\omega_n)$ is transverse:

$$p_\mu\,\Pi_{\mu\nu}(\vec{p},\omega_n) = 0\,. \tag{38}$$

Therefore the polarization tensor can be decomposed in terms of two independent transverse tensors:

$$A_{\mu\nu} = \left(\delta_{\mu 0} - \frac{p_\mu p_0}{p^2}\right)\frac{p^2}{\vec{p}^2}\left(\delta_{0\nu} - \frac{p_0 p_\nu}{p^2}\right) \tag{39}$$

and

$$B_{\mu\nu} = \delta_{\mu i}\left(\delta_{ij} - \frac{p_i p_j}{\vec{p}^2}\right)\delta_{j\nu} \ , \tag{40}$$

which satisfy:

$$A_{\mu\nu} + B_{\mu\nu} = \delta_{\mu\nu} - \frac{p_\mu p_\nu}{p^2} \ , \tag{41}$$

where $p^2 = \vec{p}\cdot\vec{p} + \omega_n^2$. Here the greek indices go over space and imaginary time, while the latin indices go only over the space. For the bosonic contribution we get:

$$\Pi^B_{\mu\nu}(\vec{p},\omega_n) = \Gamma^B(\vec{p},\omega_n)\left(\delta_{\mu\nu} - \frac{p_\mu p_\nu}{p^2}\right) \ . \tag{42}$$

and for the fermionic one:

$$\Pi^F_{\mu\nu}(\vec{p},\omega_n) = \Gamma^F_1(\vec{p},\omega_n)\, A_{\mu\nu} + \Gamma^F_2(\vec{p},\omega_n)\, B_{\mu\nu} \ , \tag{43}$$

where Γ^B, Γ^F_1, and Γ^F_2 are the coefficients obtained after the tensor decomposition of Eqs. (33) and (34).

Now the gauge field $A_\mu(\vec{p},\omega)$ has obtained its own dynamics due to fluctuations of the bosonic and fermionic degrees of freedom. Choosing the Coulomb gauge ($\vec{\nabla}\cdot\vec{A} = 0$) we obtain, using the explicit expressions for the individual contributions to the polarization tensor $\Pi_{\mu\nu}$ given by Eqs. (42) and (43), the following propagator for the gauge-field:

$$D_{00} = \frac{p^2}{\vec{p}^2}\left(\Gamma^B(\vec{p},\omega_n) + \Gamma^F_1(\vec{p},\omega_n)\right)^{-1} \ , \tag{44}$$

$$D_{0i} = 0 \ , \tag{45}$$

$$D_{ij} = \left(\delta_{ij} - \frac{p_i p_j}{\vec{p}^2}\right)\left(\Gamma^B(\vec{p},\omega_n) + \Gamma^F_2(\vec{p},\omega_n)\right)^{-1} \ . \tag{46}$$

Calculating it in detail we obtain a massive propagator for the time component due to screening by the fermions. Thus it mediates only a short-range interaction between the charges and can therefore be neglected in front of the long-range contributions. For the spatial components we obtain, after an analytical continuation to real frequencies, the following propagator:

$$D_{ij}(\vec{p},\omega) = \frac{1}{2}\left(\delta_{ij} - \frac{p_i p_j}{\vec{p}^2}\right)\frac{M}{\frac{1}{48\pi}\vec{p}^2 - i\rho\frac{M}{m}\left(\frac{\omega}{p v_F}\right)} \ , \tag{47}$$

where ρ is the fermionic density, v_F the Fermi-velocity, and m is the electron-mass as determined by the bare band-structure. This is essentially the same propagator that was found by Nagaosa and Lee [18] and Reizer [19]. A central difference, however is given by the fact that the mass of the spin-excitations determines the strength of the propagator, and hence the strength of the interaction mediated by it. In fact, after relaxing the constraint on the CP^1 variables, and according to the discussion following Eq. (37), the mass M is essentially the gap measured in neutron scattering experiments [20]. Finally it should be remarked, that the "photons" that mediate the interaction do not exist as well defined elementary excitations of the system but correspond to overdamped modes, as can be seen from the denominator of Eq. (47).

5. Discussion and Summary

The consequences of a coupling of fermions with "vector photons" was recently discussed in various contexts, ranging from purely phenomenological proposals to diverse treatments of microscopic models for high T_c superconductors. We would like to stress that the origin of the gauge-fields in the present treatment is not due to an enlargement of the Hilbert space and hence of the presence of a constraint. The gauge-fields arise by relating an $SU(2)$ rotation in spin-space and a vector on the sphere S^2. The manifold $SU(2)$ is isomorphic to S^3, however, the vector $\vec{n}$ in S^2 fixes only two of the three angles in S^3, and hence, a phase remains free. As a common feature with all other models with gauge-fields, we may expect a deviation from Fermi-liquid behavior in the normal-state as discussed by e.g. Nagaosa and Lee [18] and Reizer [19].

Since an Abelian gauge-field in $(2+1)$ dimensions is confining (attractive logarithmic potential for opposite charges), the physical spectrum of the theory contains only states with zero $\tilde{Q}$ charge (singlet states in physical terms) such as Z-Z, p-Z, and p-p bound states. As already discussed in Ref. [14], the Z-Z bound states correspond to spin-wave excitations around the antiferromagnetic wavevector, with a gap in the spectrum. The p-Z bound states are spinless charged excitations. Thus, this scenario gives an alternative way to charge-spin separation, where the bare excitations are just spin-$\frac{1}{2}$ fermions but the renormalized ones are spinless. It should be remarked here that in our case charge-spin separation results from an interaction that leads to the formation of a bound state and not by spontaneous breaking of gauge-invariance as e.g. in Ref. [18]. Finally, the p-p bound state is a singlet with a charge $2e$ leading to pairing. Hence, charge-spin separation and pairing are intimately connected in our case and result from the same interaction.

A special feature of the present theory is that a new energy scale, the mass of the magnetic excitations, appears as the strength of the propagator of the gauge-fields. This implies that confinement can be only effective in the disordered magnetic state, such that charge-spin separation and pairing should be restricted to that region. Moreover, we argue that confinement cannot be effective in one-dimension either, since in that case, a topological term is generated in the long-wavelength expansion [21] that leads to massless spin-excitations for half-integer spins [22].

The fact that confinement brings the system towards a singlet state, may explain the anomalous data in nuclear magnetic relaxation experiments [23], where a strong reduction of both the Knight-shift and the relaxation-rate on oxygen and copper is observed as a function of temperature, well above the superconducting transition, for underdoped samples of YBCO. Moreover, when charge-spin separation takes place, scattering of the charge-carriers by the gauge-fields vanishes since the new bound state has zero charge and, therefore, does not couple to the gauge-fields. Such a scenario is consistent with recent resistivity measurements in underdoped YBCO 123 [24] and 124 [25], where a reduction of the resistivity with respect to the linear temperature dependence is observed, at a temperature T^* where the nuclear relaxation rate $(1/T_1T)$ on Cu shows a maximum that is commonly associated with the spin-gap [23]. Further consequences of confinement are possibly observed in experiments probing the transfer of charge carriers in the direction perpendicular to the CuO_2 planes. In fact, anomalies in the phonon frequency (softening or hardening) are observed around T^* for infrared (IR) active phonons both in the underdoped 123 and 124 compounds, whereas the Raman active phonons show the anomaly only at T_c [26]. In the case of the IR-active phonons, a dipolar electric field is acting perpendicular to the planes, in contrast to the case of Raman active phonons. Also a suppression of the optical conductivity parallel to the c-axis was reported recently for underdoped samples of 123 [27].

We turn now our attention to the field theory (27). There we found an interaction between magnetization currents of the spin and fermionic system. According to [11] this interaction favors a spiral twist of the spin-background around the mobile hole. At the moment it is not clear whether such a deformation of the spin-background will lead to an incommensurate magnetic structure.

Finally, by transforming back to the laboratory reference frame, the physical content of the gauge-fields can be made evident. First we notice that a physical meaningful quantity is not the gauge-field by itself but the associated field-strength or "magnetic field" $B_z = \epsilon_{ij}\partial_i A_j$, where latin subindices correspond to the spatial directions. After a transformation to the laboratory reference frame, the "magnetic field" is given by $B_z = \epsilon_{ij}\vec{n} \cdot (\partial_i \vec{n} \times \partial_j \vec{n})$. This expression corresponds to chiral fluctuations of the staggered order-parameter. Therefore it is possible to probe experimentally the presence of the gauge-fields by coupling to chiral spin-fluctuations.

Summarizing, we have presented a field-theoretic treatment of the spin-fermion model from two reference-frames. In the space- and time-dependent one, gauge-fields appear that lead to charge-spin separation and pairing. In the uniform reference frame, an interaction between magnetization currents of spin and fermions takes place. Furthermore, it is seen in this reference-frame that the gauge-fields are associated to chiral spin-fluctuations.

Acknowledgements

We are grateful to A. Angelucci and C. Thomsen for instructive discussions. We acknowledge support by the Deutsche Forschungsgemeinschaft under Project No. Mu 820/5.

References

[1] V.J. Emery, Phys. Rev. Lett. **58**, 2794 (1987); C.M. Varma, Schmitt-Rink, and E. Abrahams, Solid State Commun. **62**, 681 (1987).

[2] N. Nücker, J. Fink, J.C. Fuggle, P.J. Durham, W.M. Temmerman, Phys. Rev. B**37**, 5158 (1988).

[3] G. Dopf, A. Muramatsu, and W. Hanke, Phys. Rev. Lett. **68**, 353 (1992); Europhys. Lett. **17**, 559 (1992); G. Dopf, J. Wagner, P. Dieterich, A. Muramatsu, and W. Hanke, Phys. Rev. Lett. **68**, 2082 (1992).

[4] Y. Endoh et al., Phys. Rev. B**37**, 7443 (1988).

[5] F.C. Zhang and T.M. Rice, Phys. Rev. B**37**, 3759 (1988).

[6] E. Dagotto, Int. J. Mod. Phys. B**5**, 77 (1991).

[7] P.A. Bares and G. Blatter, Phys. Rev. Lett. **64**, 2567 (1990).

[8] S-W. Cheong et al., Phys. Rev. Lett. **67**, 1791 (1991); T.E. Mason et al., Phys. Rev. Lett. **68**, 1414 (1992).

[9] R.Jackiw, Int.J.Mod.Phys. A**3**, 285 (1988).

[10] A. Angelucci and G. Jug, Int. Jour. Mod. Phys. B**3**, 1069 (1989).

[11] B.I. Shraiman and E.D. Siggia, Phys. Rev. Lett. **61**, 467 (1988); ibid. **62**, 1564 (1989); Phys. Rev. B**41**, 350 (1990); ibid. **42**, 2485 (1990).

[12] P. Prelovšek, Phys. Lett. A **126**, 287 (1988); J. Zaanen and A.M. Olés, Phys. Rev. B**37**, 4923 (1988); A. Muramatsu, R. Zeyher, and D. Schmeltzer, Europhys. Lett. **7**, 473 (1988).

[13] C. Kübert and A. Muramatsu, Phys. Rev. B**47**, 787 (1993).

[14] X.G. Wen, Phys. Rev. B**39**, 7223 (1989).

[15] R. Rajaraman, Solitons and instantons (North-Holland, Amsterdam, 1987).

[16] A. D'Adda, M. Lüscher, and P. Di Vecchia, Nucl. Phys. B**146**, 63 (1978).

[17] S. Chakravarty, B.I. Halperin, and D.R. Nelson, Phys. Rev. B**39**, 2344 (1989).

[18] N. Nagaosa and P.A. Lee, Phys. Rev. Lett. **64**, 2450 (1990).

[19] M. Reizer, Phys. Rev. B**39**, 1602 (1989);**40**, 11571 (1989).

[20] J. Rossat-Mignod et al., Physica C **185 - 189**, 86 (1991).

[21] T. Dombre and N. Read, Phys. Rev. B**38** (1988) 7181; E. Fradkin and M. Stone, Phys. Rev. B**38** (1988) 7215.

[22] F.D.M. Haldane, Phys. Lett. **93A**, 464 (1983).

[23] M. Takigawa et al., Phys. Rev. B**43**, 247 (1991).

[24] B. Bucher et al., Phys. Rev. Lett. **70**, 2012 (1993).

[25] T. Ito, K. Takenaka, and S. Uchida, Phys. Rev. Lett. **70**, 3995 (1993).

[26] A.P. Litvinchuk, C. Thomsen, and M. Cardona, Sol. St. Comm. **83**, 343 (1992); A.P. Litvinchuk et al., Zeit. f. Phys. B **92**, 9 (1993).

[27] C.C. Homes et al., Phys. Rev. Lett. **71**, 1645 (1993).

Transitions to Incompressible States

M. Imada
Institute for Solid State Physics, University of Tokyo,
Roppongi 7-22-1, Minato-ku, Tokyo 106, Japan

ABSTRACT Transitions from several different types of quantum liquid to incompressible states such as the Mott insulator, spin gap states and the fractional quantum Hall state are discussed. We analyze common features shared with these transitions as well as differences of the transitions originated from the difference in the nature of the quantum liquid.

1. Introduction

Transitions between quantum liquid and incompressible states appear in a variety of quantum many body systems. Among others, transitions from a metal to a Mott-insulator has attracted renewed interest in studies of strongly correlated electron systems. In spite of the long history of this problem, the character of the transition is far from the complete understanding. Recently, in the two-dimensional Hubbard model, Furukawa and Imada[1, 2, 3] have clarified singular divergence of the charge susceptibility at the transition point. This singularity also appears in a generalized Hubbard model on a square lattice where the next nearest neighbor transfer is finite. This observation proves that the singular divergence of the charge mass is present irrespective of its band structure. The divergence of the charge susceptibility at the transition point indicates that the transition to the Mott insulator in fermionic systems is caused not by vanishing carrier number but by the divergence of the charge mass.

The Mott insulating state appears from a variety of quantum liquid. In this paper, several different types of Mott transitions are analyzed to discuss the universality and the variety of this transition. In addition to the fundamental importance of clarifying the transition itself, it is useful to notice that the character of the transition reflects the nature of the quantum liquid near the transition point. Assuming the adiabatic continuity of the quantum liquid phase, we have a chance of probing the nature of the liquid phase through the investigation of the transition. For example, as we see below, the transitions to the Mott insulator show different properties for different choices of dimen-

Springer Series in Solid-State Sciences, Vol. 118
Correlation Effects in Low-Dimensional Electron Systems
Editors: A. Okiji and N. Kawakami

sionality, statistics of particles and interaction. In this article, the terminology of the Mott transition is generalized for the cases of the transition to a Mott insulator by the change in the particle density. We also discuss the transition to other incompressible states such as a spin gap state in a quantum spin system and the fractional quantum Hall state.

2. 1-Dimensional Systems

In one dimension, transitions to the Mott insulator show several common properties irrespective of its statistics of particles. It is related to the fact that fermions and bosons are equivalent if the hard-core repulsion exists. In the one-dimensional Hubbard model, the uniform charge susceptibility χ_c and the uniform spin susceptibility χ_s show quite different feature near the transition point. From the Bethe Ansatz analysis, it has been shown [4] that

$$\chi_c \propto \delta^{-1}, \tag{1}$$

while χ_s is smooth and remains finite near the transition point $\delta = 0$, where δ is the doping concentration from half filling i.e. the Mott insulating state. This difference may be understood from the asymptotic spin-charge separation of the Tomonaga-Luttinger liquid. Only the charge velocity v_c vanishes at the transition point while the spin velocity v_s remains finite even at $\delta = 0$. As a consequence of the vanishing charge velocity, the specific heat C at low temperatures is also singular as $C \propto \gamma T$ with $\gamma \propto \delta^{-1}$.

The singular divergence of the charge susceptibility is also seen in the one-dimensional boson system with repulsive interaction. In fact, in the hard-core boson system, $\chi_c \propto \delta^{-1}$ near the Mott transition point. The hard-core boson system is equivalent to the spin-1/2 XY model. The exact solution of the X-Y model by Katsura[5] shows

$$m = \frac{1}{\pi} \sin^{-1}(H/J), \tag{2}$$

where the magnetization density m has singular derivative at the magnetic field $H = J$ for the X-Y exchange coupling J. In case of the hard-core bosons, it leads to the relation $\chi_c = \frac{2}{\pi}\frac{1}{\sqrt{t^2-\mu^2}}$, where the chemical potential μ is equal to the transfer t at the Mott transition point. The same divergence is seen for the hard-core bosons with the nearest neighbor repulsion, which is mapped to the spin-1/2 XXZ model[6]. The exact solution of the specific heat for both the X-Y model and the Heisenberg model show the singularity[5, 6] at the saturation point of the magnetization. In the terminology of the hard-core bosons, the γ value of the specific heat behaves as $\gamma \propto \delta^{-1}$ near the Mott transition point.

Because the hard-core bosons have the particle-hole symmetry, the system at the doping concentration δ is mapped to the case of the hole concentration $1-\delta$. Therefore, the singularities of χ_c and γ at δ=0 is the same as those of the dilute boson limit $n = 1 - \delta = 0$. In the fermionic problem, in the dilute limit, the singularities are simply due to the square root singularity of the density of states. Even in the bosonic problem, the same type of singularity is due to the linear dispersion of the excitation $\varepsilon = v_c k$ at $k \ll 1$, where v_c vanishes as $v_c \propto \delta$. As is well known in the dispersion of the density wave excitations in ^{4}He, the linear dispersion $\varepsilon = v_c k$ is due to the interaction of bosons. In the case of non-interacting bosons, the dispersion becomes quadratic $\varepsilon \propto k^2$ and the specific heat is $C \propto T^{1/2}$. The charge susceptibility of noninteracting bosons is always divergent, because all the particles condense into $k = 0$ state in the ground state. The boson systems with a short-ranged interaction should approach the behavior of the non-interacting bosons in the dilute limit $n \to 0$. This is in fact seen in $\gamma \to \infty$ and $\chi_c \to \infty$ in the limit $\delta \to 0$ or $n \to 0$.

When one introduces "two-color" bosons given by the Hamiltonian

$$\mathcal{H}_1 = -t \sum_{\substack{<ij> \\ s}} (c_{is}^\dagger c_{js} + c_{js}^\dagger c_{is}), \tag{3}$$

where the Hilbert space of the boson operators $c_{is}^\dagger$ and c_{is} are restricted so as to satisfy the constraint of excluding double occupancy. The "color" degrees of freedom is represented by $s = \pm 1$. Because of the hard-core repulsion, in the one-dimensional system, physical properties in the thermodynamic limit is equivalent to the fermion Hubbard model with an infinite repulsion U. Therefore, either of (3) or the "boson t-J" Hamiltonian

$$\begin{aligned} \mathcal{H} &= \mathcal{H}_1 + \mathcal{H}_2, & (4) \\ \mathcal{H}_2 &= J \sum_{<ij>} (\mathbf{S}_i \cdot \mathbf{S}_j - \frac{1}{4} n_i n_j), & (5) \end{aligned}$$

shows the Mott transition of the same type as the t-J model in one dimension. Namely, the susceptibilities of "charge" and "color" behave quite differently at the transition. One may discuss the separation of "color" and charge in the same manner as the spin-charge separation in the fermion t-J model.

In one dimension, the transition to the Mott insulator is universally characterized by $\chi_c \propto \delta^{-1}$ and $\gamma \propto \delta^{-1}$, which is the same as the transition to a band insulator. However, the spin susceptibility (or color susceptibility) is not governed by the "band edge singularity", because the mapping between δ and $1-\delta$ is no more valid for systems with spins or "colors". The number of spins participating in the response to a magnetic field is large even at $\delta = 0$.

3. 2-Dimensional Systems

In two dimensions, the Mott transition may have a variety in contrast to the one dimensional systems. Monte Carlo studies[1]–[3] of the fermionic Hubbard model show $\chi_c \propto \delta^{-1}$ while $\chi_s \sim$ finite near the Mott transition point $\delta = 0$. In the presence of the particle-hole symmetry, the repulsive and the attractive Hubbard model are mapped each other. The antiferromagnetic order in the repulsive model or the BCS order in the attractive model open a gap at the fermi level. On the level of the Hartree-Fock approximation, the gap edge has a square root singularity in the charge density of states. Although a rigorous proof is not known, we can expect that the presence of the singularity at the gap edge is true irrespective of its dimensionality and the band structure. When the chemical potential approaches the gap edge from the inside of the gap, the physical properties should see its singularity when the chemical potential touches the gap edge. This singularity in the charge density of states strongly suggests that γ value also scales as $\gamma \propto \delta^{-1}$. These properties of χ_c, χ_s and γ are very similar to the case of the one-dimensional fermionic Hubbard model. In contrast to the case of the one-dimensional systems, the observed singularities in two dimensions are highly nontrivial.

In the hard-core boson system, the mapping condition between δ and $1-\delta$ is satisfied also in two dimensions. Therefore, the Mott transition $\delta \to 0$ is certainly equivalent to the limit of vanishing particle number $n \to 0$. In the limit of dilute density, for boson systems with short range repulsive interaction, physical properties are obtained from the low density expansion. The charge susceptibility of these interacting bosons at low density has a nonzero finite value determined from the scattering amplitude. It is in contrast to the ideal bosons where χ_c is infinity. The fact that χ_c remains finite even at the Mott transition point $\delta \to +0$ is in sharp contrast with the fermion systems with spins. The specific heat behaves as $C \propto T^2$ for hard-core bosons because the dispersion of excitation is linear while it changes to $C \propto T$ in the limit of $\delta \to 0$ as in the case of free bosons. Therefore, the coefficient of the T^2- term, i.e. η in $C \sim \eta T^2$ is singular and diverging at $\delta = 0$.

An interesting case is the "two-color" bosons defined by Eq.(3) or Eq.(4). In the case of "t-J bosons" given by Eq.(4), exact diagonalization results show that

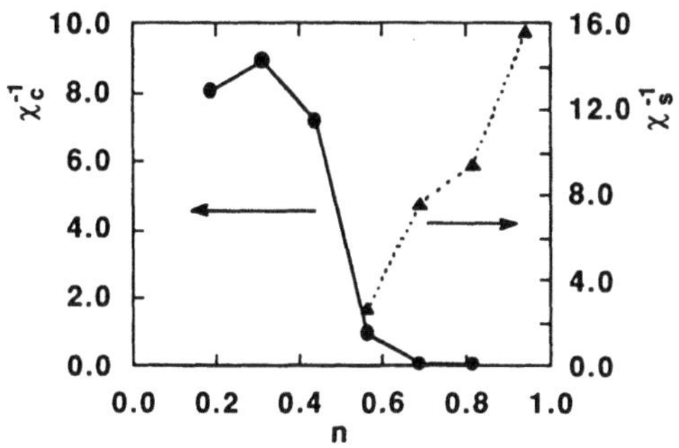

Figure 1: Filling dependence of the spin and the charge susceptibilities for the boson t-J model at $t = J$. The result is obtained from the exact diagonalization of 4×4 lattices.

the "color" and charge susceptibilities behave quite differently as in Fig.1. The charge susceptibility χ_c shows clear indication of divergent-like enhancement near $\delta = 0 (n = 1)$, while χ_s shows opposite δ dependence. The numerical data also show the formation of "antiferromagnetic" order of the "color" at $\delta = 0$. Therefore this system has substantial similarities to the fermionic Hubbard model. The "charge-color" separation seen in the boson t-J model provides useful insights on the Mott transition in the Hubbard model.

4. Discussion

In terms of the transition to an incompressible state, another interesting case is the transition to a spin-gap state. The spin compressibility of the spin-gap state is of course zero. In one dimension, a transition between a spin-gap state and a gapless state (i.e. Tomonaga Luttinger liquid) is seen in the spin-1 Haldane gap system under magnetic fields. Near the transition, the magnetization per site and the magnetic susceptibility χ_s follow the form[7] $m \propto (H - H_c)^{1/2}$ and $\chi_s \propto m^{-1}$, where the critical magnetic field is $H = H_c$. This singularity is nothing but the universal feature of the one-dimensional system discussed in §2.

In the fractional quantum Hall (FQH) state, the charge is incompressible. By changing either the magnetic field or the electron density, one can in principle see the transition to this incompressible state if the randomness can be neglected. Because of the existence of long range interaction of the quasi-particle, the compressibility may be finite. However, the quasi-particle itself proposed by Laughlin[8] does not have any dispersion. In this sense, the quasi particle mass is divergent at low doping concentration region away from FQH state within this variational approach. Its divergence at the edge of the transition has common feature with the cases discussed in §2 and §3.

In high-T_c cuprates, it has been observed in various experiments[9, 10, 11] that the carrier number decreases while the charge mass remains constant with decreasing doping. It is in sharp contrast with the properties of the two-dimensional Hubbard model discussed in §3. It suggests that the nature of carriers in the high-T_c cuprates are different from that in the Hubbard model. It has led to a speculation that a crossover from the Hubbard-like region to a pairing-dominating region may occur at high temperatures[12]. Recent experiments on the Hall coefficient at high temperatures support that the crossover in fact occurs above the room temperature[13].

It has been clarified that the character of transitions to the incompressible states may be distinguished from the singularity of the charge susceptibility at the transition point. Its difference reflects the fundamental differences of the quantum liquid phases adjacent to the incompressible states. Combining analyses for other quantities such as specific heat, we can identify the nature of the quantum liquid from the character of the transition. For example, from the singularity in the charge susceptibility and the specific heat, we can derive the charge mass divergence at the Mott transition in fermionic models irrespective of its dimensionality.

References

[1] N. Furukawa and M. Imada: J. Phys. Soc. Jpn. 61, 3331 (1992)

[2] N. Furukawa and M. Imada: J. Phys. Soc. Jpn. 62, 2557 (1993)

[3] N. Furukawa and M. Imada: these proceedings

[4] Kawakami and A. Okiji: Phys. Rev. B 40, 7066 (1989)

[5] S. Katsura: Phys. Rev. 127, 1508 (1962)

[6] M. Takahashi: Prog. Theor. Phys. 50, 1519 (1973)

[7] T. Sakai and M. Takahashi: Phys. Rev. B 43, 13383 (1991)

[8] R.B. Laughlin: Phys. Rev. Lett. 50, 1395 (1983)

[9] N.P. Ong et al.: Phys. Rev. B 35, 8807 (1987)

[10] H. Takagi et al.: Phys. Rev. B 40, 2254 (1989)

[11] T. Nishikawa and M. Sato: Physica C 209, 553 (1993)

[12] M. Imada: J. Phys. Soc. Jpn. 62, 1105 (1993)

[13] T. Nishikawa, J. Takeda and M. Sato: J. Phys. Soc. Jpn. 62, 2568 (1993)

Mass Anomaly in the Hubbard Model Near Half-Filling

N. Furukawa and M. Imada

Institute for Solid State Physics, University of Tokyo,
Roppongi 7-22-1, Minato-ku, Tokyo 106, Japan

Abstract. Ground-state properties of the two-dimensional Hubbard model is studied using the quantum Monte Carlo calculation. The numerical results show several interesting nature of the metal-insulator transition at the critical region $\delta \to 0$, where $\delta = 1 - n$ is the hole concentration. Universal nature of the metal-insulator transition in the Hubbard model is discussed.

1. Introduction

The metal-insulator (M-I) transition due to electron correlation [1] was first proposed to explain the insulating state of 3d transition metal oxides, which is difficult to reproduce through the band theory based on the Bloch state picture. Although the insulating state at half-filling seems to be easily understood, the critical properties at the transition point have been made clear only in some special cases. The Hubbard model is considered as one of the canonical models to explain the M-I transition due to the strong correlation effect. In spite of its extremely simplified form, this model in general cases is still far from complete understanding due to the lack of reliable analytical treatments.

In this paper, we numerically study the M-I transition of the Hubbard model. One of the purposes of this paper is to clarify the critical phenomena at the M-I transition. We calculate the ground state properties of the two-dimensional Hubbard model using the quantum Monte Carlo method with the projection algorithm. Combined with the results obtained for the one- and infinite-dimensional systems, we clarify universal properties of the M-I transition of the Hubbard model at half-filling.

2. Numerical Results in Two Dimensions

We investigate square-lattice models with nearest-neighbor hopping $t = 1$ and the interaction $U = 4$. The square-lattice model with only the nearest-neighbor transfer has some special properties due to its band structure, such as the perfect nesting, the particle-hole symmetry, and the van Hove singularity. In order to examine the relevance of the properties given above, we have also examined the Hubbard model with both the nearest-neighbor transfer t and the second-neighbor transfer t'.

Springer Series in Solid-State Sciences, Vol. 118
Correlation Effects in Low-Dimensional Electron Systems
Editors: A. Okiji and N. Kawakami

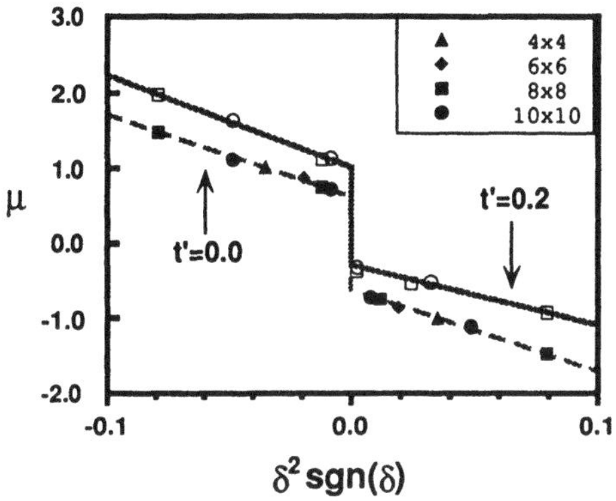

Figure 1: Doping dependence of the chemical potential for $t' = 0$ (open symbols) and $t' = 0.2$ (filled symbols).

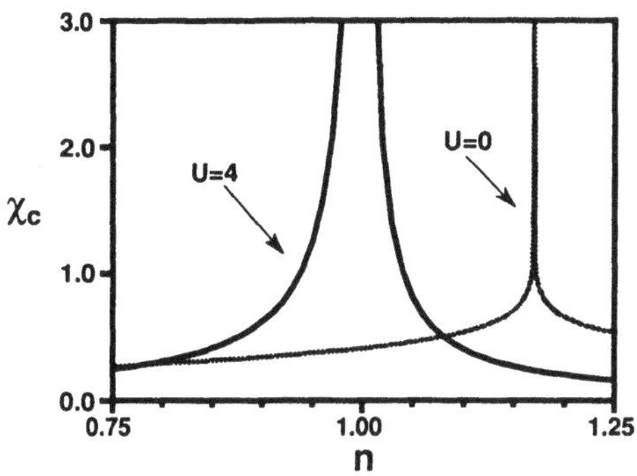

Figure 2: Doping dependence of the charge susceptibility at $t' = 0.2$, for $U = 4$ and $U = 0$.

We show doping dependence of the chemical potential at both $t' = 0$ and $t' = 0.2$ in Fig. 1. The data well fit the form

$$|\mu(n) - \mu(n \to 1)| \propto \delta^2, \tag{1}$$

where $\delta = 1 - n$ is the doping concentration [2, 3, 4]. From eq. (1), the charge susceptibility diverges as

$$\chi_c = \left(\frac{\partial \mu}{\partial n}\right)^{-1} \propto |\delta|^{-1}, \tag{2}$$

which seems to be a universal property in this M-I transition, irrespective of its band structure. In Fig. 2, the singular nature of χ_c at $t' = 0.2$ for the interacting system $U = 4$ is compared with that of the noninteracting system $U = 0$. The $|\delta|^{-1}$ divergence of χ_c in the interacting system is clearly distinguished from the logarithmic van Hove singularity in the noninteracting system.

Although the charge susceptibility diverges at half-filling, the uniform spin susceptibility does not seem to diverge [3]. These numerical results pose serious constraints on theoretical analyses. If we assume that the spin-charge separation takes place as in one dimension the singularity in the charge susceptibility comes from vanishing of the charge velocity v_c as the system approaches the transition point. If, on the other hand, the system belongs to the Fermi liquid fixed point, the observed data may be interpreted as the divergence of the effective mass $m^*/m \propto |\delta|^{-1}$ accompanied with the anomaly in the Landau parameter F_0^a [3]. In either cases, the anomaly in the charge susceptibility is due to the divergence of the charge mass.

Now we consider the nature of the spin correlation, which may be related to the anomaly in the mass. The numerical results at $U = 4$ and $t' = 0$ show that the spin structure factor diverges as

$$S(Q) \propto (\delta - \delta_c)^{-1}, \tag{3}$$

with $\delta_c \lesssim 0.01$. This critical nature is explained as follows, in analogy with the one-dimensional system [5]. Let us assume that the "mean hole-hole distance" in two dimensions $\xi \propto \delta^{-1/2}$ gives the relevant scale of the length, and the spin correlation behaves as the Heisenberg model in the short-distance part $r \ll \xi$, while the metallic behavior appears at $r \gg \xi$. If the singular part of the spin structure factor is dominated by the Heisenberg-like spin correlation in the short distance part, the spin structure factor is given by

$$S(Q) \simeq \int_0^{\xi} \mathrm{d}^2 r\, |\langle S_0 \cdot S_r \rangle_{\mathrm{Heis}}| \propto \xi^2 \propto \frac{1}{\delta}, \tag{4}$$

which reproduces the numerical result with $\delta_c = 0$. The singular nature of $S(Q)$ is determined from the behavior of the cut-off of the short-ranged part, which gets longer and longer as the doping concentration is reduced.

Although it is not obvious, the above property implies that the low-energy responses of the system is determined by the metallic system while the high-energy part is described by the localized spins. As the system becomes closer to the M-I transition, the band width of the metallic part becomes smaller. This is consistent with the mass anomaly, at least, qualitatively.

3. Discussion

A straightforward way to study the nature of the M-I transition is to investigate the dc conductivity, which is related to the charge stiffness D_c as

$$\sigma(\omega) \sim D_c \delta(\omega) \qquad \text{at} \qquad \omega \sim 0. \tag{5}$$

The charge stiffness is given by

$$D_c \propto \frac{n_{\mathrm{eff}}}{m^*} \tag{6}$$

in the framework of the quasi-particle picture. At the M-I transition point, either of $n_{\text{eff}} \to 0$ or $m^* \to \infty$ should take place.

Within the Gutzwiller approximation, Brinkman and Rice [6] have shown that there exists the singularity in the effective mass $m^* \to \infty$ at the transition point. In the Hubbard approximation, however, vanishing of the effective carrier number $n_{\text{eff}} = \delta \equiv 1 - n$ causes the M-I transition so that the effective mass m^* does not show anomaly. The above contradiction is solved to some extent by recent developments in both analytical and numerical calculations in one-, two- and infinite-dimensional Hubbard model.

In the one-dimensional system [7, 8, 9], the M-I transition at $n \to 1$ takes place as $v_c \propto |\delta|$, where v_c is the charge velocity. This leads to the divergence of the charge susceptibility $\chi_c \propto |\delta|^{-1}$ [10]. The specific heat coefficient also diverges as $\gamma \propto |\delta|^{-1}$, which may be interpreted as the divergence of the charge mass.

In two dimensions, our Monte Carlo data have made it clear that the critical behavior at the M-I transition is analogous to that of the one-dimensional system. It has been numerically shown that the Luttinger sum rule is satisfied at finite U [3, 11], so that the number of particles that can contribute to the low-energy transport does not vanish at $n \to 1$. Considering these results, our conclusion is that the Mott transition in the two-dimensional Hubbard model is driven by the charge mass singularity $m^* \to \infty$ with the carrier number being kept finite.

In the infinite-dimensional system, the mass anomalously increases in the metallic phase as in the case of Brinkman and Rice when the interaction U is adiabatically increased to the critical value $U \to U_c$ from below [12, 13]. The renormalization parameter at the Fermi level seems to scale as $z \propto 1 - U/U_c$. As the Coulomb interaction U approaches the critical value, the reduction of the band width in the metallic part is observed.

The singular nature in the charge mass described above, which leads to charge localization in the insulator phase, may be a fundamental property of the M-I transition due to the strong correlation effect, irrespective of its dimensionality.

Finally, let us mention experimental results. Effective mass anomalies near the M-I transition in 3d transition metal oxides such as (La,Sr)TiO_3 [14, 15, 16] are observed. It seems to be a good example of the M-I transition driven by the mass anomaly, although this system has some complexity due to the orbital degeneracy and a small charge gap. We also note that, in realistic systems, effects of the randomness should make the system insulating at $\delta \neq 0$ before the mass divergence. Formation of the three-dimensional antiferromagnetic order may become another reason for the saturation of the mass.

To summarize, we have found a singular nature of the mass in the critical region of the M-I transition in the two-dimensional Hubbard model. This anomaly seems to be similar to that of the one-dimensional systems. Although the situation may be slightly different, singular nature of the effective mass seems to be present also in the Hubbard models at larger dimensions where the Fermi liquid theory is applicable to the paramagnetic metal phase.

This work was supported by a Grant-in-Aid for Scientific Research on Priority Areas, "Computational Physics as a New Frontier in Condensed Matter Research" and "Science of High-T_c Superconductivity" from the Ministry of Education, Science and Culture.

References

[1] N. Mott, Proc. Phys. Soc. **A62**, 416 (1949); *Metal Insulator Transitions* (Taylor and Francis, London, 1990).

[2] N. Furukawa and M. Imada, J. Phys. Soc. Jpn. **60**, 3604 (1991).

[3] N. Furukawa and M. Imada, J. Phys. Soc. Jpn. **61**, 3331 (1992).

[4] N. Furukawa and M. Imada, J. Phys. Soc. Jpn. **62**, 2557 (1993).

[5] M. Imada, N. Furukawa, and T. M. Rice, J. Phys. Soc. Jpn. **61**, 3861 (1992).

[6] W. F. Brinkman and T. M. Rice, Phys. Rev. **B2**, 4302 (1970).

[7] H. J. Schulz, Phys. Rev. Lett. **64**, 2831 (1990).

[8] N. Kawakami and S. K. Yang, Phys. Lett. **A148**, 359 (1990).

[9] H. Frahm and V. E. Korepin, Phys. Rev. **B42**, 10553 (1990)

[10] T. Usuki, N. Kawakami, and A. Okiji, Phys. Lett. **A135**, 476 (1989); N. Kawakami and A. Okiji, Phys. Rev. **B40**, 7066 (1989)

[11] A. Moreo, D. J. Scalapino, R. L. Sugar, and S. R. White, Phys. Rev. **B41**, 2313 (1990).

[12] X. Y. Zhang, M. J. Rosenberg, and G. Kotliar, Phys. Rev. Lett. **70**, 1666 (1993).

[13] A. Georges and W. Krauth, Phys. Rev. **B48**, 7167 (1993).

[14] Y. Fujishima, Y. Tokura, T. Arima, and S. Uchida, Phys. Rev. **B46**, 11167 (1992).

[15] Y. Tokura, Y. Taguchi, Y. Okada, Y. Fujishima, T. Arima, K. Kumagai, and Y. Iye, Phys. Rev. Lett. **70**, 2126 (1993).

[16] K. Kumagai, T. Suzuki, Y. Taguchi, Y. Okada, Y. Fujishima, and Y. Tokura, Phys. Rev. **B48**, 7636 (1993).

Index of Contributors

Springer Series in Solid-State Sciences

Editors: M. Cardona P. Fulde K. von Klitzing H.-J. Queisser

1 **Principles of Magnetic Resonance** 3rd Edition 2nd Printing By C. P. Slichter
2 **Introduction to Solid-State Theory** 2nd Printing By O. Madelung
3 **Dynamical Scattering of X-Rays in Crystals** By Z. G. Pinsker
4 **Inelastic Electron Tunneling Spectroscopy** Editor: T. Wolfram
5 **Fundamentals of Crystal Growth I** Macroscopic Equilibrium and Transport Concepts 2nd Printing By F. E. Rosenberger
6 **Magnetic Flux Structures in Superconductors** By R. P. Huebener
7 **Green's Functions in Quantum Physics** 2nd Edition 2nd Printing By E. N. Economou
8 **Solitons and Condensed Matter Physics** 2nd Printing Editors: A. R. Bishop and T. Schneider
9 **Photoferroelectrics** By V. M. Fridkin
10 **Phonon Dispersion Relations in Insulators** By H. Bilz and W. Kress
11 **Electron Transport in Compound Semiconductors** By B. R. Nag
12 **The Physics of Elementary Excitations** By S. Nakajima, Y. Toyozawa, and R. Abe
13 **The Physics of Selenium and Tellurium** Editors: E. Gerlach and P. Grosse
14 **Magnetic Bubble Technology** 2nd Edition By A. H. Eschenfelder
15 **Modern Crystallography I** Symmetry of Crystals. Methods of Structural Crystallography 2nd Edition By B. K. Vainshtein
16 **Organic Molecular Crystals** Their Electronic States By E. A. Silinsh
17 **The Theory of Magnetism I** Statics and Dynamics 2nd Printing By D. C. Mattis
18 **Relaxation of Elementary Excitations** Editors: R. Kubo and E. Hanamura
19 **Solitons** Mathematical Methods for Physicists 2nd Printing By. G. Eilenberger
20 **Theory of Nonlinear Lattices** 2nd Edition By M. Toda
21 **Modern Crystallography II** Structure of Crystals 2nd Edition By B. K. Vainshtein, V. L. Indenbom, and V. M. Fridkin
22 **Point Defects in Semiconductors I** Theoretical Aspects By M. Lannoo and J. Bourgoin
23 **Physics in One Dimension** Editors: J. Bernasconi and T. Schneider
24 **Physics in High Magnetics Fields** Editors: S. Chikazumi and N. Miura
25 **Fundamental Physics of Amorphous Semiconductors** Editor: F. Yonezawa
26 **Elastic Media with Microstructure I** One-Dimensional Models By I. A. Kunin
27 **Superconductivity of Transition Metals** Their Alloys and Compounds By S. V. Vonsovsky, Yu. A. Izyumov, and E. Z. Kurmaev
28 **The Structure and Properties of Matter** Editor: T. Matsubara
29 **Electron Correlation and Magnetism in Narrow-Band Systems** Editor: T. Moriya
30 **Statistical Physics I** Equilibrium Statistical Mechanics 2nd Edition By M. Toda, R. Kubo, N. Saito
31 **Statistical Physics II** Nonequilibrium Statistical Mechanics 2nd Edition By R. Kubo, M. Toda, N. Hashitsume
32 **Quantum Theory of Magnetism** 2nd Edition By R. M. White
33 **Mixed Crystals** By A. I. Kitaigorodsky
34 **Phonons: Theory and Experiments I** Lattice Dynamics and Models of Interatomic Forces By P. Brüesch
35 **Point Defects in Semiconductors II** Experimental Aspects By J. Bourgoin and M. Lannoo
36 **Modern Crystallography III** Crystal Growth 2nd Edition By A. A. Chernov
37 **Modern Chrystallography IV** Physical Properties of Crystals Editor: L. A. Shuvalov
38 **Physics of Intercalation Compounds** Editors: L. Pietronero and E. Tosatti
39 **Anderson Localization** Editors: Y. Nagaoka and H. Fukuyama
40 **Semiconductor Physics** An Introduction 5th Edition By K. Seeger
41 **The LMTO Method** Muffin-Tin Orbitals and Electronic Structure By H. L. Skriver
42 **Crystal Optics with Spatial Dispersion, and Excitons** 2nd Edition By V. M. Agranovich and V. L. Ginzburg
43 **Structure Analysis of Point Defects in Solids** An Introduction to Multiple Magnetic Resonance Spectroscopy By J.-M. Spaeth, J. R. Niklas, and R. H. Bartram
44 **Elastic Media with Microstructure II** Three-Dimensional Models By I. A. Kunin
45 **Electronic Properties of Doped Semiconductors** By B. I. Shklovskii and A. L. Efros
46 **Topological Disorder in Condensed Matter** Editors: F. Yonezawa and T. Ninomiya
47 **Statics and Dynamics of Nonlinear Systems** Editors: G. Benedek, H. Bilz, and R. Zeyher

Springer Series in Solid-State Sciences

Editors: M. Cardona P. Fulde K. von Klitzing H.-J. Queisser

48 **Magnetic Phase Transitions**
Editors: M. Ausloos and R. J. Elliott

49 **Organic Molecular Aggregates**
Electronic Excitation and Interaction Processes
Editors: P. Reineker, H. Haken, and H. C. Wolf

50 **Multiple Diffraction of X-Rays in Crystals**
By Shih-Lin Chang

51 **Phonon Scattering in Condensed Matter**
Editors: W. Eisenmenger, K. Laßmann, and S. Döttinger

52 **Superconductivity in Magnetic and Exotic Materials** Editors: T. Matsubara and A. Kotani

53 **Two-Dimensional Systems, Heterostructures, and Superlattices**
Editors: G. Bauer, F. Kuchar, and H. Heinrich

54 **Magnetic Excitations and Fluctuations**
Editors: S. W. Lovesey, U. Balucani, F. Borsa, and V. Tognetti

55 **The Theory of Magnetism II** Thermodynamics and Statistical Mechanics By D. C. Mattis

56 **Spin Fluctuations in Itinerant Electron Magnetism** By T. Moriya

57 **Polycrystalline Semiconductors**
Physical Properties and Applications
Editor: G. Harbeke

58 **The Recursion Method and Its Applications**
Editors: D. G. Pettifor and D. L. Weaire

59 **Dynamical Processes and Ordering on Solid Surfaces** Editors: A. Yoshimori and M. Tsukada

60 **Excitonic Processes in Solids**
By M. Ueta, H. Kanzaki, K. Kobayashi, Y. Toyozawa, and E. Hanamura

61 **Localization, Interaction, and Transport Phenomena** Editors: B. Kramer, G. Bergmann, and Y. Bruynseraede

62 **Theory of Heavy Fermions and Valence Fluctuations** Editors: T. Kasuya and T. Saso

63 **Electronic Properties of Polymers and Related Compounds**
Editors: H. Kuzmany, M. Mehring, and S. Roth

64 **Symmetries in Physics** Group Theory Applied to Physical Problems
By W. Ludwig and C. Falter

65 **Phonons: Theory and Experiments II**
Experiments and Interpretation of Experimental Results By P. Brüesch

66 **Phonons: Theory and Experiments III**
Phenomena Related to Phonons
By P. Brüesch

67 **Two-Dimensional Systems: Physics and New Devices**
Editors: G. Bauer, F. Kuchar, and H. Heinrich

68 **Phonon Scattering in Condensed Matter V**
Editors: A. C. Anderson and J. P. Wolfe

69 **Nonlinearity in Condensed Matter**
Editors: A. R. Bishop, D. K. Campbell, P. Kumar, and S. E. Trullinger

70 **From Hamiltonians to Phase Diagrams**
The Electronic and Statistical-Mechanical Theory of sp-Bonded Metals and Alloys By J. Hafner

71 **High Magnetic Fields in Semiconductor Physics**
Editor: G. Landwehr

72 **One-Dimensional Conductors**
By S. Kagoshima, H. Nagasawa, and T. Sambongi

73 **Quantum Solid-State Physics**
Editors: S. V. Vonsovsky and M. I. Katsnelson

74 **Quantum Monte Carlo Methods in Equilibrium and Nonequilibrium Systems** Editor: M. Suzuki

75 **Electronic Structure and Optical Properties of Semiconductors** 2nd Edition
By M. L. Cohen and J. R. Chelikowsky

76 **Electronic Properties of Conjugated Polymers**
Editors: H. Kuzmany, M. Mehring, and S. Roth

77 **Fermi Surface Effects**
Editors: J. Kondo and A. Yoshimori

78 **Group Theory and Its Applications in Physics**
By T. Inui, Y. Tanabe, and Y. Onodera

79 **Elementary Excitations in Quantum Fluids**
Editors: K. Ohbayashi and M. Watabe

80 **Monte Carlo Simulation in Statistical Physics**
An Introduction 2nd Edition
By K. Binder and D. W. Heermann

81 **Core-Level Spectroscopy in Condensed Systems**
Editors: J. Kanamori and A. Kotani

82 **Photoelectron Spectroscopy**
Principle and Applications
By S. Hüfner

83 **Physics and Technology of Submicron Structures**
Editors: H. Heinrich, G. Bauer, and F. Kuchar

84 **Beyond the Crystalline State** An Emerging Perspective By G. Venkataraman, D. Sahoo, and V. Balakrishnan

85 **The Fractional Quantum Hall Effect**
Properties of an Incompressible Quantum Fluid
By T. Chakraborty and P. Pietiläinen

86 **The Quantum Statistics of Dynamic Processes**
By E. Fick and G. Sauermann

87 **High Magnetic Fields in Semiconductor Physics II**
Transport and Optics Editor: G. Landwehr

88 **Organic Superconductors**
By T. Ishiguro and K. Yamaji

89 **Strong Correlation and Superconductivity**
Editors: H. Fukuyama, S. Maekawa, and A. P. Malozemoff

MIX
Papier aus verantwortungsvollen Quellen
Paper from responsible sources
FSC® C105338

If you have any concerns about our products,
you can contact us on
ProductSafety@springernature.com

In case Publisher is established outside the EU,
the EU authorized representative is:
Springer Nature Customer Service Center GmbH
Europaplatz 3, 69115 Heidelberg, Germany

Printed by Libri Plureos GmbH
in Hamburg, Germany